Innovations in Chemical Physics and Mesoscopy

NANOSCIENCE AND NANOENGINEERING

Novel Applications

Edited by

V. B. Dement'ev, DSc
A. K. Haghi, PhD
V. I. Kodolov, DSc

Apple Academic Press Inc.
3333 Mistwell Crescent
Oakville, ON L6L 0A2 Canada

Apple Academic Press Inc.
9 Spinnaker Way
Waretown, NJ 08758 USA

© 2019 by Apple Academic Press, Inc.
Exclusive worldwide distribution by CRC Press, a member of Taylor & Francis Group
No claim to original U.S. Government works
International Standard Book Number-13: 978-1-77188-696-3 (Hardcover)
International Standard Book Number-13: 978-1-351-13878-9 (eBook)

All rights reserved. No part of this work may be reprinted or reproduced or utilized in any form or by any electric, mechanical or other means, now known or hereafter invented, including photocopying and recording, or in any information storage or retrieval system, without permission in writing from the publisher or its distributor, except in the case of brief excerpts or quotations for use in reviews or critical articles.

This book contains information obtained from authentic and highly regarded sources. Reprinted material is quoted with permission and sources are indicated. Copyright for individual articles remains with the authors as indicated. A wide variety of references are listed. Reasonable efforts have been made to publish reliable data and information, but the authors, editors, and the publisher cannot assume responsibility for the validity of all materials or the consequences of their use. The authors, editors, and the publisher have attempted to trace the copyright holders of all material reproduced in this publication and apologize to copyright holders if permission to publish in this form has not been obtained. If any copyright material has not been acknowledged, please write and let us know so we may rectify in any future reprint.

Trademark Notice: Registered trademark of products or corporate names are used only for explanation and identification without intent to infringe.

Library and Archives Canada Cataloguing in Publication

Nanoscience and nanoengineering : novel applications / edited by V.B. Dement'ev, DSc, A.K. Haghi, PhD, V.I. Kodolov, DSc.

(Innovations in chemical physics and mesoscopy)
This book contains papers based on reports presented at the the 6th International Conference "From Nanostructures, Nanomaterials and Nanotechnologies to Nanoindustry--NanoIzh 2017," which was held in Izhevsk, Russia, in April 2017, as well as several papers previously published in Russian in the Russian journal Chemical Physics & Mesoscopy (Institute of Mechanics, Ural Division, Russian Academy of Sciences, Izhevsk, Russia).
Includes bibliographical references and index.
Issued in print and electronic formats.
ISBN 978-1-77188-696-3 (hardcover).--ISBN 978-1-351-13878-9 (PDF)

1. Nanotechnology. I. Dement'ev, V. B. (Viacheslav Borisovich), editor II. Haghi, A. K., editor III. Kodolov, Vladimir Ivanovich, editor IV. Series: Innovations in chemical physics and mesoscopy

T174.7.N25 2018	620'.5	C2018-904586-8	C2018-904587-6

Library of Congress Cataloging-in-Publication Data

Names: Dement'ev, V. B. (Viacheslav Borisovich), editor. | Haghi, A. K., editor. | Kodolov, Vladimir I. (Vladimir Ivanovich), editor.

Title: Nanoscience and nanoengineering : novel applications / editors, V.B. Dement'ev, A.K. Haghi, V.I. Kodolov.

Other titles: Nanoscience and nanoengineering (Dement'ev)

Description: Toronto ; New Jersey : Apple Academic Press, 2019. | Includes bibliographical references and index.

Identifiers: LCCN 2018037191 (print) | LCCN 2018038122 (ebook) | ISBN 9781351138789 (ebook) | ISBN 9781771886963 (hardcover : alk. paper)

Subjects: | MESH: Nanotechnology--methods | Nanostructures--analysis

Classification: LCC R857.N34 (ebook) | LCC R857.N34 (print) | NLM QT 36.5 | DDC 610.28--dc23

LC record available at https://lccn.loc.gov/2018037191

Apple Academic Press also publishes its books in a variety of electronic formats. Some content that appears in print may not be available in electronic format. For information about Apple Academic Press products, visit our website at **www.appleacademicpress.com** and the CRC Press website at **www.crcpress.com**

ABOUT THE EDITORS

Vjacheslav B. Dement'ev, DSc

Vjacheslav B. Dement'ev, DSc, is a Professor and the Director of the Institute of Mechanics, Ural Division, of the Russian Academy of Sciences. He has published several articles and books. He has been named an Honorable Scientist of the Russian Federation.

A. K. Haghi, PhD

A. K. Haghi is the author and editor of 165 books, as well as 1000 published papers in various journals and conference proceedings. Dr. Haghi has received several grants, consulted for a number of major corporations, and is a frequent speaker to national and international audiences. Since 1983, he served as professor at several universities. He is currently Editor-in-Chief of the *International Journal of Chemoinformatics and Chemical Engineering* and *Polymers Research Journal* and on the editorial boards of many international journals. He is also a member of the Canadian Research and Development Center of Sciences and Cultures (CRDCSC), Montreal, Quebec, Canada. He holds a BSc in urban and environmental engineering from the University of North Carolina (USA), an MSc in mechanical engineering from North Carolina A&T State University (USA), a DEA in applied mechanics, acoustics and materials from the Université de Technologie de Compiègne (France), and a PhD in engineering sciences from Université de Franche-Comté (France).

Vladimir I. Kodolov, DSc

Vladimir I. Kodolov is Professor and Head of the Department of Chemistry and Chemical Technology at M. T. Kalashnikov Izhevsk State Technical University in Izhevsk, Russia, as well as Chief of Basic Research at the High Educational Center of Chemical Physics and Mesoscopy at the Udmurt Scientific Center, Ural Division at the Russian Academy of Sciences. He is also the Scientific Head of Innovation Center at the Izhevsk Electromechanical Plant in Izhevsk, Russia.

He is Vice Editor-in-Chief of the Russian journal *Chemical Physics and Mesoscopy* and is a member of the editorial boards of several Russian journals. He is the Honorable Professor of the M. T. Kalashnikov Izhevsk State Technical University, Honored Scientist of the Udmurt Republic, Honored Scientific Worker of the Russian Federation, Honorary Worker of Russian Education, and also Honorable Academician of the International Academic Society.

ABOUT THE SERIES INNOVATIONS IN CHEMICAL PHYSICS AND MESOSCOPY

The Innovations in Chemical Physics and Mesoscopy book series publishes books containing original papers and reviews. These volumes report on research developments in the following fields: nanochemistry, mesoscopic physics, computer modeling, and technical engineering, including chemical engineering. The books in this series will prove very useful for academic institutes and industrial sectors round the world interested in advanced research.

EDITORS-IN-CHIEF:

Vjacheslav B. Dement'ev, DSc
Professor and Director, Institute of Mechanics, Ural Division,
Russian Academy of Sciences
E-mail: demen@udman.ru

Vladimir Ivanovitch Kodolov, DSc
Professor and Head of Chemistry and Chemical Technology Department at
M. T. Kalashnikov Izhevsk State Technical University; Director of BRHE
center of Chemical Physics and Mesoscopy, Udmurt Scientific Centre,
Russian Academy of Sciences
E-mail: kodol@istu.ru; vkodol.av@mail.ru

EDITORIAL BOARD:

A. K. Haghi, PhD
Professor, Associate Member of University of Ottawa, Canada;
Member of Canadian Research and Development Center of Science and Culture
E-mail: akhaghi@yahoo.com

Victor Manuel de Matos Lobo, PhD
Professor, Coimbra University, Coimbra, Portugal

Richard A. Pethrick, PhD, DSc
Research Professor and Professor Emeritus, Department of Pure and Applied
Chemistry, University of Strathclyde, Glasgow, UK

Eli M. Pearce, PhD
Former President, American Chemical Society; Former Dean, Faculty of Science
and Art, Brooklyn Polytechnic University, New York, USA

Mikhail A. Korepanov, DSc
Research Senior of Institute of Mechanics, Ural Division, Russian Academy of Sciences

Alexey M. Lipanov, DSc
Professor and Head, Udmurt Scientific Center, Russian Academy of Sciences;
Editor-in-Chief, *Chemical Physics & Mesoscopy* (journal)

Gennady E. Zaikov, DSc
Professor and Head of the Polymer Division at the N. M. Emanuel Institute of
Biochemical Physics, Russian Academy of Sciences

BOOKS IN THE SERIES
Multifunctional Materials and Modeling
Editors: Mikhail. A. Korepanov, DSc, and Alexey M. Lipanov, DSc
Reviewers and Advisory Board Members: Gennady E. Zaikov, DSc,
and A. K. Haghi, PhD

Mathematical Modeling and Numerical Methods in Chemical Physics and Mechanics
Ali V. Aliev, DSc, and Olga V. Mishchenkova, PhD
Editor: Alexey M. Lipanov, DSc

Applied Mathematical Models and Experimental Approaches in Chemical Science
Editors: Vladimir I. Kodolov, DSc, and Mikhail. A. Korepanov, DSc

Computational Multiscale Modeling of Multiphase Nanosystems: Theory and Applications
Alexander V. Vakhrushev, DSc

Nanoscience and Nanoengineering: Novel Applications
Editors: V. B. Dement'ev, DSc, A. K. Haghi, PhD, and V. I. Kodolov, DSc

CONTENTS

Contributors..*xiii*

Abbreviations...*xxiii*

Preface...*xxv*

**PART I: Development of Chemical Mesoscopics &
Nanoengineering**...1

1. **Chemical Mesoscopics: New Scientific Trend** 3
   V. I. Kodolov and V. V. Trineeva

2. **Progress on Development of Nanotechnology**............................19
   M. R. Moskalenko

3. **Development of the Prognostic Apparatus in the
   Study of Nanobiobodies**..23
   E. M. Basarygina, T. A. Putilova, and M. V. Barashkov

4. **Non-Stationary Thermodynamic Processes in Rocket Engines
   of Solid Fuels: Chemical Equilibrium of Combustion Products**...............31
   A. V. Aliev, O. A. Voevodina, and E. S. Pushina

5. **Corpuscular-Wave Processes**..43
   G. A. Korablev

6. **Spatial-Energy Interactions and Entropy's Curves** 57
   G. A. Korablev, V. I. Kodolov, G. E. Zaikov, Yu. G. Vasil'ev, and N. G. Petrova

7. **Functional Dependencies in the Equations of Motion of the Planets**........97
   G. A. Korablev

8. **Exchange Spatial-Energy Interactions** 107
   G. A. Korablev and G. E. Zaikov

PART II: Nanosystems Formation Processes 131

9. **Investigation of the Formation of Nanofilms on the
   Substrates of Porous Aluminum Oxide**..133
   A. V. Vakhrushev, A. Yu. Fedotov, A. V. Severyukhin, and R. G. Valeev

Contents

10. Metal/Carbon Nanocomposites and Their Modified Analogues: Theory and Practice 143

V. I. Kodolov and V. V. Trineeva

11. The Metal/Carbon Nanocomposites Modification with Use of Ammonium Polyphosphate for the Application as Nanomodifier of Epoxy Resins 161

R. V. Mustakimov, V. I. Kodolov, I. N. Shabanova, and N. S. Terebova

PART III: New Insights and Developments 175

12. Efficiency of Metal/Carbon Nanocomposite Application in Lily Growing Under Protected Soil Conditions 177

V. M. Merzlyakova, A. A. Lapin, and V. I. Kodolov

13. Biocidal Activity of Phthalocyanine–Polymer Complexes 189

A. V. Lobanov, A. B. Kononenko, D. A. Bannikova, S. V. Britova, E. P. Savinova, O. A. Zhunina, S. M. Vasilev, V. N. Gorshenev, G. E. Zaikov, and S. D. Varfolomeev

14. Solutions and Films of Silver Nanoparticles Produced by Photochemical Method and Their Bactericidal Activity 199

A. B. Kononenko, D. A. Bannikova, S. V. Britova, E. P. Savinova, O. A. Zhunina, A. V. Lobanov, S. M. Vasilev, V. N. Gorshenev, G. E. Zaikov, and S. D. Varfolomeev

15. Polymer Composites with Gradient of Electric and Magnetic Properties 207

Jimsher N. Aneli, L. Nadareshvili, A. Akhalkatsi, M. Bolotashvili, and G. Basilaia

16. Modern Aspects of Technologies of Atomic Force Microscopy and Scanning Spectroscopy for Nanomaterials and Nanostructures Investigations and Characterizations 217

Victor A. Bykov, Arseny Kalinin, Vyatcheslav Polyakov, and Artem Shelaev

17. X-Ray Photoelectron Investigation of the Chemical Structure of Polymer Materials Modified with Carbon Metal-Containing Nanostructures 225

I. N. Shabanova, V. I. Kodolov, and N. S. Terebova

18. Small-Sized Technological Electron Spectrometer with Magnetic Energy-Analyzer 237

Yu. G. Manakov, I. N. Shabanova, V. A. Trapeznikov, and Ye. A. Morozov

PART IV: Structure and Properties of Nanostructures and Nanosystems 245

19. NMR ^{1}H, ^{13}C, and ^{17}O Spectroscopy of the *Tert*-Butyl Hydroperoxide 247

N. A. Turovskij, Yu. V. Berestneva, E. V. Raksha, and G. E. Zaikov

Contents   xi

20. On Processes of Charge Transfer in Electric Conducting Polymeric Composites 259

Tamaz A. Marsagishvili and Jimsher N. Aneli

21. Routes to Chemical Modification of the Surface of Nanosized Fluorides 273

A. V. Safronikhin, H. V. Ehrlich, and G. V. Lisichkin

22. Prospects for the Nanosilica Powder Aquaspersions in Feed for Fish 283

A. A. Lapin, M. L. Kalayda, V. V. Potapov, and V. N. Zelenkov

PART V: Selected Short Notes, Abstracts, and Conference Communications 295

23. Carbon Nanoparticles Based on Graphite Nitrate Cointercalation Compounds 297

E. V. Raksha, Yu. V. Berestneva, O. M. Padun, A. N. Vdovichenko, Savoskin M. V., and G. E. Zaikov

24. Selected Communications, Short Notes, and Abstracts 301

D. S. Vokhmyanin, Y. V. Pershin, N. M. Karavaeva, V. V. Trineeva, V. I. Kodolov, Yu. Tolchkov, T. Panina, Z. Mikhaleva, E. Galunin, N. Memetov, A. Tkachev, A. I. Khayrullina, E. V. Petrova, A. F. Dresvyannikov, L. R. Khayrullina, I. O. Grigoryeva, V. B. Dement'yev, P. G. Ovcharenko, A. Y. Leshchev, K. E. Chekmyshev, S. M. Reshetnikov, V. Ya. Kogai, K. G. Mikheev, G. M. Mikheev, E. V. Raksha, Yu. V. Berestneva, V. Yu. Vishnevskij, A. A. Maydanik, V. A. Glazunova, V. V. Burkhovetskij, A. N. Vdovichenko, Savoskin M. V., M. R. Moskalenko, D. A. Romanov, V. E. Gromov, M. A. Stepicov, E. A. Gaevoi, E. A. Grigorieva, V. O. Apanina, I. Sh. Fatykhov, A. I. Kadyrova, V. G. Kolesnikova, T. N. Ryabova, A. S. Saushin, E. V. Popova, A. N. Latyshev, O. V. Ovchinnikov, A. A. Gurov, S. E. Porozova, A. M. Khanov, O. Yu. Kamenschikov, O. A. Shulyatnikova, D. V. Volykhin, V. G. Klyuev, E. V. Aleksandrovich, R. A. Gaisin, A. A. Lapin, S. D. Filippov, R. Y. Krivenkov, E. M. Khusnutdinov, T. N. Mogileva, I. P. Angelov, R. G. Zonov, T. M. Makhneva, A. A. Sukhikh, I. V. Pavlov, A. Yu. Fedotov, E. M. Borisova, O. R. Bakieva, F. Z. Gilmutdinov, V. Y. Bayankin, M. Yu Vasilchenko, V. A. Bazhenov, K. P. Shirobokov, R. V. Mustakimov, S. D. Litvinov, I. I. Markov, Sabu Thomas, D. L. Bulatov, Yu. A. Ostrouh, I. N. Shabanova, N. S. Terebova, and G. V. Sapozhnikov

Index 353

CONTRIBUTORS

A. Akhalkatsi
I. Javakhishvili Tbilisi State University, 3, I. Chavchavadze Ave. Tbilisi 0176, Georgia

E. V. Aleksandrovich
Laser Laboratory, Institute of Mechanics of the Ural Branch of the Russian Academy of Sciences, T. Baramzinoy 34, Izhevsk, Russia

A. V. Aliev
Federal State Budgetary Educational Institution of Higher Education
Kalashnikov Izhevsk State Technical University, Izhevsk 426069, Studencheskaya St 7, Russia

Jimsher N. Aneli
Institute of Machine Mechanics, Mindeli St. 10, Tbilisi 0186, Georgia

I. P. Angelov
Institute of Organic Chemistry with Centre of Phytochemistry, BAS, Sofia 1113, Bulgaria

V. O. Apanina
Siberian State Industrial University, Novokuznetsk, Russia

O. R. Bakieva
Physical–Technical Institute, Udmurt Fedral Research Center, Ural Division, Russian Academy of Science, Kirova St. 132, Izhevsk 426000, Russia

D. A. Bannikova
All-Russian Research Institute of Veterinary Sanitation, Hygiene and Ecology, Moscow, Russia

M. V. Barashkov
LLC, Agrocomplex "Churilovo," Chelyabinsk, Russia

E. M. Basarygina
Department of Mathematical and Natural Sciences Disciplines, Chelyabinsk State Academy of Agricultural Engineering, Lenin Prospect 75, Chelyabinsk, Russia

G. Basilaia
Institute of Machine Mechanics, Mindeli St. 10, Tbilisi 0186, Georgia

V. Y. Bayankin
Physical–Technical Institute, Udmurt Fedral Research Center, Ural Division, Russian Academy of Science, Kirova St. 132, Izhevsk 426000, Russia

V. A. Bazhenov
Izhevsk State Agricultural Academy, Izhevsk, Russia

Yu. V. Berestneva
L. M. Litvinenko Institute of Physical Organic and Coal Chemistry, R. Luxemburg St. 24, Donetsk 83114, Ukraine

M. Bolotashvili
Institute of Machine Mechanics, Mindeli St. 10, Tbilisi 0186, Georgia

E. M. Borisova
Department of Physical and Organic Chemistry, Udmurt State University, Universitetskaya 1, Izhevsk, Russia

S. V. Britova
All-Russian Research Institute of Veterinary Sanitation, Hygiene and Ecology, Moscow, Russia

D. L. Bulatov
Laser Laboratory, Institute of Mechanics of the Ural Branch of the Russian Academy of Sciences, T. Baramzinoy 34, Izhevsk, Russia

V. V. Burkhovetskij
Public Institution Donetsk Institute for Physics and Engineering Named after A. A. Galkin, Donetsk, Ukraine

Victor A. Bykov
NT-MDT-Spectral Instruments Companies Group, Moscow 124460, Russia
Moscow Institute of Physics and Technology, Zhukovsky, Russia

K. E. Chekmyshev
Institute of Mechanics, Udmurt Federal Research Center, Ural Branch of the Russian Academy of Sciences, Laser Laboratory, T. Baramzinoy 34, Izhevsk, Russia

V. B. Dement'ev
Laser Laboratory, Institute of Mechanics, Udmurt Federal Research Center, Ural Branch of the Russian Academy of Sciences, T. Baramzinoy 34, Izhevsk, Russia

A. F. Dresvyannikov
Department of Analytical Chemistry, Kazan National Research Technological University, Kazan, Russia

H. V. Ehrlich
Lomonosov Moscow State University, Moscow, Russia

I. Sh. Fatykhov
Department of Plant-Growing, Izhevsk State Agricultural Academy, Izhevsk, Russia

A. Yu. Fedotov
Department of Mechanics of Nanostructure, Institute of Mechanics, Udmurt Federal Research Center, Ural Branch of the Russian Academy of Science, T. Baramzinoy 34, Izhevsk, Russia

S. D. Filippov
Akvalon Ltd, Chelyabinsk, Russia

E. A. Gaevoi
Siberian State Industrial University, Novokuznetsk, Russia

R. A. Gaisin
Kalashnikov Izhevsk State Technical University, Studencheskaya St. 7, Izhevsk, Russia

E. Galunin
Department of Technology and Methods of Nanoproducts Manufacturing, Tambov State Technical University, Tambov, Russia

F. Z. Gilmutdinov
Physical–Technical Institute, Udmurt Fedral Research Center, Ural Division, Russian Academy of Science, Kirova St. 132, Izhevsk 426000, Russia

V. A. Glazunova
Public Institution Donetsk Institute for Physics and Engineering Donetsk, Ukraine

Contributors

V. N. Gorshenev
Department of Bio-Chemistry, N. M. Emanuel Institute of Biochemical Physics, Russian Academy of Sciences, Kosygina St. 4, Moscow, Russia

E. A. Grigorieva
Siberian State Industrial University, Novokuznetsk, Russia

I. O. Grigoryeva
Department of Analytical Chemistry, Kazan National Research Technological University, Kazan, Russia

V. E. Gromov
Siberian State Industrial University, Novokuznetsk, Russia

A. A. Gurov
Faculty of Mechanical Engineering, Perm National Research Polytechnic University, Perm Krai, Russia

A. I. Kadyrova
Department of Plant-Growing, Izhevsk State Agricultural Academy, Izhevsk, Russia

M. L. Kalayda
Department of Water Resource, Kazan State Power Engineering University, Krasnoselskaya St., 51, Kazan, The Republic of Tatarstan, Russia

Arseny Kalinin
NT-MDT-Spectral Instruments Companies Group, Moscow 124460, Russia
Moscow Institute of Physics and Technology, Zhukovsky, Russia

O. Yu. Kamenschikov
Perm State National Research University, Perm, Russia

N. M. Karavaeva
Nanostructures Laboratory, Izhevsk Electromechanical Plant "KUPOL", Izhevsk, Russia

A. M. Khanov
Faculty of Mechanical Engineering, Perm National Research Polytechnic University, Perm Krai, Russia

A. I. Khayrullina
Department of Analytical Chemistry, Kazan National Research Technological University, Kazan, Russia

L. R. Khayrullina
Department of Analytical Chemistry, Kazan National Research Technological University, Kazan, Russia

E. M. Khusnutdinov
Laser Laboratory, Institute of Mechanics, Udmurt Federal Research Center, Ural Branch of the Russian Academy of Sciences, T. Baramzinoy 34, Izhevsk, Russia

V. G. Klyuev
Optics and Spectroscopy Department, Voronezh State University, Voronezh, Russia

V. I. Kodolov
Department of Chemistry and Chemical Technology, Kalashnikov Izhevsk State Technical University, Studencheskaya St. 7, Izhevsk, Russia
Basic Research-High Educational Centre of Chemical Physics and Mesoscopy, Ural Division of the Russian Academy of Science, Izhevsk, Udmurt Republic, Russia

V. Ya. Kogai
Laser Laboratory, Institute of Mechanics, Ural Branch of the Russian Academy of Sciences, T. Baramzinoy 34, Izhevsk, Russia

V. G. Kolesnikova
Department of Plant-Growing, Izhevsk State Agricultural Academy, Izhevsk, Russia

A. B. Kononenko
All-Russian Research Institute of Veterinary Sanitation, Hygiene and Ecology, Moscow, Russia
Moscow Institute of Physics and Technology, Zhukovsky, Russia

G. A. Korablev
Department of Physics, Izhevsk State Agricultural Academy, Studencheskaya St. 11, Izhevsk, Russia
Basic Research-High Educational Centre of Chemical Physics and Mesoscopy, Ural Division of
Russian Academy of Sciences, Izhevsk, Udmurt Republic, Russia

R. Y. Krivenkov
Laser Laboratory, Institute of Mechanics, Udmurt Federal Research Center, Ural Branch of the Russian
Academy of Sciences, T. Baramzinoy 34, Izhevsk, Russia

A. A. Lapin
Department of Water Resource, Kazan State Power Engineering University, Krasnoselskaya St. 51,
Kazan, The Republic of Tatarstan, Russia

A. N. Latyshev
Faculty of Physics, Voronezh State University, Voronezh, Russia

A. Y. Leshchev
Institute of Mechanics, Udmurt Federal Research Center, Ural Branch of the Russian Academy of
Sciences, Laser Laboratory, T. Baramzinoy 34, Izhevsk, Russia

G. V. Lisichkin
Lomonosov Moscow State University, Moscow, Russia

S. D. Litvinov
Medical University "Reaviz", Tchapaevskaya St. 227, Samara, Russia

A. V. Lobanov
Department of Chemical and Biological Processes Dynamics, Semenov Institute of Chemical Physics
of Russian Academy of Sciences, Moscow, Russia
Plekhanov Russian University of Economics, Moscow, Russia

T. M. Makhneva
Laser Laboratory, Institute of Mechanics, Ural Branch of the Russian Academy of Sciences,
T. Baramzinoy 34, Izhevsk, Russia

Yu. G. Manakov
Udmurt State University, Universitetskaya St., Izhevsk, 426034, Russia

I. I. Markov
Medical University "Reaviz", Tchapaevskaya St. 227, Samara, Russia

Tamaz A. Marsagishvili
Institute of the Inorganic Chemistry and Electrochemistry, Mindeli str.11, Tbilisi 0186, Georgia

A. A. Maydanik
L. M. Litvinenko Institute of Physical Organic and Coal Chemistry, R. Luxemburg St. 24,
Donetsk 83114, Ukraine

N. Memetov
Department of Technology and Methods of Nanoproducts Manufacturing, Tambov State Technical
University, Tambov, Russia

Contributors xvii

V. M. Merzlyakova
Izhevskaya State Agricultural Academy, Izhevsk, Russia

Z. Mikhaleva
Department of Technology and Methods of Nanoproducts Manufacturing, Tambov State Technical University, Tambov, Russia

G. M. Mikheev
Laser Laboratory, Institute of Mechanics, Udmurt Federal Research Center, Ural Branch of the Russian Academy of Sciences, T. Baramzinoy 34, Izhevsk, Russia

K. G. Mikheev
Laser Laboratory, Institute of Mechanics of the Ural Branch of the Russian Academy of Sciences, T. Baramzinoy 34, Izhevsk, Russia

T. N. Mogileva
Laser Laboratory, Institute of Mechanics, Udmurt Federal Research Center, Ural Branch of the Russian Academy of Sciences, T. Baramzinoy 34, Izhevsk, Russia

Ye. A. Morozov
Izhevsk State Technical University, 7, Studentcheskaya St., Izhevsk, Russia

M. R. Moskalenko
History of Science and Technology Department, Institute of Humanities and Arts, Ural Federal University, Ekaterinburg, Russia

R. V. Mustakimov
Department of Chemistry and Chemical Technology, Kalashnikov Izhevsk State Technical University, Studencheskaya St. 7, Izhevsk, Russia
Basic Research-High Educational Centre of Chemical Physics and Mesoscopy, Ural Branch of the Russian Academy of Science, Izhevsk, Udmurt Republic, Russia

L. Nadareshvili
Institute of Cybernetics of Georgian Technical University, 5, S.Euli str. Tbilisi 0186, Georgia

Yu. A. Ostrouh
Laser Laboratory, Institute of Mechanics, Ural Branch of the Russian Academy of Sciences, T. Baramzinoy 34, Izhevsk, Russia

P. G. Ovcharenko
Institute of Mechanics, Udmurt Federal Research Center, Ural Branch of the Russian Academy of Sciences, Laser Laboratory, T. Baramzinoy 34, Izhevsk, Russia

O. V. Ovchinnikov
Faculty of Physics, Voronezh State University, Voronezh, Russia

O. M. Padun
Supramolecular Chemistry Department, L. M. Litvinenko Institute of Physical Organic and Coal Chemistry, R. Luxemburg St. 24, Donetsk 83114, Ukraine

T. Panina
Department of Technology and Methods of Nanoproducts Manufacturing, Tambov State Technical University, Tambov, Russia

I. V. Pavlov
Kalashnikov Izhevsk State Technical University, Studencheskaya St. 7, Izhevsk, Russia

xviii

Y. V. Pershin
Nanostructures Laboratory, Izhevsk Electromechanical Plant "KUPOL", Izhevsk, Russia
Department of Chemistry and Chemical Technology, Kalashnikov Izhevsk State Technical University, Studencheskaya St. 7, Izhevsk, Russia

E. V. Petrova
Department of Analytical Chemistry, Kazan National Research Technological University, Kazan, Russia

N. G. Petrova
Department of Security and Communications, Agency of Informatization and Communication, Udmurt Republic, Russia

Vyatcheslav Polyakov
NT-MDT-Spectral Instruments Companies Group, Moscow 124460, Russia

E. V. Popova
Faculty of Physics, Voronezh State University, Voronezh, Russia

S. E. Porozova
Faculty of Mechanical Engineering, Perm National Research Polytechnic University, Perm Krai, Russia

V. V. Potapov
Scientific-Research Geotechnological Center of Russian Academy of Sciences, North East Highway, 30, Petropavlovsk-Kamchatsky, Russia

E. S. Pushina
Federal State Budgetary Educational Institution of Higher Education, Kalashnikov Izhevsk State Technical University, Studencheskaya St. 7, Izhevsk 426069, Russia

T. A. Putilova
Department of Mathematical and Natural Sciences Disciplines, Chelyabinsk State Academy of Agricultural Engineering, Lenin Prospect 75, Chelyabinsk, Russia

E. V. Raksha
L. M. Litvinenko Institute of Physical Organic and Coal Chemistry, R. Luxemburg St. 24, Donetsk 83114, Ukraine

S. M. Reshetnikov
Department of Physical and Organic Chemistry, Udmurt State University, Universitetskaya 1, Izhevsk, Russia

D. A. Romanov
Siberian State Industrial University, Novokuznetsk, Russia

T. N. Ryabova
Department of Plant-Growing, Izhevsk State Agricultural Academy, Izhevsk, Russia

A. V. Safronikhin
Lomonosov Moscow State University, Moscow, Russia

G. V. Sapozhnikov
Laboratory of Nanostructures, Physico-Technical Institute, Udmurt Federal Research Center, Ural Branch of the Russian Academy of Sciences, Kirov St. 132, Izhevsk 426000, Russia
FGBOU GOU VPO "Udmurtski Gosudarstvenni Universitet", Universitetskaya St. 1, Izhevsk 426034, Russia

A. S. Saushin
Laser Laboratory, Institute of Mechanics, Udmurt Federal Research Center, Ural Branch of the Russian Academy of Sciences, T. Baramzinoy 34, Izhevsk, Russia

Contributors

E. P. Savinova
All-Russian Research Institute of Veterinary Sanitation, Hygiene and Ecology, Moscow, Russia

M. V. Savoskin
L. M. Litvinenko Institute of Physical Organic and Coal Chemistry, R. Luxemburg St. 24, Donetsk 83114, Ukraine

A. V. Severyukhin
Department of Mechanics of Nanostructure, Institute of Mechanics, Ural Division, Russian Academy of Science, T. Baramzinoy 34, Izhevsk, Udmurt Republic, Russia

I. N. Shabanova
Laboratory of Nanostructures, Physico-Technical Institute, Udmurt Federal Research Center, Ural Branch of the Russian Academy of Sciences, Kirov St. 132, Izhevsk 426000, Russia
Department of Chemistry and Chemical Technology, Kalashnikov Izhevsk State Technical University, Studencheskaya St. 7, Izhevsk, Russia

Artem Shelaev
NT-MDT-Spectral Instruments Companies Group, Moscow 124460, Russia
Moscow Institute of Physics and Technology, Zhukovsky, Russia

K. P. Shirobokov
Department of Chemistry and Chemical Technology, Kalashnikov Izhevsk State Technical University, Studencheskaya, St. 7, Izhevsk, Russia

O. A. Shulyatnikova
Perm State Medical Academy Academy, Perm, Russia

M. A. Stepicov
Siberian State Industrial University, Novokuznetsk, Russia

A. A. Sukhikh
Laser Laboratory, Institute of Mechanics, Ural Branch of the Russian Academy of Sciences, T. Baramzinoy 34, Izhevsk, Russia

N. S. Terebova
Laboratory of Nanostructures, Physico-Technical Institute, Udmurt Federal Research Center, Ural Branch of the Russian Academy of Sciences, Kirov St. 132, Izhevsk 426000, Russia
Deparment of Chemistry and Chemical Technology, Kalashnikov Izhevsk State Technical University, Studencheskaya St. 7, Izhevsk, Russia

Sabu Thomas
School of Chemical Sciences, International and Inter University Centre for Nanoscience and Nanotechnology, Mahatma Gandhi University, Priyadarshini Hills P.O., Kottayam 686560, Kerala, India

A. Tkachev
Department of Technology and Methods of Nanoproducts Manufacturing, Tambov State Technical University, Tambov, Russia

Yu. Tolchkov
Department of Technology and Methods of Nanoproducts Manufacturing, Tambov State Technical University, Tambov, Russia

V. A. Trapeznikov
Physico-Technical Institute of the Ural Branch of the Russian Academy of Sciences, Izhevsk, Russia

V. V. Trineeva
Institute of Mechanics, Ural Division, Russian Academy of Sciences, Izhevsk, Russia
Department of Chemistry and Chemical Technology, Kalashnikov Izhevsk State Technical University, Studencheskaya St. 7, Izhevsk, Russia

N. A. Turovskij
Physical Chemistry Department, Donetsk National University, Universitetskaya St. 24, Donetsk 83001, Ukraine

A. V. Vakhrushev
Department of Mechanics of Nanostructure, Institute of Mechanics, Ural Division, Russian Academy of Sciences, T. Baramzinoy 34, Izhevsk, Russia
Department of Nanotechnology and Microsystems, Kalashnikov Izhevsk State Technical University, Studencheskaya St. 7, Izhevsk, Udmurt Republic, Russia

R. G. Valeev
Department of Physics and Chemistry of Surface, Physical-Technical Institute, Ural Branch of the Russian Academy of Science, Kirova St. 132, Izhevsk, Udmurt Republic, Russia

S. D. Varfolomeev
Department of Bio-Chemistry, N. M. Emanuel Institute of Biochemical Physics of Russian Academy of Sciences, Moscow, Russia

M. Yu Vasilchenko
Izhevsk State Agricultural Academy, Izhevsk, Russia

Yu. M. Vasilchenko
Department of Chemistry and Chemical Technology, Kalashnikov Izhevsk State Technical University, Studencheskaya St. 7, Izhevsk, Russia

Yu. G. Vasil'ev
Department of Veterinary, Izhevsk State Agricultural Academy, Studencheskaya St. 11, Izhevsk, Russia

S. M. Vasilev
Institute of Modern Standards, Moscow, Russia

A. N. Vdovichenko
L. M. Litvinenko Institute of Physical Organic and Coal Chemistry, R. Luxemburg St. 24, Donetsk 83114, Ukraine

Vishnevskij V. Yu.
L. M. Litvinenko Institute of Physical Organic and Coal Chemistry, R. Luxemburg St. 24, Donetsk 83114, Ukraine

O. A. Voevodina
Federal State Budgetary Educational Institution of Higher Education Kalashnikov Izhevsk State Technical University, Izhevsk 426069, Studencheskaya St 7, Russia

D. S. Vokhmyanin
Faculty of Mechanical Engineering, Perm National Research Polytechnic University, Perm Krai, Russia

D. V. Volykhin
Optics and Spectroscopy Department, Voronezh State University, Voronezh, Russia

G. E. Zaikov
Department of Bio-Chemistry, N. M. Emanuel Institute of Biochemical Physics, Russian Academy of Sciences, Kosygina St. 4, Moscow 119991, Russia

Contributors

V. N. Zelenkov
FGBNU Institute of Vegetable Crops, Vereya, Ramensky District, Moscow, Russia

O. A. Zhunina
Russian Scientific Center of Molecular Diagnostics and Therapy, Moscow, Russia

R. G. Zonov
Laser Laboratory, Institute of Mechanics, Ural Branch of the Russian Academy of Sciences, T. Baramzinoy 34, Izhevsk, Russia

ABBREVIATIONS

AAO	anodic aluminum oxide
Acac	acetylacetone
AFM	atomic force microscopy
APP	ammonium polyphosphate
Asp	aspartic acid
BET	Brunauer–Emmett–Teller
C/M	carbon metal-containing
Citr	citric acid
CNTs	carbon nanotubes
CuSe	copper selenide
Dbm	dibenzoylmethane
DFT	density functional theory
DMSO	dimethylsulfoxide
DS	dispersion system
ECPC	electrical conducting polymer composites
EPA	environmental protection agency
FePc	iron phthalocyanine
FTIR	Fourier-transform infrared spectroscopy
GFs	green temperature functions
GIAO	gauge-including atomic orbital
Gly	glycine
HDS	high-dispersive silicon dioxide
LAMMPS	large-scale atomic/molecular massively parallel simulator
MD method	molecular dynamics method
MEAM	modified method of the immersed atom
MgPc	magnesium phthalocyanine
MHC	major histocompatibility complex
MnPc	manganese phthalocyanine
MRI	magnetic resonance imaging
NCs	nanocomposites
NDS	nano-dispersive silica
NM	nanomaterials
NP	nanoparticles
NT	nanotechnology

OBD	optical beam deflection
OL	optical limiting
PC	polycarbonate
Pc	phthalocyanines
PEG	polyethylene glycol
Phen	o-phenanthroline
PL	photoluminescence
PLE	photoluminescence excitation
PMMA	polymethyl methacrylate
PVA	polyvinyl alcohol
PVP	poly-N-vinylpyrrolidone
PXRD	powder X-ray diffraction
RE	rare earth
RS	Raman scattering
SDS	sodium dodecyl sulfate
SEI	spatial-energy exchange interactions
SEM	scanning electron microscopy
SEP	spatial-energy parameter
SPM	scanning probe microscopy
TEM	transmission electron microscopy
TERS	tip-enhanced Raman scattering
Trp	tryptophan
USSR	Union of Soviet Socialist Republics
XPS	X-ray photoelectron spectroscopy
XRD	X-ray diffraction
ZnPc	zinc phthalocyanine

PREFACE

This book *Nanoscience and Nanoengineering: Novel Applications* is the fifth volume of the book series Innovation in Chemical Physics and Mesoscopy.

This volume compiles the most important information about the new scientific trend chemical mesoscopics as well as new information in the science about nanomaterials and processes of nanochemistry and nano-engineering. This book contains papers based on reports presented at the the 6th International Conference "From Nanostructures, Nanomaterials and Nanotechnologies to Nanoindustry–NanoIzh 2017," which was held in Izhevsk, Russia, in April 2017, as well as several papers previously published in Russian in the Russian journal *Chemical Physics & Mesoscopy* (Institute of Mechanics, Ural Division, Russian Academy of Sciences, Izhevsk, Russia). The volume covers the fundamentals of chemical mesoscopics along with investigations in the field of nanotechnology and nanomaterials engineering.

This book is unique and important because the new trends such as chemical mesoscopics are discussed. Researchers, professors, postgraduate students, and other scientists will find much interesting information here and will gain new knowledge on the new trends of chemical mesoscopics, nanotechnology, and nanoengineering.

This book will be useful for researchers, professors/instructors (for teaching specific courses), students, and postgraduates and also for personal re-qualification, university/college libraries, and bookstores.

The Editors

PART I

Development of Chemical Mesoscopics & Nanoengineering

CHAPTER 1

CHEMICAL MESOSCOPICS: NEW SCIENTIFIC TREND

V. I. KODOLOV[1,2] and V. V. TRINEEVA[1,3]

[1]*Basic Research-High Educational Centre of Chemical Physics and Mesoscopy, Ural Division of Russian Academy of Sciences, Izhevsk, Udmurt Republic, Russia*

[2]*Department of Chemistry and Chemical Technology, Kalashnikov Izhevsk State Technical University, Studencheskaya St. 7, Izhevsk, Russia*

[3]*Institute of Mechanics, Ural Division, Russian Academy of Sciences, T. Baramzinoy 34, Izhevsk, Russia*

Corresponding author. E-mail: kodol@istu.ru

ABSTRACT

The definition and theoretical fundamentals of a new scientific trend—chemical mesoscopics—are considered in the example of a new class of nanostructures: metal/carbon nanocomposites (NCs). The theoretical and experimental methods for the processes direction prognosis as well as for the chemical systems reactivity estimation are discussed. The hypothesis of mesoscopic metal-containing cluster creation at mechanic–chemical formation of metal/carbon NCs is proposed. The electron structures of carbon shells for metal-containing clusters are determined. The metal nature influence of metallic phase cluster on the carbon shell is shown. The media electron structure changes are possible under the influence of even minute quantities of NCs, and it is confirmed by the X-ray photoelectron spectroscopic investigations and also is explained by the Chemical Mesoscopics principles. It is shown that the orientation processes, in which mesoscopic particles (NCs) participate, lead to the changes of sub-molecular structures of polymeric compositions.

1.1 INTRODUCTION

At present, we observe the creation of a new synthetic science on the basis of synergetic, fractal theory, chemistry within nanoreactors, and mesoscopic physics. These scientific trends consider the objects of nano-sized level (0.1–1000 nm) and investigate properties of mesoparticles. Their phenomena take place at definite conditions (size of phase coherence must be less than 1000 nm).

From the comparison of the notions and fundamentals of abovementioned scientific trends, it is possible to speak about the creation of a new scientific trend or new science—the chemical mesoscopics.

Thus, the definition of chemical mesoscopics can be given as follows. Chemical mesoscopics is a scientific trend (new science) concerning the formation of nanostructures including the synthesis within nanoreactors as well as the behavior of these nanostructures in different media and compositions. Theoretical fundamentals of this new science give the explanations for peculiarities of processes within nanoreactors and also the influence of nanostructures' minute quantities on media and compositions.

Some fundamentals of this scientific trend are considered in the example of metal/carbon nanocomposites (NCs), with a special focus on its application for the modification of polymeric compositions.

1.2 CONDITIONS FOR PROCESSES IN CHEMICAL MESOSCOPICS

The restrictions for processes defined by the limits of chemical mesoscopics are the following:

- The mesoscopic particle, which may be presented as a big molecule or a linked molecules group, is found in active interaction with medium.
- In this case, the size of phase coherency is located in limits up to 1000 nm.
- Then, the phenomena such as interference, spectrum quantization, and charge quantization appear.
- In other words, there is the source of quant radiation which activates the certain functional (active) groups in medium when nanosized particles and also confined sp ace take place.

The aforesaid restrictions are at the metal/carbon NCs obtaining. Let us consider the obtaining of copper/and nickel/carbon NCs from metal oxides and polyvinyl alcohol (PVA). At the grinding of reagents on the first stage, when the reagents relation is given as 1:4 (1 part—metal oxide and 4 parts—PVA), the decreasing of metal-containing phase sizes is compared by RedOx process. Metals are reduced with the change of system electro structure that is, confirmed by spectroscopic and diffracted investigations. During the second (thermal) stage, which is realized at the temperature increasing to 400°C (no more), the result is fixed.[1–4]

Chemical mesoscopics explain these processes by the nanoreactor formation, in which the metal oxide cluster formed is coordinated with hydroxyl groups of PVA and promotes to its separation as well as the hydrogen separation from polymeric chain. Then in this processes, the electron flow from nanoreactor walls across the cluster appeared that leads to the change of metal oxidation state. As the result of RedOx processes, the metal clusters are formed within shells which contain fragments of polyacetylene and carbine chains. Those shells actively interact with d metals and initiate d electrons transition on more high energy levels that promote the increase of metal magnetic characteristics.

The diffract grams of NCs obtained are given for the confirmation of metal phase formation (Figs. 1.1 and 1.2 and Tables 1.1 and 1.2).[5]

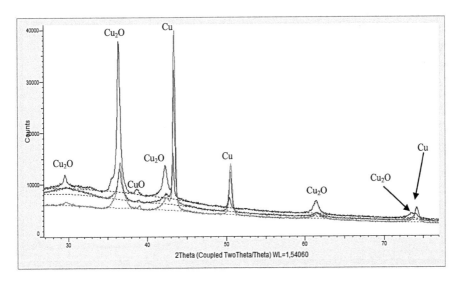

FIGURE 1.1 (See color insert.) Diffract grams of copper/carbon nanocomposites obtained from copper oxide within matrices for different marks of PVA: Cu/C NC (red)—PVA mark BF-14, NC (blue)—BF-17, and NC (dark)—BF-24.

TABLE 1.1 The Influence of PVA Marks on Process Completion.

Marks of PVA	The line color on diffract grams of copper/carbon nanocomposite	I_{1350}/I_{1580}
PVA mark BF-17	Blue	0.77
PVA mark BF-14	Red	0.82
PVA mark BF-24	Dark	0.72

Table 1.1 shows the dependence of process result from the nature of polymeric matrix or marks of PVA.

TABLE 1.2 The Influence of Process Temperature on the Quantity Metallic Nickel in Nanocomposite.

The curve color on diffract grams of nickel/carbon nanocomposite	I(NiO)/I(Ni)	The maximal temperature, °C
Dark	0.94	420
Red	0.62	400

Table 1.2 illustrates the influence of process temperature on the RedOx process quality (perfect). According to data of diffract grams (Fig. 1.2), the increase in temperature above 400°C leads to the decreasing of the metallic nickel at the obtaining of nickel/carbon NC. This fact can be explained by the disturbance of metal coordination with the nanoreactors walls.

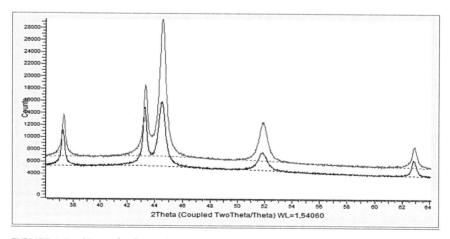

FIGURE 1.2 (See color insert.) Diffract grams of nickel/carbon nanocomposites obtained from nickel oxide within matrices. PVA at different temperature of thermals.

Chemical Mesoscopics: New Scientific Trend

The comparison of these diffract grams shows that the RedOx processes of obtaining of copper/carbon NCs proceed more perfect than analogous processes for nickel/carbon NC obtaining.

The comparison of Figures 1.1 and 1.2 shows that the copper oxide completely participates in reduction process (Fig. 1.1). At the same time, the nickel oxide is reduced only on 65% (at 400°C). This fact is confirmed by X-ray photoelectron spectroscopy investigations. In C1s spectra, there are lines correspond to the C1s energy for CH groups. According to C1s spectra for nickel/carbon NC, the content of CH groups is equal to 35%.

It is necessary that the absence of progressive and rotatory motions and the decreasing of vibration mode possibilities for the creation of processes realization conditions according to principles of chemical mesoscopics. When these conditions are secured, the transport of "quantized electrons" takes place across the nano (meso) particles, what in order to this process promotes to proceeding of RedOx processes. These conditions may be possible only at the correspondent sizes and forms of nanoreactors and also at the interactions of nanoreactors walls with clusters created within these nanoreactors.

1.3 ABOUT THE NANOREACTORS FORMATION IN MEDIA

1.3.1 WHAT ARE NANOREACTORS AND HOW ARE THE CORRESPONDENT PROCESSES REALIZED WITHIN NANOREACTORS?

What are nanoreactors? Let us give the definition of this notion.

Nanoreactors are defined as specific nanostructures and can be nanosized cavity in polymeric matrices or space bounded part, in which reactant chemical particles are orientated with the creation of transitional state before pre-determined product formation.

According to Buchachenko[6], the following classification of nanoreactors is known.

1.3.1.1 ONE-DIMENSIONAL NANOREACTORS

(1) *In the clearance between probe and surface;* (2) *Crystal canals;* (3) *Complexes;* (4) *Crystal solvates;* (5) *Macromolecules;* (6) *Micelles;* (7)*Vesicles;* and (8) *Pores.*

1.3.1.2 TWO-DIMENSIONAL NANOREACTORS

(1) *The double electrical layer;* (2) *Monomolecular layers on surface;* (3) *Membranes;* (4) *Interface layers (boundaries);* and (5) *Adsorption layers.*

1.3.1.3 CLUSTER NANOREACTORS—METAL CLUSTERS

Thus, the great variety of nanoreactors formed at different conditions takes place. Then, it is necessary to note that the formation of correspondent activated complex, which protects the process direction, is relieved within nanoreactors. The charge quantization into medium promotes the process evolution during the certain time interval. Also, it is noted that the formation of nanostructures within nanoreactors are determined by the nature of reactants, which participate in synthesis, and by the energetic and geometric characteristics of nanoreactors.

The analysis of modern scientific data and our experiments show that the following peculiarities of nanostructures formation within nanoreactors may be noted:

1. *The principal peculiarity—the decreasing of collateral parallel processes and the process direction to special product side.*
2. *The low energetic expenses and the high rates of processes.*
3. *The dependence of obtained nanostructures properties from the energetic and geometric characteristics of nanoreactors.*

What basic parameters and equations can be used for the nanostructures formation within nanoreactors and what matrices are possible?

First, the coordination number of element or elements group is necessary to take into account the definition of process direction. In this case, it is possible for the application of multiplate theory.

Second, the chemical potential difference is a basic factor for the self-organization of process to start.

Third, the growth and form of nanostructures can be determined by using Kolmogorov–Avrami equations.

As the bright example of the coordination number influence on nano-structures formation can be presented[7] by the Nan rotors formation based on zinc oxide systems $(ZnO)_m$ with the addition of indium oxide (In_2O_3). The zinc coordination number is equal to 4. Therefore, the ribbon formation can be equated and explained. The indium (In) coordination number is equal to 6. Then, the possibility of six rabbets contained at rotor formation takes place when In is included in the chain of ZnO system (Fig. 1.3).

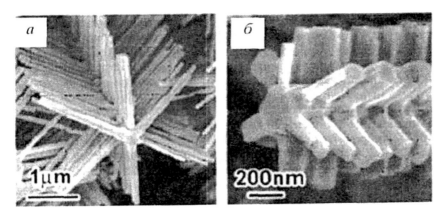

FIGURE 1.3 Nan rotor with six rabbets on base of ZnO and additive In_2O_3.

However, let us return to metal/carbon NCs in which metals can be copper or nickel. The coordination number of copper may be 2 or 4 in the dependence of metal oxidation state (for +1—coordination number is 2 and for +2—coordination number is 4). At the same time, the nickel coordination number may be 4 or 6 depending on ligand nature. The formation of carbon shell from PVA chains starts off coordination process of four chain fragments with the metal-containing cluster. When the dehydration and dehydrogenization proceed, the connected fragments of polyacetylene and carbine appear. The fragments formed are coordinated on the metal. The electron structure of carbon fibers formed is investigated by X-ray photoelectron spectroscopy.

According to C1s spectra, three types of satellites (sp, sp^2, and sp^3 hybridization) with different relation between them are found. These satellites also take place in C1s spectrum. The intensities relation as I_{sp}^2/I_{sp}^3 for Cu/C NC corresponds to 1 and 7 and this parameter for Ni/C NC is equal to 1. It is known[8] that the satellites intensities relation also equals 1. According to PEM data the nickel/carbon NC contains tubular nanostructures. The development of the nanostructures self-organization process is estimated by means of the equation based on Kolmogorov–Avrami equation.[9] In this equation, the RedOx potentials are used because of the metal reduction that occurs in process of metal/carbon nanocomposite formation.

$$W = 1 - k \exp\left[-\tau^n \exp\left(zF\Delta\varphi/RT\right)\right], \qquad (1.1)$$

where k is the proportionality coefficient considering the temperature factor, τ is duration of process, n is the fractal dimension, z is the number of electrons participating in the process, $\Delta\varphi$ is the difference of potentials on the border "nanoreactor wall–reaction mixture," F is the Faraday number, R is the universal gas constant, and T is the temperature. The calculations are made when the process duration changing takes with a half-hour interval. It was accepted for calculations that n equals 2 (two-dimensional growth), potential of Redox process during the metal reduction equals 0.34 V, temperature equals 473 K, Faraday number corresponds to 26.81 Ah/mol, R equals 2.31 Watt·hour/mol·degree. Calculation results are presented in Table 1.3.

TABLE 1.3 The Calculations Results of Process Duration for Copper/Carbon Nanocomposite Obtaining.

Duration (h)	0.5	1.0	1.5	2.0	2.5
Content of product (%)	22.5	63.8	89.4	98.3	99.8

The calculation results practically coincide with experimental data.

The coordination processes lead to the changes of metal electron structure with unpaired electrons formation. That is accompanied by the metal atomic magnetic moment increasing, and also the appearance of unpaired electrons on the carbon shell surface (Table 1.4).

TABLE 1.4 Experimental EPR Data and Atomic Magnetic Moments (μ_B) for Copper and Nickel/Carbon Nanocomposites.

Type of metal/carbon nanocomposite	g-factor	Number of unpaired electrons, spin/g	Atomic magnetic moment, μ_B metal/carbon nanocomposite/massive sample
Copper/carbon nanocomposite	2.0036	1.2×10^{14}	1.3/–
Nickel/carbon nanocomposite	2.0036	10^{22}	1.8/0.5

Source: Adapted from Ref. [10].

Thus, the metal/carbon NC, obtained within PVA nanoreactors from metal oxides, represent active nanoparticles, which can be source of electron quants for media and compositions. According to chemical mesoscopics, minute quantities of these nanoparticles (with size less 100 nm) in essence can improve the materials properties.

1.4 THE INFLUENCE OF NANOSTRUCTURES ON MEDIA AND COMPOSITIONS

When the additive sizes are decreased to nanometer sizes, the phenomena of mesoscopic particles appear. According to Imri[11] and Mosklets,[12] the phenomena such as interference, spectrum quantization, and charge quantization occur when the mesoparticles have the limitation in motions or in the energetic possibilities realization. In this case, the mesoparticles can only vibrate, and also the electron transport is possible across them.

Polymeric compositions properties change under the mesoparticles influence can be achieved at equal distribution of these particles in composition volume and at its coagulation absence. Last is possible at the following conditions:

- certain polarity and dielectric constant of medium;
- minute concentration of mesoparticles;
- ultrasound action on the correspondent suspension for the proportional distribution of mesoparticles.

The assignment of active nanostructures (mesoparticles) during the compositions modification is concluded in the activation of matrices self-organization in needful direction. For the realization of this goal, the determination of organized phase part is necessary. The Johnson–Mehl–Avrami–Kolmogorov equation is applied for the organized phase determination.

$$W = 1 - \exp(-k\tau^n), \qquad (1.2)$$

where W is the part of organized phase, k is the parameter defined the rate of organized phase growth, τ is the duration of organized phase growth, and n is the fractal dimension.

The consideration of metal/carbon NC reactivity is interesting in processes of interaction with different substances and media.

The presence of double bonds and unpaired electrons promotes the interactions of phosphorus, silicon, and sulfur substances with NC. As the result of this interaction, the reduction of correspondent elements takes place accompanied by the growth of metal atomic magnetic moments for NC and the increasing of their polarization action on media. For instance, after the silicon introduction into copper/carbon nanocomposite its properties are changed:

- Copper atomic magnetic moment grows to 3 μ_B.
- Antioxidant activity increases in 11 times.
- Dynamic viscosity of liquid glass, which contains 0.001% of Cu/C nanocomposite with silicon is increased by 21%.

These experimental results show the possibility of NC interaction with active media and substances. The illustrations of that possibility realization can be the investigations of NC fine dispersed suspensions.

The quantum charge wave expansion leads to the functional group's polarization (dipole moments) change as well as the extinction growth. It determines the increasing of peaks intensities in IR spectra. The individual peaks growth effects in IR spectra are observed at the introduction of NC minute quantities.

According to Kodolov and Trineeva[13] and Kodolov et al.,[14,15] the positive results on materials properties improvement are presented when the minute quantities of metal/carbon NC are introduced in these materials. According to Khokhriakov et al.,[17] the hypothesis about nanostructures influence transmission on macromolecules of polymeric matrices is proposed. This hypothesis complies with chemical mesoscopics principles which consider quantum effects at the certain conditions of nano meso particle existence. In this case, the peak intensity growth in IR spectra is observed when the quantity of introduced nanocomposite is decreased. This fact complies with fundamental principle, when there is excess of vibration above rotation motion. The nanocomposite action efficacy estimation in finely dispersed suspension on medium is carried out by means of the relation comparison of peak intensity to its half width ($I/a_{0.5}$).

The relation ($I/a_{0.5}$) reflects according to Herzberg–Kondrat'ev equation excess of vibration above rotation motion.

$$v = \omega + B\,[J'(J'+1) - J(J+1)], \qquad (1.3)$$

where v is the vibration frequency of correspondent group, ω is the value of vibration motion lot, B is the value of rotation motion lot, and J' and J are the rotation quantum numbers. If this relation is increased, then the rotation motion lot will be less in system.

The second significant factor for active interaction characteristics into media appears as the change of special model films electron structure. These films are obtained on the base of linear functional polymers such as PVA, polymethyl methacrylate (PMMA), and polycarbonate (PC). The polarity of polymers increases in row from PVA to PC. The "stamp" of NC is characterized by sp, sp^2, and sp^3 hybridization satellites and it appears in C1s spectra of polymeric films at concentration intervals of NC in the following order: for PVA—at 10^{-3}% NC; for PMMA—interval 10^{-2} to 10^{-4}% NC; and for PC—interval 10^{-2} to 10^{-5}% NC. Besides, it is noted[8] that the changes for relations I_{sp}^2/I_{sp}^3 take place. In these experiments, the illustration of bond between electron structure and sub-molecular structure of polymeric substances is presented. The analysis of adduced examples gives the basis for hypothesis about the propagation electron quants as direct flow or diffuse radiation in the dependence on the NC concentration and on the nature of NC, and also medium nature (magnetic and electric characteristics).

It is interesting to observe the direction of carbon fibers in comparison with the direction of sub-molecular structures orientation in the nanostructured polycarbonate surface layers.

Thus, it is possible that the wave which initiates self-organization process in polymeric composition is expensed from these fibers associated with metal cluster, and then stimulates the correspondent orientation of sub-molecular structures in nanostructured composite surface layers.

The self-organization mechanism for polymeric compositions modified by the metal/carbon NC minute quantities is concluded in the conditions created for composition polarization, which brings the great change of electron and sub-molecular structures of materials. Certainly, these structures change influence on the modified materials properties. The sufficient quantity of experimental works carried out on the modification of different materials with the using of metal/carbon NC. These experiments confirm the improvement of material properties on account of their modification by the minute (less 0.01%) quantities of metal/carbon NC.

The examples of results for modification with using of above-considered suspensions or NC concentrates will be presented below.

For example, it is noted the decreasing of thermal conductivity in 1.5 times for polycarbonate, modified by 0.01% of copper/carbon NC as well as the increasing the light radiation transmission in the field 400–500 nm at the decreasing in the field 560–760 nm.

In the following example, it is shown that polyvinyl chloride film, modified by $8 \cdot 10^{-4}$%, has electrostatic properties.

In other instance, the increasing of adhesive durability more than 50% is defined for glues nanostructured by $1-3 \cdot 10^{-4}$% of copper or nickel/carbon NC.

For the current conductive glues and pastes, the increase of adhesion durability and the decreasing of electric resistance are observed (Table 1.5).[13–16]

TABLE 1.5 The Comparison of Electric Resistance for Modified Glues and Pastes and Their Modified Analogous.

Parameters	The current conductive paste	The current conductive glue
Specific volume resistance for non-modified sample, Ohm cm	2.4×10^{-4}	3.6×10^{-4}
Specific volume resistance for modified sample, Ohm cm	2.2×10^{-5}	3.3×10^{-5}

The introduction of metal/carbon NCs in quantity equaled 0.005% in epoxy resins leads to the increase of thermostability on 75–100°C.

For reinforced glass plastics, the introduction of 0.02% copper/carbon NC leads the increase of durability on 32.2%.

There are many examples of the materials properties improvement owing to the use of the metal/carbon NC minute (or super small) quantity.

1.5 PERSPECTIVES OF CHEMICAL MESOSCOPICS DEVELOPMENT

Some fundamental points of new scientific trend development appear the creation of mathematical prognostic program, in which there is place for the theory of formation and application of nanoreactors which promote to processes proceeding of useful nanostructures or nanosystems obtaining. In this theory, it is necessary to determine the correspondent conditions for the formation of nanoreactors with the same geometric and energetic characteristics.

Certainly, theoretical apparatus for kinetics of processes within nanoreactor requires development, as well as the determination of peculiarities in the dependence on nanoreactor type. The separate and very important problem is the mechanism of the quant radiation formation. In other words, what conditions are necessary for the creation of the charge (electron) quantization, and also the interference of quant waves with the formation of new oscillators which will promote the wave propagation into media.

With the development of different theories, the correspondent trends in technological fields take place, especially technologies of nanostructures (nanosystems) producing as well as the manufactory of nanostructured materials, obtained owing to the modification of compositions by nanoproducts including metal/carbon NC.

The proposed technologies on the basis of chemical mechanics, electrochemistry, chemistry on surface, or processes on the phase boundaries are needed in the development of theoretical apparatus for results prognosis. At the same time, it is necessary to develop experimental methods of processes control, and also the estimation of results obtained during the processes of nanoproducts obtaining.

The development of this new scientific trend will be more at universal studies of nanostructures (nanosystems) properties such as catalytic and stimulator of processes, synergetic, quantum nanogenerator, etc. Simultaneously, it is necessary to increase the quantity of various (on content, structures, and properties) nanoproducts including finely dispersed suspensions and nanostructures concentrators.

1.6 CONCLUSION

The unification of four scientific trends such as synergetic, fractal theory, chemistry within nanoreactors, and mesoscopic physics, is proposed on the base of its comparison on investigated objects and observed phenomena. In this case, the new science could be named as chemical mesoscopics. Some peculiarities and regularities of this new scientific trend on the instance of creation, characteristics investigations, and application of nanostructure new class such as metal/carbon NCs are discussed. On the basis of discussion, it may be concluded that chemical mesoscopics consider the nano- or meso-particles reactivity, and also the processes with these particles participation.

The development of chemical mesoscopics supposes the theory of creation for mechanisms of nanosystems formation as well as mechanisms of synergetic and self-organization in media and compositions.

Certainly, the new scientific trend will help the creation of new methods for the processes investigations and also for studies of nanostructures properties.

The extensive applications of new science results will be expected in practice.

KEYWORDS

- **chemical mesoscopics**
- **self-organization**
- **RedOx synthesis**
- **charge quantization**
- **nanostructures**

REFERENCES

1. Kodolov, V. I.; Didik, A. A.; Volkov, A. Yu.; Volkova, E. G. Pat. RF N 2221744, 2004.
2. Kodolov, V. I; Didik, A. A.; Shayakhmetova, Ye. Sh.; Kuznetsov, A. P.; Volkov, A. Yu.; Volkova, E. G. Pat. RF N 2223218, 2004.
3. Kodolov, V. I.; Nikolaeva, O. A.; Zakharova, G. S.; Shayakhmetova, Ye. Sh.; Volkova, E. G.; Volkov, A. Yu.; Makarova, L. G. Pat. RF N 2225835, 2004.
4. Kodolov, V. I.; Trineeva, V. V.; Blagodatskikh, I. I.; Vasil'chenko Yu, M.; Vakhrushina, M. A.; Bondar, A. Yu. The Nanostructures Obtaining and the Synthesis of Metal/Carbon Nanocomposites in Nanoreactors. In *Nanostructure, Nanosystems, and Nanostructured Materials. Theory, Production and Development*; Apple Academic Press: Toronto/Point Pleasant, NJ, 2013; pp 558, 101–145.
5. Trineeva, V. V. The Technology of Metal/Carbon Nanocomposites Production and Its Application for the Polymeric Materials Modification. Thesis on DSc– Kazan, KNRTU, 2015, p 300.
6. Buchachenko, A. L. Chemistry on the Border Centuries – Achievement and Prospects. *Russ. Chem. Rev.* **1999,** *68* (2), 85–102.
7. Zakharova, G. S.;Volkov, V. L.; Ivanovskaya, V. V.; Ivanovskii, A. L. *Nanotubes and Related Nanostructures of Metal Oxides;* UB RAS: Ekaterinburg, Russia, 2005; p 241.
8. Shabanova, I. N.; Kodolov, V. I.; Terebova, N. S.; Trineeva, V. V. *X-ray Photoelectron Spectroscopy in the Investigation Metal/Carbon Nanosystems and Nanostructured Materials;* M. Izhevsk: Publ. Udmurt University: Izhevsk, Russia, 2012; p 252.
9. Kodolov, V. I.; Trineeva, V. V.; Vasil'chenko, Yu. M. The Calculating Experiments for Metal/Carbon Nanocomposites Synthesis in Polymeric Matrices with the Application of Avrami Equations. In *Nanostructures, Nanomaterials and Nanotechnologies to Nanoindustry;* Apple Academic Press: Toronto, NJ, 2015; pp 105–118.
10. Lipanov, A. M.; Kodolov, V. I.; Mel'nikov, Ya. M.; Trineeva, V. V.; Pergushov, V. I. *The Influence of Minute Quantities of Metal/Carbon Nanocomposites on the Polymeric Materials Properties;* DAN (Doklady Akademy Nauk): Russia, 2016; Vol. 466, pp 15–17.
11. Imri, Yi. *Introduction in Mesoscopic Physics;* Fizmatlit: Moscow, 2004; p 301.
12. Mosklets, M. V. *Fundamentals of Mesoscopic Physics;* NTU KhPI: Khar'kov, Ukraine, 2010; p 180.
13. Kodolov, V. I.; Trineeva, V. V. *How Mesoscopic Physics Explains RedOx Synthesis of Metal/carbon Nanocomposites within Nanoreactors of Functional Polymers.* Chemical Physics & Mesoscopy; 2015; Vol. 17, No. 4, pp 580–587.
14. Kodolov, V. I.; Trineeva, V. V.; Khokhriakov, N. V. Synthesis and Application of Metal/ Carbon and Metal/Polymeric Nanocomposites: Theory, Experiment and Production. In *The Problems of Nanochemistry for the Creation of New Materials;* EPMD: Torun, Poland, 2012; Vol. 250, pp 7–15.
15. Kodolov, V. I.; Khokhriakov, N. V.; Trineeva, V. V.; Blagodatskikh, I. I. Problems of Nanostructures Activity Estimation, Nanostructures Directed Production and Application. In *Nanomaterials Yearbook – 2009. From Nanostructures, Nanomaterials and Nanotechnologies to Nanoindustry;* Nova Science Publishers Inc.: Hauppauge, NY, 2009; Vol. 383, pp 1–17.
16. Kodolov, V. I.; Trineeva, V. V. Fundamental Definitions for Domain Nanostructures and Metal/carbon Nanocomposites. In *Nanostructure, Nanosystems, and Nanostructured*

Materials. Theory, Production and Development; Apple Academic Press: Toronto, NJ, 2013; Vol. 558, pp 1–42.

17. Khokhriakov, N. V.; Kodolov, V. I.; Korablev, G. A.; Trineeva, V. V.; Zaikov, G. E. Prognostic Investigations of Metal or Carbon Nanocomposites and Nanostructures Synthesis Processes Characterization. In *Nanostructure, Nanosystems, and Nanostructured Materials. Theory, Production and Development;* Apple Academic Press: Toronto, NJ, 2013; Vol. 558, pp 43–99.

CHAPTER 2

PROGRESS ON DEVELOPMENT OF NANOTECHNOLOGY

M. R. MOSKALENKO[*]

History of Science and Technology Department, Institute of Humanities and Arts, Ural Federal University, Ekaterinburg, Russia

[*]E-mail: max.rus.76@mail.ru

ABSTRACT

This chapter discusses the scientific and technological forecasting. Characteristics are studied for scientific and technological forecasts applicable to the development of nanotechnology.

2.1 INTRODUCTION

Nowadays, the intermittent development of the sphere of scientific researches, designing, and engineering on an atomic scale, which is called nanotechnology, can be observed. Of course, tempestuous development of nanotechnology gives life to the most incredible and bold forecasts of its usage in everyday life and industry, which are stirred up by sensation-seeking journalists. That is why it is necessary to know the general specifications of scientific-technical forecasts in order to estimate all the perspectives of the development of nanotechnology.

Usually, these standard methods are used in scientific-technical forecasts: analysis of trends and tendencies of development; extrapolation of existing tendencies; elicitation of the processes' cyclicality; studying of principles and regularities of the process or phenomenon; and making of scenarios. Sometimes science fiction is related to a specific type of forecasting where the method of intuitional foreseeing and artistic imagination of artists are used. Several issues should be considered when we speak about scientific-technical forecasting in general and especially when we speak about forecasting of nanotechnology.

2.2 HOW DO SOCIAL CONDITIONS OF ANY SPECIFIC SOCIETY CONTRIBUTE TO INNOVATIONS?

In other words, how do social conditions of any specific society contribute to the development of high-tech economics and integration of knowledge? For example, well-developed system of education and the cult of scientific knowledge in the late *Union of Soviet Socialist Republics* (USSR) contributed to the creation of many original development projects and inventions.

However, the economic system was not able to realize them; no required mechanisms and institutions were made to provide the integration. Also, the phenomenon of "resource curse" can also take place—when the excessive amount of natural resources and corresponding export profits turn all other branches of economics (including high-tech ones) to be secondary and subordinated to the import.[1-4]

2.3 ABOUT VALUE OF NEW BREAKTHROUGH TECHNOLOGIES AND INVENTIONS

It can be quite difficult to value new breakthrough technologies and inventions. For example, airplane appeared later then dirigibles did, despite the fact that before the First World War dirigibles had been predicted to be the main striking air force; the attitude toward first machine-guns was skeptical among soldiers; first cars and vacuum cleaners often caused nothing but skeptical smile, etc.

There could be some engineering designs that seem to be quite realistic and perspective, but then they appeared being hardly possible to build in practice: nuclear aircrafts in 1950s; enormous cargo and passenger seaplanes (some development projects could weigh up to 1000 tons); hovercraft tanks; etc.

Tempestuous development of any technology can crucially change industry and everyday life within a short period of time (automotive industry in 1900s, personal computers in 1980–1990s, and cellular networks in 1990 to early 2000s, etc.).

2.4 NONLINEAR DEVELOPMENT OF TECHNOLOGY AND TECHNIQUES

The development of technology and technique is often nonlinear: tempestuous, intermittent development can slow down and become more fluent (classic example—the history of tanks and aviation).

Tempestuous development of any sphere of science and technology always goes with excessively optimistic forecasts, like how this thing will change the face of humanity; then comes a period of more deliberated valuation and normally forecasts become more restrained.

2.5 CONCLUSION

These peculiarities of scientific-technical forecasting should be considered while forecasting and planning the development of nanotechnology in various branches of industry.

KEYWORDS

- **scientific forecasting**
- **technological forecasting**
- **nanotechnology**
- **nonlinear**
- **development of technology**
- **designing**
- **social conditions**

REFERENCES

1. Canton J. The Strategic Impact of Nanotechnology on the Future of Business and Economics. http://www.globalfuturist.com/dr-james-canton/insights-and-future-forecasts/stratigic-impact-of-nanotechnology-onbusiness-and-economics.html.
2. Nanotechnology: A Technology Forecast. http://www.tfi.com/pressroom/pr/nano.html.
3. Toffler, A. *Future Shock;* Random House: New York, NY, 1970; p 505.
4. Zigunenko, S. N. *Hundred-of Great Records for Military Technique;* Veche: Anenii Noi, Moldova, 2012; p 432.

CHAPTER 3

DEVELOPMENT OF THE PROGNOSTIC APPARATUS IN THE STUDY OF NANOBIOBODIES

E. M. BASARYGINA[1*], T. A. PUTILOVA[1], and M. V. BARASHKOV[2]

[1]*Department of Mathematical and Natural Sciences Disciplines, Chelyabinsk State Academy of Agricultural Engineering, Lenin Prospect 75, Chelyabinsk, Russia*

[2]*LLC, Agrocomplex "Churilovo," Chelyabinsk, Russia*

Corresponding author. E-mail: b_e_m@mail.ru

ABSTRACT

In order to develop a prognostic device, applied for nano- and bioobjector study, the technique of rapid assessment of photosynthetic pigments, which can detect stress conditions in plants, is considered. In accordance with the established methodology, a rapid assessment of the photosynthetic apparatus pigment systems requires consistent implementation of the following stages: Registration of the absorption spectrum of the leaves (with the help of a photometric measurement equipment); definition of indicators pigment systems; and comparison of the values. The proposed express assessment is designed for production conditions of greenhouse complexes, does not require sample preparation, precedes physical and chemical analysis and complements the tissue diagnostics.

The developed technique allowed us to determine recommended values of optical density for pigment systems of the photosynthetic apparatus, which are characteristic for actively growing and developing cucumber plants having high yield. Timely detection of stressful conditions and carrying out compensatory measures help to prevent crop losses, ensure environmental cleanliness and product quality.

3.1 INTRODUCTION

The development of the prognostic apparatus in the study of the pigments of chloroplasts belonging to the nanobiobacteria is of great practical importance, since it allows us to identify the stress conditions of plants. Timely diagnostics of stress helps to select technological regimes that promote the removal of plants from this state and reduce its negative consequences, that is, decrease in productivity.[1–3]

Many biological materials, including photosynthesis pigments (chlorophylls), are classified as nanomaterials (Table 3.1).[4–7] The main purpose of photosynthetic pigments is the absorption of light energy and its conversion into chemical energy.[8–10]

TABLE 3.1 Typical Sizes of Biological Objects in the Nanometer Range.

Type	Substance	Size, nm
Amino acid	Glycine	0.42
	Tryptophan	0.67
Nucleotide	Cytosine	0.81
	Guanine phosphate	0.86
	Adenosine triphosphate	0.95
Acid	Stearic acid $C_{17}H_{35}CO_2H$	0.87
Molecules	Chlorophyll plants	1.1
Protein	Insulin	2.2
	Hemoglobin	7
	Elastin	5
	Fibrinogen	50
	Lipoprotein	20
Virus	Influenza virus	60
	Tobacco mosaic virus (length)	120

Source: Adapted from Ref. [4].

Photosynthesis occupies a central place in the energy of the cell, as it serves as the primary source of all energy used by living organisms. The process of photosynthesis, which is inseparably linked with the reactions of energy and plastic metabolism, forms the basis of the metabolism of plant cells.[8–10]

At the level of the whole plant, photosynthesis acts as the main donor of metabolites for such processes as respiration, absorption, and assimilation of mineral substances, growth, and development.[8–10]

The process of photosynthesis, located in the center of donor–acceptor bonds, plays a key role in shaping the overall and economic productivity of plants. Photosynthesis forms the basis of the primary natural productivity of ecosystems and determines the formation of crops of agricultural plants.[8–10] By means of photosynthesis, about 95% of the organic mass of the crop is created and all the energy accumulated in the plant organism is accumulated.[1–3]

To carry out phytomonitoring and adjust the nutrition of plants under protected soil conditions, the deficiency of nutrient elements is diagnosed in soil (substrate) and plant (Fig. 3.1).[1–3]

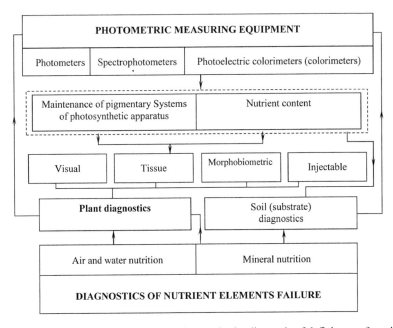

FIGURE 3.1 Photometric measuring equipment in the diagnosis of deficiency of nutrients.

3.2 SOIL (SUBSTRATE), DIAGNOSTICS IN ENVIRONMENT, AND PLANT DIAGNOSTICS

Some types of plant diagnostics such as injectable, morpho-biometric, visual, and tissue are well known. Tissue and substrate diagnostics are carried out by physicochemical analysis, the results of which are established with established norms or recommended values.[1-3] The physicochemical analysis of the root environment and plants is based on the use of photometric measuring equipment such as photometers, photoelectrocolorimeters, spectrophotometers, etc.[11-12] Photometric equipment (particularly, spectrophotometers) can also be used to register the absorption spectra of leaves for the subsequent evaluation of the state of the pigment systems of the photosynthetic apparatus.[11-12]

To assess the pigment systems of the photosynthetic apparatus in FGBOU VO South Ural GAU, a special methodology was developed, approbation of which was carried out jointly with LLC "Agrocomplex" Churilovo (Chelyabinsk). The proposed express assessment is designed for production conditions of greenhouse complexes, does not require sample preparation, precedes physical and chemical analysis, and complements the tissue diagnostics.

At present, any research in the field of photosynthesis, in particular, the study of the influence of growth conditions, nutrition, and light intensity on the development and photosynthetic activity of plants requires more or less detailed characterization of pigment systems. The characteristics of the pigment systems of the plant leaf apparatus include a quantitative evaluation:[10]

- The content of chlorophyll a, chlorophyll b;
- The amount $(a + b)$;
- Relationships (a/b);
- Content of carotene kp and xanthophylls kc;
- The amount $(kc + kp)$;
- Relationships (kc/kp);
- The ratio of green pigments to yellow $(a + b)/(kc/kp)$.

To quantify the pigment content, a spectrophotometric method is used, which is based on the characteristic property of plant pigments—the ability to absorb rays of light of a certain wavelength. An important condition for the quantitative determination of pigments is complete extraction from plant tissue.[10-12] Calculations related to the determination of the concentration of the substance in solution use the molar absorption coefficient (ε) or the specific absorption coefficient (d) of the substance and take into account the nature of the solvent (Tables 3.2 and 3.3).[10]

Development of the Prognostic Apparatus in the Study 27

TABLE 3.2 Absorption Maxima and Molar Absorption Coefficients for Chlorophyll in Various Solvents.

Solvent	λ Cope, nm	λ, nm	$\varepsilon \cdot 10^{-4}$, L mol^{-1} cm^{-1}
Methanol	432.0	665.7	6.66
Ethanol	431.5	664.7	6.94
Ethyl ether	428.8	660.6	8.51
Acetone	430.1	662.0	7.66
Isopropanol	431.6	664.2	7.44
Benzene	432.8	665.6	7.82

Source: Adapted from Ref. [10].

TABLE 3.3 The Specific Absorption Coefficients of Chlorophylls *a* and *b* in Various Solvents.

Solvent	Wavelength, λ, nm	Specific absorption coefficients, d, L g^{-1} cm^{-1}	
		Chlorophyll *a*	Chlorophyll *b*
Ethyl ether	660	102.0	4.5
	642.5	16.3	57.6
Acetone (water 80%)	663	82.04	9.27
	645	16.75	456
Methanol	665.2	70.5	–
	6524	–	3527

Source: Adapted from Ref. [10].

In accordance with the developed method, rapid assessment of pigment systems of the photosynthetic apparatus assumes the following stages:

- registration of the absorption spectrum of leaves (using photometric measuring equipment);
- determination of indicators of pigment systems;
- comparison of the obtained values of the indicators with the recommended values, which are established as a result of observations and depend on the culture, variety, and vegetation period.

3.3 METHODS FOR ESTIMATION AND PROGNOSIS OF NANOBIOBODIES DEVELOPMENT

The main indicator of pigment systems is the optical density, which is calculated by the expression:[13]

$$A = log\frac{I_0}{I},\tag{3.1}$$

where I is the intensity of incident light; I_0 is the transmitted light intensity.

The characteristics of the pigment systems included an estimate of the uptake of chlorophylls a and b (A_a and A_b), their sums ($A_a + A_b$), ratios (A_a/A_b); carotene kp (A_{kp}), xanthophylls kc (A_{kc}), the sum of carotenoids $kc + kp$ ($A_{kc} + A_{kp}$), their ratios kc/kp (A_{kc}/A_{kp}); the ratio of green pigments to yellow (($A_a + A_b$)/($A_{kc} + A_{kp}$)). The main absorption maxima of the pigments of the photosynthetic apparatus were taken in accordance with publications.[10–12]

The use of the obtained values of the indices makes it possible, according to the recommendations of Nobel,[13] to approximately estimate the pigment content on the basis of the Lambert–Beer law:[10,13]

$$A = \varepsilon cx,\tag{3.2}$$

where ε is the molar absorption coefficient; c is the concentration; and x is the layer thickness.

Approbation of the developed methodology was carried out in LLC Agrocomplex "Churilovo." Experimental studies were conducted with cucumber plants ("Meva" variety) grown by the method of light culture on a mineral-like substrate. The absorption spectrum of the leaves (lower tier) was taken with a UV 1800 Shimadzu spectrophotometer. The sample size was 20 plants. The processing of experimental data was carried out according to standard methods of biological (variational) statistics.[14,15] The results of the experimental studies are presented in Table 3.4.

The developed technique allowed to determine the recommended values of the optical density for the pigment systems of the photosynthetic apparatus: $a/b = 2.81 \pm 0.02$, $kc/kp = 3.85 \pm 0.14$; and $(a + b)/(kc + kp) = 5.47 \pm 0.03$, which are characteristic for actively growing and developing cucumber plants with high yield.

Deviations from the recommended values were observed in plants in a stressful state, manifested a slowdown in growth and development and showed a decrease in productivity. Timely detection of stressful conditions and implementation of compensatory measures help to prevent crop losses, ensure environmental cleanliness and product quality.

Table 3.4 shows the approximate content of pigment systems in cucumber leaves, calculated using the proposed indicators.

Development of the Prognostic Apparatus in the Study 29

TABLE 3.4 Content of Pigment Systems in Cucumber Leaves (%Pigment Content, Wet Mass).

Chlorophylls		Carotenoids	
a	b	kc	kp
0.17	0.06	0.03	0.01

3.4 CONCLUSION

Analysis of the results obtained shows that they agree well with the data of Taylor et al.[10] and Nobel[13] which confirms the possibility of using the proposed indicators in calculations related to the content of pigment systems of the photosynthetic apparatus. The developed technique contributes to the development of the prognostic apparatus used in the study of nanobiobodies.

KEYWORDS

- **prognostic apparatus**
- **nanobioobjects**
- **photosynthesis pigments**
- **photometric equipment**
- **plant diagnostics**
- **absorption spectrum**
- **optical density**

REFERENCES

1. Autko, A. *Vegetable Growing of Protected Soil*; Publishing House VVERER: St Paul, MN, 2006; p 320.
2. Belogubova, E. N; Vasiliev, A. M; Gil, L. S. *Modern Vegetable Growing of Open and Closed Ground*; PE Ruta: Zhitomir, Ukraine, 2007; p 532.
3. Vasko, V. T. *Theoretical Bases of Plant Growing*; PROFI-INFORM: Saint Petersburg, Russia, 2004; p 200.
4. Fedorenko, V. F. *Nanotechnologies and Nanomaterials in the Agro-industrial Complex*; Scientific. Analyte Overview. FGNU Rosinformagrotekh: Moscow, 2007; p 96.
5. Roco, M. K.; William, R. S.; Alivistaus, P. *Nanotechnology in the Next Decade. Forecast Direction of Development (Trans. in English);* Mir: Moscow, 2002; p 292.

6. Poole, C.; Owens, F. *World of Materials and Technologies. Nanotechnology* (Trans. in English); The Technosphere: Bengaluru, India, 2005; p 336.
7. Andrievsky, R. A.; Ragulya, R. A. *Nanostructured Materials*; Publishing Center Academy: Moscow, 2005; p 192.
8. Medvedev, S. S. *Plant Physiology*; BHV-Petersburg: Saint Petersburg, Russia, 2013; p 512.
9. Taylor, D.; Green, N.; Stout, U. Biology. In *3 Tons of Transl*; Soper, R., Ed.; Mir: Moscow, 2001–2002; p 1341.
10. Gavrilenko, V. F.; Zhigalova, T. V. *A Large Workshop on Photosynthesis*; Publishing Center Academy: Moscow, 2003; p 256.
11. Schmidt, V. *Optical Spectroscopy for Chemists and Biologists. (Trans. in English);* Ivanovskaya, N. P., Savilov, S. V., Eds.; Technosphere: Mallathahalli, India, 2007; p 368.
12. *Methods of Spectrophotometry in UV and Visible Regions for Inorganic Analysis;* Marchenko, Z., Balcerzak, M., Eds.; (Trans. in Polish). BINOM Laboratory of Knowledge: Moscow, 2009; p 711.
13. Nobel, P. *Physiology of the Plant Cell (Physicochemical Approach)/Per. (Trans in English)*; Rapanowicz, I. I., Prof. Gunara, I. I., Eds.; Mir: Moscow, 1973; p 288.
14. Rokitsky, P. F. *Biological Statistics;* VSh: Moscow, 1973; p 320.
15. Armor, B. A. *Methodology of Field Experience (with the Basics of Statistical Processing of Research Results);* ID Alliance: Moscow, 2011; p 352.

CHAPTER 4

NON-STATIONARY THERMODYNAMIC PROCESSES IN ROCKET ENGINES OF SOLID FUELS: CHEMICAL EQUILIBRIUM OF COMBUSTION PRODUCTS

A. V. ALIEV[*], O. A. VOEVODINA[*], and E. S. PUSHINA[*]

Federal State Budgetary Educational Institution of Higher Education, Kalashnikov Izhevsk State Technical University, Studencheskaya St. 7, Izhevsk 426069, Russia

[*]Corresponding author. E-mail: aliev@istu.ru; mien@istu.ru; meh_mod@istu.ru

ABSTRACT

The method of calculation of non-stationary intra chamber processes in the rocket solid propellant engines based on two approaches is considered. The first approach assumes consideration of products of combustion as mechanical mix, other approach—as the mix which is in chemical equilibrium. For an increase of reliability of the solution of intra ballistic tasks in which the assumption of chemical equilibrium of products of combustion is used, computing algorithms of calculation of structure of products of combustion are changed. In algorithm of the decision system of the nonlinear equations of chemical equilibrium when determining iterative amendments instead of a method of Gauss, the orthogonal QR-method is used. Besides, possibility of application in a task about structure of products of combustion of genetic algorithms is considered.

It is shown that in the tasks connected with forecasting of non-stationary intra ballistic characteristics in rocket solid propellant engines, application of models of mechanical mix and chemically equilibrium structure of products of combustion leads to qualitatively and quantitatively coinciding results. The maximum difference of parameters does not exceed 5–10%. In tasks about an exit to the mode of mid-flight rocket engines of solid propellant with high-temperature products of combustion difference in results is more essential, and can reach 20% and more.

4.1 INTRODUCTION

In modern techniques for calculating the problems of internal ballistics of solid propellant rocket engines, the main assumption adopted in the construction of mathematical models of processes is that combustion products in the combustion chamber are assumed to be chemically inert with a mechanical mixture (frozen mixture) of the gas that originally filled the chamber (e.g., air), the combustion products of the ignition charge and the products of solid fuel combustion.[1,2] Nevertheless, one cannot exclude from consideration kinetic effects due to the chemical interaction of individual components of combustion products, the number of which can be more than a hundred.[3–5] However, the consideration of kinetic effects in solving the design problems of solid propellant is very laborious because of the limited use of computer resources and because of the lack of reliable data on the kinetic parameters taken into account in the calculations of chemical reactions.

According to Alemasov and Tishin[6] and Alemasov and Kryukov[7], in the solution of thermodynamic problems in chemical reactors (these include rocket solid propellant engines), it is assumed that in the combustion chamber almost instantly established chemical equilibrium. This assumption, like the assumption for frozen, is not fully true, because the speed of some chemical reactions can be relatively low. In this regard, the real thermodynamic properties of the combustion products in terms of the rocket solid propellant engines will have values located in the interval between the characteristics of the combustion products, obtained in chemically inert compound and the chemical conditions of full thermodynamic equilibrium. Two solutions as asymptotes set the boundaries of the actual values of the parameters that can be obtained using current models of chemically reacting gas mixtures, however, the required computational resources become acceptable with the task of designing solid propellant.

It is of interest to develop methods for calculating the internal ballistics of solid propellant engines, taking into account the chemical equilibrium in the combustion chamber of the engine. The availability of such methods allows us to assess the impact of processes of chemical kinetics on the accuracy of the forecasting performance of the SRB. It should be noted that the issues related to the calculation of equilibrium composition of combustion products that are relevant to the tasks associated with the disposal of SRM,[8] as they allow to establish the presence in the products of processing hazardous and toxic substances.

There are a large number of software products that enables the calculation of chemical equilibrium composition of combustion products, are referenced, for example, by Alemasov and Kryukov[7] and Thermodynamics Program.[9] However, the application of them to products and software calculation of internal ballistics of solid propellant motors for different types is difficult. This is because the computational algorithms to determine the chemical equilibrium composition is taken as the basis for creating software products that can accumulate errors that can lead to abnormal termination of the task of internal ballistics (the problem about chemical equilibrium composition of combustion products is solved at each step of time integration for every point in space within the combustion chamber of the SRB).

The problem of the composition of the products of combustion of solid fuel according to Alemasov and Tishin[6] and Alemasov and Kryukov[7] can be reduced to, for example, the following system of nonlinear equations for the unknown $\gamma_i = \ln p_i$, $i = 1, I + J$ and M_T (where I, J, p_i—accordingly, the number of elementary substances in the combustion products, the number of connections and the partial pressure of elemental substances and compounds).

$$\gamma_j - \sum_{i=1}^{I} a_{ij} \cdot \gamma_j + \ln K_j^P = 0, \quad j = 1, J; \tag{4.1}$$

$$\ln(\sum_{j=1}^{q} a_{ij} \cdot \exp(\gamma_j)) - \ln M_T - \ln b_{iT} = 0, \quad i = 1, I; \tag{4.2}$$

$$\ln(\sum_{l=1}^{I+J} \exp(\gamma_j)) - \ln p = 0. \tag{4.3}$$

In this system, recorded J equations for chemical reactions, the I equations of conservation of mass for the source of the substances included in the conditional formula of the solid fuel, and Dalton's equation for partial pressures of each of the $J + I$ substances. Here and below the following notation:

- K^P—the constant of chemical equilibrium;
- M_T—the number of moles of the starting materials;
- a_{ij}—the number of atoms of the i chemical element in the j residue;

34 Nanoscience and Nanoengineering: Novel Applications

- b_{iT}—factors that determine a fuel composition;
- p—the pressure of the combustion products.

In accordance with Alemasov and Tishin,[6] the equations are solved by Newton's method the values of unknown γ_i and M_T established iterative algorithm (n is the iteration number).

$$\gamma_i^{(n+1)} = \gamma_i^{(n)} + \Delta_i, \quad i = 1, I + J,$$

$$M_{\Delta}^{(n+1)} = M_{\Delta}^{(n)} + \Delta_{I+J+1},$$

Amendments are established by solving the system of linear equations of the form $A\Delta = B$. Here, A is the matrix—Jacobian, constructed for eqs 4.1–4.3, and B is the vector of right parts. The elements of the matrix A and vector B are recalculated at each iteration. Solution of system of linear equations at each iteration for the unknown Δ_i by Alemasov and Tishin[6] and Thermodynamics Program[9] is the Gauss method.[10] However, it is known that the error of the method of Gauss increases with the number of unknowns in the system of linear equations. Moreover, due to the change of the Jacobian at each iteration a high probability of degeneracy of the matrix A (degeneration correspond to zero value of the determinant of the matrix A). These features, in some cases, lead to abnormal termination of the Newton method when solving the system of eqs 4.1–4.3, which is unacceptable in the solution of unsteady internal ballistics. However, there are orthogonal methods for solving systems of linear equations,[10,11] a feature of which is their resistance to the accumulation of computational errors. One such method is the matrix representation of A in the product of $A = Q \cdot R$, where Q is an orthogonal matrix ($Q \cdot Q^T = E$), and R is a lower triangular matrix. The amended definition of Δ_i in this case is determined by the decision matrix equations of the $Q \cdot R \cdot \Delta_i = B$. Due to the nature of matrices Q and R solution of this equation is straightforward.[12] In general, the complexity of the QR-method is several times the complexity of the method of Gauss, but the price for it is the absolute reliability of convergence of the Newton method applied to the system of eqs 4.1–4.3. Note that according to Belov,[13] it is noted the effectiveness of the application for solving problems of equilibrium thermodynamics methods of factorization, which is essentially the application of the QR-decomposition of a square matrix.

Another way to solve systems of nonlinear eqs 4.1–4.3 is to view the problem of chemical equilibrium composition of combustion products as a problem of mathematical programing. This approach was proposed by Sorkin.[14] If you consider the values of variables γ_i and M_T, there is no need

Non-Stationary Thermodynamic Processes in Rocket Engines   35

to establish any restrictions, the mathematical programing problem can be formulated as a problem of unconstrained optimization. The problem unconstrained optimization of the objective function Φ $(\gamma_1, \gamma_2,..., \gamma_{I+J}, M_I)$ write, for example, in the form,

$$\min \Phi(\gamma_1, \gamma_2,..., \gamma_I, M_T) = \delta_1 \times \sum_{j=1}^{J}(\gamma_j - \sum_{i=1}^{I} a_{ij} \times \gamma_j + \ln K_j^p)^2 +$$

$$+\delta_2 \times \sum_{i=1}^{I} (\ln(\sum_{j=1}^{q} a_{ij} \times \exp(\gamma_j)) - \ln M_T - \ln b_{iT})^2 + \delta_3 \times (\ln(\sum_{l=1}^{I+J} \exp(\gamma_j)) - \ln p)^2. \quad (4.4)$$

In the formula for the objective function whose minimum you want to ensure δ_1, δ_2, δ_3—scale coefficients, which values can be taken, such as $\delta_1 = \delta_2 = \delta_3 = 1$.

In connection with appearance of new effective methods of solution of problems of mathematical programing (optimization problems), this statement deserves attention, in particular, is a highly effective method based on genetic algorithms.[15]

The sequence of solution of the equilibrium composition of products of burning fuel when using genetic algorithm is established in the following order:

- Set the genotype (the set of independent arguments of the objective function consisting of values of partial pressures p_i and M_m). Each of the variables is a particular gene. A chromosome ϑ is a vector containing specific values of the variables p_i and M_m. The beginning of solving the problem corresponds to a population consisting of N individuals, each of which has a unique chromosome—$\vartheta_n = \vartheta(\gamma_1^{(n)}, \gamma_2^{(n)},...,\gamma_{I+J}^{(n)}, M_m^{(n)})$. The population of N individuals is updated crossings.
- For each individual, calculated functionality of Φ $(\gamma_1, \gamma_2,..., \gamma_{I+J}, M_T)$. Individual, corresponding to the smallest functionality is moved to a new generation.
- If the value of the Φ $(\gamma_1, \gamma_2,..., \gamma_{I+J}, M_T)$ exceeds the preset precision, the decision on the need for a new iteration and the calculation of new generation.
- Upgrades designed population due to interbreeding. At the next crossing of a chromosome ϑ_n is recalculated by the algorithm, called the crossover. The crossover is designed to consider the importance of the chromosomes involved in the hybridization, and the chromosomes corresponding to these values of the objective functions. At the next stage of the algorithm, the structure of the

population is reviewed. Of the weak chromosomes are removed and added again received a new and strong (from the point of view of objective function value) chromosome. A new gene, included in the newly calculated unique chromosome $\mathcal{G}'_n = \mathcal{G}(\gamma_1^{(n)}, \gamma_2^{(n)}, \ldots, \gamma_{1+J}^{(n)}, M_m^{(n)})$, the next iteration is calculated as a linear combination of genes in chromosome numbers s and t,

$$\mathcal{G}'_n = a \cdot \mathcal{G}'_s + b \cdot \mathcal{G}'_t.$$

The values of coefficients a and b are calculated by the formulas

$$a = 1 + \alpha - u(1 + 2\alpha), \quad b = u(1 + 2\alpha) - \alpha; \quad \alpha \in [0,1], \quad u \in (0,1).$$

Calculations on the recorded algorithm run as long as the value of the objective function for an individual inside the chromosome will not be less than the specified terms of accuracy. It should be noted that the solution of system of nonlinear eqs 4.1–4.3 using genetic algorithms is provided by almost any prescribed accuracy.[16] The disadvantages of these algorithms should include a large number of required for the solution iterations.

4.2 EXPERIMENTAL

Figure 4.1 is based on showing the convergence of the solution of the nonlinear eqs 4.1–4.3 QR-method (Fig. 4.1a) and genetic method (Fig. 4.1b). The calculation is performed for a metal-free fuel composition, conditional formula which has the form $C_h H_m O_r N_t Cl_q$.

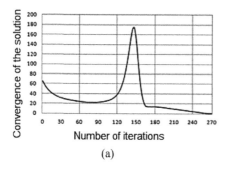
(a)

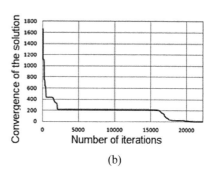
(b)

FIGURE 4.1 Convergence of methods of solving the problem of a chemical equilibrium composition of the products. (a) The Newton method in combination with QR-method and (b) genetic algorithm.

The high reliability of the calculation algorithms to the problem of chemical equilibrium composition of the combustion products allows you to incorporate the algorithms into software products providing calculations of the internal ballistics of rocket solid propellant engines of standard designs, including adjustable.[17]

Note that when known products of combustion values of specific heats c_p, c_v for the mixture of combustion products are set by the ratios

$$c_v = \sum_{i=1}^{I+J} \alpha_i c_{v_i}; \quad c_p = \sum_{i=1}^{I+J} \alpha_i c_{p_i}. \tag{4.5}$$

In a model of chemically inert mixture, specific heat capacity is determined by the formulas

$$c_v = c_{v6}\alpha_6 + c_{vm}\alpha_m + c_{v0}\alpha_0;$$

$$c_p = c_{p6}\alpha_6 + c_{pm}\alpha_m + c_{p0}\alpha_0.$$

In the last equations, α_0, α_6, α_m is the mass concentration in the combustion products of the air originally filling the combustion chamber of the rocket solid propellant engines, the products of combustion of sample igniter composition and combustion products of solid fuel. Specific heat capacity included in these equations corresponds to the same components.

The following are the results of the analysis of parameters of internal ballistics for the two types of rocket engines solid fuel-regulated (Fig. 4.2a) and propulsion (Fig. 4.2b).

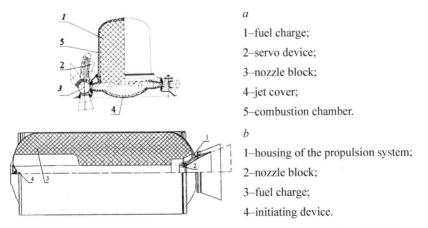

a
1–fuel charge;
2–servo device;
3–nozzle block;
4–jet cover;
5–combustion chamber.

b
1–housing of the propulsion system;
2–nozzle block;
3–fuel charge;
4–initiating device.

FIGURE 4.2 Structural diagram of the rocket solid propellant engines. (a) Adjustable engine and (b) cruise engine.

Mathematical model of processes in the regulated rocket solid propellant engines is given by Solomonov et al.[18] and Aliev et al.[19] The results of the calculations of the model engine weighing up to 50 kg and work time up to 35 presented in Figure 4.3. The calculations performed in the embodiment of chemical equilibrium composition of combustion products marked with one in a chemically inert composition (Fig. 4.2). In RDTT, as the igniter composition is used tubular piece booked ends, is made of fuels with a conditional formula $H_aO_bC_cN_dS_fK_g$ and solid fuel mixed metal-free low-temperature conditional formula which has the form $H_aO_bC_cN_dS_fK_g$. Program mode Figure 4.3 denotes the number 3 and corresponds to engine operation at two pressure levels—4.0 and 7.0 MPa.

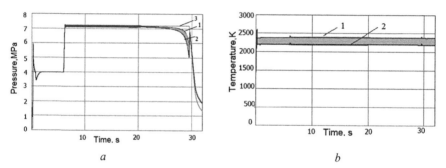

FIGURE 4.3 Dependence of pressure and temperature in the combustion chamber of a controlled rocket solid propellant engines. (a) Pressure combustion products and (b) the temperature in the chamber. 1—equilibrium flow, 2—frozen mix, and 3—software pressure.

The analysis presented in Figure 4.3 of the results allows us to draw the following conclusions:

- On the output stage of the engine on mode visible differences in the curves $p(t)$ for pressure in the combustion chamber of the engine, obtained using both models, is not observed.
- Control system provides pressures in the combustion chamber at the software level (curve 3). Noticeable difference of pressures (up to 5% or more) is observed only for transient conditions in the final stage rocket solid propellant engines. Time transient during which the reduction of pressure occurs in the combustion chamber of the engine is regulated to the programmed mode does not exceed 0.7–0.8 s.
- The difference between the temperature values obtained by the solution of the problem of internal ballistics model nonreactive and

chemically equilibrium mixture practically significant. The analysis shows that the temperature difference in the compared calculations can reach 10% or more.

A mathematical model of intrachamber processes in solid propellant sustainer engine (Fig. 4.2b) is significantly different from the model of the controlled engine. In particular, the in-cell process should be considered in the gas dynamic formulation (below, we use a one-dimensional space distribution of the intra ballistic parameters), there is no need to adjust the operating pressure level in the chamber, solid propellant engine, etc., are presented below the results on mathematical models.[2] In the calculations, it was supposed that the initial pressure in the combustion chamber solid propellant engines is 0.7 MPa. The length of the Central channel 3 of the engine is m, and diameter 1.5 m. The analysis is performed for fuel, temperature of combustion products which is 3200 K. The calculations were performed partitioning of the channel charge on the estimated 200 volumes, the solution of gas-dynamic tasks were performed by the method of large particles taking into account the modifications proposed by Lipanov et al.[2] and Aliev and Mishchenkova.[10] Calculation of chemical equilibrium composition of combustion products was carried out at each time step and in each of the selected volumes. On PC the average performance of such a calculation was performed over 50 h.

Figure 4.4 shows the results of pressure calculations as a function of time in the forward volume of the combustion chamber.

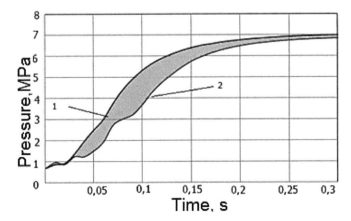

FIGURE 4.4 Dependence of the pressure in the combustion chamber from the time. 1—chemical equilibrium model and 2—model of chemically inert products of combustion.

Curve 1 in Figure 4.4 corresponds to the calculation of initial phase of solid propellant engine (corresponds to the time from 0 to 0.3) in the chemical equilibrium composition of combustion products, and curve 2— calculation of a chemically inert composition. As expected, the output of the solid propellant engine at quasi-steady-state operation when using models of chemical equilibrium composition of products of combustion occurs more rapidly and in less time. A series of calculations show that the difference between curves 1 and 2 is greater than the higher temperature products of combustion of solid fuel. It should be further noted that for the considered version of the sustainer rocket engine solid fuel 2 curve to more closely match the experimental results.

4.3 CONCLUSION

In conclusion, we can draw the following conclusions:

- Application of models of chemical equilibrium and chemically inert composition of combustion products, is advisable when solving problems of internal ballistics as it allows you to set boundaries changes in the basic characteristics of the solid propellant rocket engines, which are placed inside their actual values. The exact values of characteristics can be installed, solving the problem of internal ballistics as a problem of chemical kinetics; in problems on the work of the controlled solid propellant rocket engines, it is advisable to use a model of chemically inert products of combustion, and this is due to the relatively low-temperature products of combustion of solid fuel (up to 2800 K).
- The task of entering the mode propulsion solid propellant rocket engines the difference between the results obtained by models of inert and chemically equilibrium mixture of combustion products, increases markedly with increasing temperature of the combustion products of solid fuel.
- An account of the problem of internal ballistics with the definition of chemical equilibrium composition of combustion products is very time consuming. To reduce the computation time when using such a formulation can be pre-counted thermodynamic characteristics for the values of pressure and mass concentrations of the composition of the combustion products in the chamber, and then, propositional their polynomials, for example, first or second order.

KEYWORDS

- **rocket engine**
- **propellant**
- **chemical equilibrium**
- **internal ballistics**
- **QR-method**
- **genetic algorithm**

REFERENCES

1. Aliev A. V., Lipanov, A. M., Milekhin, Yu. M. *Internal Ballistics of Solid-Propellant Rocket Engines*, Mechanical Engineering, 2007; p 504.
2. Lipanov, A. M.; Bobryshev, V. P.; Aliev, A. V. *Numerical Experiment in the Theory of RSPRE*; Science: Ekaterinburg, Russia, 1994; p 303.
3. Varnatts, Yu.; Maas, A. T.; Dibbl, R. *Burning Physical and Chemical Aspects, Modeling, Experiments, Formation of the Polluting Substances/Lane from English*; Agafonov, G. L., Vlasov, P. A., Eds.; Publishing House Fizmatlit: Moscow, 2003; p 352.
4. Cherny G. G.; Losev S. A. *Physical and Chemical Processes in Gas Dynamics*; Publishing House Lomonosov: Moscow, 1995.
5. Bochkov, M. V.; Lovachev, L. A.; Chetverushkin, B. N. Chemical Kinetics of NOx Formation in Methane-air Combustion. *Mathem. Mod.* **1992**, *4*, 3–36.
6. Alemasov, V. E.; Dregalin, A. F.; Tishin, A. P. *Theory of Rocket Engines*; Mashinostroenie: Moscow, 1980; p 464.
7. Alemasov, V. E.; Dregalin, A. F.; Kryukov, V. G.; Naumov, V. I. *Mathematical Modeling of High-temperature Processes in Power Power Installations;* Nauka Publications: Moscow, 1989; p 285.
8. Burdyugov, S. I.; Korepanov, M. A.; Kuznetsov, N. P., et al. *Utilization of Solid Propellant Rocket Engines;* Izhevsk Institute Computer Researches: Izhevsk, Russia, 2008; p 512.
9. Thermodynamics Program. *Catalogue of Innovative Development of the Izhevsk State Technical University. 2nd Prod., Additional and Reslave;* Publishing House Izhevsk State Technical University: Izhevsk, Russia, 2001; p 95.
10. Aliev, A. V.; Mishchenkova, O. V. *Mathematical Modeling in Equipment;* Izhevsk Institute of Computer Researches: Izhevsk, Russia, 2012; p 476.
11. George A., Liu Dzh. *Computer Solution of Large Sparse Positif Definite Systems;* Publishing House MIR: Moscow, 1984; p 333.
12. Voevodina O. A.; Mishchenkova O. V. The Program for Calculation of Equilibrium Structure of Products of Combustion of Organic Fuel with Application of LU-of a Method and QR-of a Method//the Certificate on the State Registration of the Computer Program No. 2014618276 Granted by Federal Service for Intellectual Property. It Is Registered in the Register of the Computer Programs 13.08.2014.

13. Belov, G. V. *Thermodynamic Modeling: Methods, Algorithms, Programs;* Nauchnyj Mir: Moscow, 2002; p 184.
14. Sorkin R. E. *Gazodinamik of Rocket Engines on Solid - Propellant;* Science: Moscow, 1967; p 368.
15. Tenenev, V. A.; Yakimovich, B. A. *Genetic Algorithms in Modeling of Systems;* Publishing House Izhevsk State Technical University: Izhevsk, Russia, 2010; p 308.
16. Voevodina, O. A.; Mishchenkova, O. V. The Program for Calculation of Equilibrium Structure of Products of Combustion of Organic Fuel with Application of Genetic Algorithm//Certificate About the State Registration of the Computer Program No. 2014618275, Granted by Federal Service for Intellectual Property. It Is Registered in the Register of the Computer Programs 13.08.2014.
17. Aliev, A. V. *Applied Programs "Solid - propellant Rocket Engine";* Izhevsk State Technical University: Izhevsk, Russia, 2001; p 24.
18. Solomonov, Yu. S.; Lipanov, A. M.; Aliev, A. V.; Dorofeev, A. A.; Cherepov, V. I. *Solid Propellant Regulable Rocket Engine/RARAN*; Mechanical Engineering: Moscow, 2011; p 416.
19. Aliev, A. V.; Loshkarev, A. N.; Cherepov, V. I. *Mathematical Model of Work of Regulable Solid-propellant Rocket Engine//Chemical Physics and Mezoskopiya;* 2006; pp 311–320.

CHAPTER 5

CORPUSCULAR-WAVE PROCESSES

G. A. KORABLEV[1,2*]

[1]*Department of Physics, Izhevsk State Agricultural Academy, Studencheskaya St. 11, Izhevsk, Russia*

[2]*Basic Research-High Educational Centre of Chemical Physics and Mesoscopy, Ural Division of Russian Academy of Sciences, Izhevsk, Udmurt Republic, Russia*

E-mail: korablevga@mail.ru

ABSTRACT

Two principles of adding energy characteristics of structural interactions are fulfilled if the process flows either along the potential gradient or against it. Transforming these rules onto the corpuscular-wave dualism, we can assume that corpuscular interactions flow along the potential gradient (principle of adding reciprocals of energies), and wave processes—against the potential gradient (principle of algebraic addition of energies). Such approach is confirmed by the empiric equation, in which the act of quantum action is narrowed to the energy redistribution in the system "particle-wave."

It is demonstrated that the angular vector of rotational–translation motion of electrons at quantum transitions changes in compliance with the quantum number of the square tangent of this angle.

5.1 INTRODUCTION

The problem of quantum-wave dualism was mainly solved in the period of quantum mechanics development. Thus, the application of de Broglie equation allows defining the borders of such phenomena. But the predominating property depends on the process conditions. And it is quite complicated to find out which of them will operate in each particular case, although it is known that the wave pattern more often takes place at low energies, and corpuscular—at high ones.

One of the founders of quantum mechanics Max Born commented on that: "Each process can be interpreted either from the corpuscular or wave point. However, the proof that we are actually dealing with particles or waves is beyond our capabilities since we are not able to define all characteristic properties of the process. Therefore, we can only say that wave and corpuscular descriptions should be considered as two ways of considering one and the same objective process complementing each other."[1]

Thus, these problematic issues of quantum-wave dualism need to be further investigated and discussed. In this chapter, the attempt is made to clarify them from the point of the notions of space–energy interactions.

5.2 ON TWO PRINCIPLES OF ADDING ENERGY CHARACTERISTICS OF INTERACTIONS

The analysis of kinetics of various physical and chemical processes shows that in many cases the reciprocals of velocities, kinetic, or energy characteristics of the corresponding interactions are added.

In particular, such supposition is confirmed by the formula of electron transport possibility (W_∞) due to the overlapping of wave functions 1 and 2 (in steady state) during electron–conformation interactions:

$$W_\infty = \frac{1}{2} \frac{W_1 W_2}{W_1 + W_2} \qquad (5.1)$$

Equation 5.1 is used when evaluating the characteristics of diffusion processes followed by non-radiating transport of electrons in proteins.[2]

And also "From classical mechanics it is known that the relative motion of two particles with the interaction energy $U(r)$ takes place as the motion of material point with the reduced mass μ:

$$\frac{1}{\mu} = \frac{1}{m_1} + \frac{1}{m_2} \qquad (5.2)$$

in the field of central force $U(r)$, and general translational motion—as a free motion of material point with the mass:

$$m = m_1 + m_2 \qquad (5.3)$$

Corpuscular-Wave Processes

Such things take place in quantum mechanics as well."[3]

For moving thermodynamic systems, the first commencement of thermo-dynamics is as follows:

$$\delta E = d\left(U + \frac{mv^2}{2}\right) \pm \delta A, \tag{5.4}$$

where δA is the amount of energy transferred to the system;

element $d\left(U + \frac{mv^2}{2}\right)$ characterize the changes in internal and kinetic energies of the system;

$+ \delta A$ is the work performed by the system and

$- \delta A$ is the work performed with the system.

As the work value numerically equals the change in the potential energy, then:

$$+ \delta A = - \Delta U \tag{5.5}$$

and

$$- \delta A = + \Delta U \tag{5.6}$$

It is probable that not only in thermodynamic but in many other processes in the dynamics of moving particles interaction not only the value of potential energy is critical, but its change as well. Therefore, the following should be fulfilled for two-particle interactions:

$$\delta E = d\left(\frac{m_1 v_1^2}{2} + \frac{m_2 v_2^2}{2}\right) \pm \Delta U \tag{5.7}$$

Here

$$\Delta U = U_2 - U_1, \tag{5.8}$$

where U_2 and U_1 are the potential energies of the system in final and initial states.

The character of the change in the potential energy value (ΔU) was analyzed by its sign for various potential fields and the results are given in Table 5.1.

From the table, it is seen that the values $-\Delta U$ and accordingly $+ \delta A$ (positive work) correspond to the interactions taking place along the potential gradient, ΔU, and $-\delta A$ (negative work) occur during the interactions against the potential gradient.

TABLE 5.1 Directedness of the Interaction Processes.

No.	Systems	Type of potential field	Process	U	$\dfrac{r_2}{r_1}$ $\left(\dfrac{x_2}{x_1}\right)$	$\dfrac{U_2}{U_1}$	Sign ΔU	Sign δA	Process directedness in potential field
1	Opposite electrical charges	Electrostatic	Attraction	$-k\dfrac{q_1 q_2}{r}$	$r_2 < r_1$	$U_2 > U_1$	−	+	Along the gradient
			Repulsion	$-k\dfrac{q_1 q_2}{r}$	$r_2 > r_1$	$U_2 < U_1$	+	−	Against the gradient
2	Similar electrical charges	Electrostatic	Attraction	$-k\dfrac{q_1 q_2}{r}$	$r_2 < r_1$	$U_2 > U_1$	+	−	Against the gradient
			Repulsion	$-k\dfrac{q_1 q_2}{r}$	$r_2 > r_1$	$U_2 < U_1$	−	+	Along the gradient
3	Elementary masses m_1 and m_2	Gravitational	Attraction	$-\gamma\dfrac{m_1 m_2}{r}$	$r_2 < r_1$	$U_2 > U_1$	−	+	Along the gradient
			Repulsion	$-\gamma\dfrac{m_1 m_2}{r}$	$r_2 > r_1$	$U_2 < U_1$	+	−	Against the gradient
4	Spring deformation	Field of elastic forces	Compression	$k\dfrac{\Delta x^2}{2}$	$x_2 < x_1$	$U_2 > U_1$	+	−	Against the gradient
			Extension	$k\dfrac{\Delta x^2}{2}$	$x_2 > x_1$	$U_2 > U_1$	+	−	Against the gradient
5	Photo effect	Electrostatic	Repulsion	$-k\dfrac{q_1 q_2}{r}$	$r_2 > r_1$	$U_2 < U_1$	−	+	Along the gradient

Corpuscular-Wave Processes

The solution of two-particle task of the interaction of two material points with masses m_1 and m_2 obtained under the condition of the absence of external forces, corresponds to the interactions flowing along the gradient, the positive work is performed by the system (similar to the attraction process in the gravitation field).

The solution of this equation via the reduced mass (μ) is the Lagrange equation for the relative motion of the isolated system of two interacting material points with masses m_1 and m_2, which in coordinate x is as follows:

$$\mu \times x'' = -\frac{\partial U}{\partial x}; \quad \frac{1}{\mu} = \frac{1}{m_1} + \frac{1}{m_2}.$$

Here, U is the mutual potential energy of material points and μ is the reduced mass. At the same time, $x'' = a$ (feature of the system acceleration).

For elementary portions of the interactions Δx can be taken as follows:

$$\frac{\partial U}{\partial x} \approx \frac{\Delta U}{\Delta x}$$

That is $\quad \mu a \Delta x = -\Delta U.$

Then

$$\frac{1}{1/(a\Delta x)} \frac{1}{(1/m_1 + 1/m_2)} \approx -\Delta U \, ;$$

$$\frac{1}{(1/(m_1 a\Delta x)) + 1/(m_2 a\Delta x)} \approx -\Delta U \qquad (5.9)$$

or

$$\frac{1}{\Delta U} \approx \frac{1}{\Delta U_1} + \frac{1}{\Delta U_2} \qquad (5.10)$$

where ΔU_1 and ΔU_2 are potential energies of material points on the elementary portion of interactions, ΔU is the resulting (mutual) potential energy of this interactions.

Thus:

1. In the systems in which the interactions proceed along the potential gradient (positive performance) the resulting potential energy is found based on the principle of adding reciprocals of the corresponding energies of subsystems.[4] Similarly, the reduced mass for the relative motion of two-particle system is calculated.

2. In the systems in which the interactions proceed against the potential gradient (negative performance) the algebraic addition of their masses as well as the corresponding energies of subsystems are performed (by the analogy with Hamiltonian).

From the eq 5.10, it is seen that the resulting energy characteristic of the system of two material points interaction is found based on the principle of adding reciprocals of initial energies of interacting subsystems.

"Electron with the mass m moving near the proton with the mass M is equivalent to the particle with the mass: $\mu = \dfrac{mM}{m+M}$."[5]

Therefore, when modifying eq 5.10, we can assume that the energy of atom valence orbitals (responsible for interatomic interactions) can be calculated[6] by the principle of adding reciprocals of some initial energy components based on the following equations:

$$\frac{1}{q^2/r_i} + \frac{1}{W_i n_i} = \frac{1}{P_e} \tag{5.11}$$

or

$$\frac{1}{P_0} = \frac{1}{q^2} + \frac{1}{(Wrn)_i} \tag{5.12}$$

$$P_e = P_0/r_i \tag{5.13}$$

Here, W_i is the electron orbital energy;[7] r_i is the orbital radius of i;[8] $q = Z*/n*$;[9] n_i is the number of electrons of the given orbital; $Z*$ and $n*$ are nucleus effective charge and effective main quantum number, and r is the bond dimensional characteristics. For a free electron, $P = P_e = Wr$, where $W = 0.510034$ and MeV $= 0.81872 \times 10^{-13}$ J. As the dimensional characteristic, we used the value of electron classic radius $r = 2.81794 \times 10^{-15}$ m and, therefore, $P_e = 2.30712 \times 10^{-28}$ Jm.

5.3 ACT OF QUANTUM ACTION

The formalism of eqs 5.10–5.12 is not principally new. Already in 1924, the following equation was obtained based on Compton's effect:

$$\frac{1}{hv'} = \frac{1}{hv} + \frac{1-\cos\theta}{mc^2} \tag{5.14}$$

Corpuscular-Wave Processes

Here hv' is the energy of scattered photon, hv is the energy of incident photon, mc^2 is the own energy of electron, and Θ is the scattering angle. At the same time, the energy of photons decreases by the value additionally obtained by the electron. In this way, the act of quantum action takes place, resulting in the energy redistribution between the corpuscular, and wave properties of the interacting systems.

It is even easier, if the action proceeds during the interaction of the pair of similar particles. During the interaction along the potential gradient (corpuscular mechanism) the resultant energy, $W_k = \dfrac{W}{2}$. If this process goes against the gradient (wave movement) and the total energy $W_w = 2W$. The ratio between them $\dfrac{W_w}{W_k} = 4$.

Electric current is the motion of electrons along the potential gradient. If we assume that the magnetic field generated by them is the wave process, the ratio between the electric and magnetic constants needs to contain this digit 4, which is confirmed by the following empirical equation:

$$h = \left(\frac{4 + 2\alpha}{2\pi} \right)^2 p_e \frac{\varepsilon}{\mu} \tag{5.15}$$

Here ε is electric constant, μ is magnetic constant, h is Plank's constant, α is fine structure constant—parameter characterizing the interactions of quantum electron-positron and electromagnetic fields. Number π is determined by the ratio between the rotational motion (circle perimeter) and translational motion (length of diameter). The percentage error of calculations in this equation is about 0.06%.

The proportionality coefficient in eq 5.15 has the velocity dimensionality (m/sec) for the ratio (F/Hn), that is, in such way the rate of energy redistribution in the system "particle-wave" is characterized.

Therefore, the act of quantum action expressed via Plank's constant is narrowed to the energy equilibrium-exchange redistribution between the corpuscular and wave processes.

Generalizing the formalism of eqs 5.10 and 5.15 onto all other interactions flowing along the potential gradient, we can make the conclusion that corpuscular processes take place in these cases, and wave dualism corresponds to the interactions against the potential gradient.

5.4 ANGLE OF ELECTRON WINDING

It is known that a particle can have three main motions: translational, rotational, and oscillatory. But quantum mechanics does not consider the issue of electron trajectory as we can speak only of the possibility of its location in the given point in space.

But an electron also moves if this translational motion goes along the potential gradient, then it corresponds to corpuscular process, and rotational motion—to wave one. The correlation of these energy redistribution acts depends on the values of initial energy criteria of the subsystems. During quantum transitions, these can be orbital bond energies of the corresponding levels.

Thus, the main parameters of quantum transitions are as follows:

1. Energy of electromagnetic wave of quantum transition following Plank's equation $E = h\nu$, where ν is the electromagnetic wave frequency. In such way, the oscillatory motion demonstrates itself in quantum transitions, since the electromagnetic wave itself is the process of distribution of the corresponding oscillations.
2. Difference of bond energies of electrons on different energy level of transition: $\Delta W = W_2 - W_1$.
3. Resultant energy of their corpuscular interaction:

$$\frac{1}{W_k} = \frac{1}{W_1} + \frac{1}{W_2} \tag{5.16}$$

Let us consider some macro processes important in this case. The silkworm winds the natural (organic) silk thread only at a definite rotation angle. In cosmonautics, the cellulose-viscose thread is wound around the metal cylinder of the spaceship following the special technology, and, what is important, at the same winding angle as the silkworm. The spaceship becomes most durable, more technologically high-quality and lighter.[10] We can also speak of other examples of such phenomenon.

This angle (mainly as applicable to organic systems) was called the geodesic angle: $\varphi_g = 54.73° = 54°44'$.

In a general case, the winding angle (Θ) is the angle between the geodesic line and vector of rotational motion. The geodesic line is the shortest distance between two points in a geometric figure of rotation. Besides, planets are also rotating around the sun along the geodesic line. For five primary planets, the angle between the axis of rotation and orbit equals from 62° to 66.5°. The earth Θ, apparently taking into account also the moon influence, is 66°33'.

Corpuscular-Wave Processes

The sun has the same value Θ. In astronomic terms: obliquity of the sum ecliptic and obliquity of the earth equator to the orbit are numerically the same and equal to $22°27'$. Is not it the reason for special efficiency of solar action on the earth biophysical processes?

Nitrogen, oxygen, hydrogen and, most important, carbon are the main elements of organic materials. Carbon is a specific element, capable of easier hybridization of atomic orbitals with the quantum transition 2s-2p. Therefore, when temperature and pressure rise, the conditions for such hybridization of carbon atoms are formed in organic materials, and this, apparently, takes place in the winding technique in spaceships. And in the silkworm, the same way as in many other natural processes, the corresponding fermentative reactions take place, on which we are still learning how to work.

To calculate Θ and φ_g, we use the formalism of Compton eq 5.14, modifying it as applicable to quantum transitions:

$$\frac{1}{hv} - \frac{1}{W_k K} = \frac{1 - \cos \Theta}{\Delta W} \tag{5.17}$$

By this equation, the difference of energies of wave and corpuscular processes numerically equals the difference of bond energies of electrons on the corresponding orbitals, but when implementing the addition principles (in this case—deduction) of reciprocals of these parameters and taking into account the quantum geometry of transitions. In accordance with the law of energy conservation, this is the process of its redistribution during the quantum action. Angle Θ is the angular vector of electron movement, which is quantized by an integer number (K) via the square tangent of this angle: $tg^2\varphi_g = 2; tg^2 60° = 3; tg^2 45° = 1$.

The calculations by eq 5.17 are given in Tables 5.2 and 5.3. At the same time, the values of the angle Θ are mainly correlated with the value $\varphi = \dfrac{hv}{W_k}$ in compliance with Table 5.3.

The notions of breaking stress in the process of plastics stretching by its winding pitch are used in chapters:[10,12,13] σ_α is axial, σ_β is circumferential stress, which are replaced by the value N_α is axial "effort" and N_β is circumferential "effort" proportional to them. At the same time, the following equation is fulfilled:

$$\frac{\sigma_\beta}{\sigma_\alpha} = \frac{N_\beta}{N_\alpha} = tg^2\varphi_g = 2 \tag{5.18}$$

TABLE 5.2 Energies of Quantum Transitions.

Atom	Transition	W_1 (eV)	W_2 (eV)	ΔW (J)	W_k (J)	λ (A^0)[11]	hv (J)
C (IV)	2s-2p	19.201	11.792	11.871	11.705	1549	12.824
N (V)	2s-2p	25.724	15.445	16.469	15.462	1238	16.046
O (VI)	2s-2p	33.859	17.195	26.699	18.267	1031	19.267
Al (III)	3s-3p	10.706	5.7130	7.9997	5.9886	1854	10.7145
Si (IV)	3s-3p	14.690	8.0848	10.583	8.3554	1393	14.260
C (III)	2s²-2s2p	19.201·2	19.201 + 11.792	11.871	27.480	977	20.332
N (IV)	2s²-2s2p	25.724·2	25.724 + 15.445	16.469	36.638	765	25.967
Si (III)	3s²-3s3p	14.690·2	14.690 + 8.0848	10.583	20.557	1206	16.4715
Al (II)	3s²-3s3p	10.706·2	10.706 + 5.7130	7.9997	14.889	1670	11.895

TABLE 5.3 Quantization of the Geometry of Structural Transitions.

Atom	Transition	$\varphi = \dfrac{h\nu}{W_k}$	$\langle\varphi\rangle$	K	$\theta°$	$\langle\theta\rangle$		Functions of square tangent (k)
C (IV)	2s-2p	1.0956	60.9°	2	54.45°	60.02°	–	$tg^2\varphi_r = 2$
N (V)	2s-2p	1.0377		2	59.67°			
O (VI)	2s-2p	1.0547		2	65.93°			
Al (III)	3s-3p	1.7951	$\varphi_z° + 45.47° = 100.2°$	3 = 2 + 1	45.45°	46.2°	61.6°	$tg^260° = 3$
Si (IV)	3s-3p	1.7067		3 = 2 + 1	47.02°			
C (III)	2s²-2s2p	0.7399	43.1°	1	31.97°	31.7°	42.27°	$tg^245° = 1$
N (IV)	2s²-2s2p	0.7087		1	35.38°			
Si (III)	3s²-3s3p	0.8013		1	29.27°			
Al (II)	3s²-3s3p	0.7589		1	30.17°			

"This condition allows obtaining the equally tensioned system of threads with the minimal item weight."[10]

The quantum functions of square tangent $k = 1, 2$, and 3 numerically determine the ratios of two triangle legs, whose values characterize energy dependencies via axial and circumferential stresses in the system with quantum and wave processes.

From Table 5.3, it is seen that quantum transitions of 2s-2p type for carbon atom, as distinguished from all other elements, are not accompanied with the changes in geodesic angle and coefficient k. Obviously, this property predetermines the unique features of the winding geodesic angle influence on the biosystems stability. Besides, in all transitions (except for 2s-2p) the correlation $\varphi \approx \frac{4}{3}\theta$ is fulfilled, which proves that such coefficient mainly compensates structural features of more complex transitions.

Some difference between the values of the angles φ and θ or φ and is obviously determined by the effect of particle scattering around the main coordinate axes. Similarly, at the conformation of cellular structures the particles are statistically concentrated along the coordinate axes of hexagons with the deviations by 2.6°, 4.4°, and 7.9° (Fig. 5.1).[14]

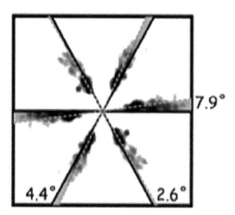

FIGURE 5.1 Statistic distribution of the cell number along the coordinate axes.[14]

The average number of such deviations equaled to 4.97 approximately corresponds to the difference $60° - \varphi_g° = 5.27°$.

The dynamics of hexagonal formation of cellular systems are in compliance with the established[15] condition of approximate equilibrium of spatial-energy characteristics of the subsystems by all bond lines. This is also

Corpuscular-Wave Processes

facilitated by the fact that biosystems with elements of the second period in their structure produce the winding angular vector (Θ) of 60°.

5.5 CONCLUSIONS

1. Two principles of adding energy characteristics of structural interactions can be transformed into the processes of corpuscular-wave dualism.
2. It is assumed that in the process of rotational–translation motion of the electron the energies redistribute in the system "particle-wave," which is demonstrated via the angular vector of such motion (winding angle).
3. The dependencies obtained give the possibility to consider in a unified way a lot of structural-dynamic processes different by nature and scale. For example, the characteristic of spin-orbital interaction— the constant of fine structure $= \dfrac{r}{\lambda}$, where r is the electron classical radius and λ is its Compton wavelength.
4. Formally, but similarly: the interaction force of two long conductors with current is proportional to the ratio $\dfrac{l}{2\pi r}$, where l is the length of the conductors and r is the distance between them.
5. In these examples, as in many others, this approach allows evaluating structural interactions based on the ratios of corpuscular and wave spatial-energy parameters in each action.
6. In material science, as well as in physical–chemical investigations, when calculating the electron winding angle in some structures, it is possible to give certain recommendations on innovative technologies, for instance, when manufacturing parts (in mechanical engineering) or winding on the linen flax base (in agriculture).
7. The difference in phases of electric and magnetic oscillations in electromagnetic wave is $\dfrac{\pi}{2}$. Entering value into the coefficient of eq 5.15, we have the equation for Plank constant with the accuracy close to that of initial data:

$$h = \left(\frac{4}{\pi^2} + a\right) P_e \frac{\varepsilon}{\mu} \tag{5.19}$$

where $a = 0.0023293$ and experimental quantum correction to the spin is the factor of the electron.

KEYWORDS

- **potential gradient**
- **quantum transitions**
- **corpuscular-wave dualism**
- **geodesic angle**
- **potential energy**

REFERENCES

1. Marison, J. B. *General Physics with Biological Examples;* Vysshaya Shkola: Moscow, 1986; p 623.
2. Rubin, A. B. *Biophysics. Book 1. Theoretical Biophysics;* Vysshaya Shkola: Moscow, 1987; p 319.
3. Blokhintsev, D. I. *Basics of Quantum Mechanics;* Vysshaya Shkola: Moscow, 1961; p 512.
4. Korablev, G. A.; Zaikov, G. E. *Quantum and Wave Characteristics of Spatial Energy Interactions. Bioscience Methodologies in Physical Chemistry;* Apple Academic Press: Waretown, NJ, 2013; pp 130–143.
5. Eyring, G.; Walter, J.; Kimball, G. *Quantum Chemistry;* F. L.: Moscow, 1948; p 528.
6. Korablev, G. A. *Spatial-Energy Principles of Complex Structures Formation, Netherlands;* Brill Academic Publishers and VSP: Leiden, Netherlands, 2005; p 426. (Monograph).
7. Fischer, C. F. *Average-energy of Configuration Hartree-Fock Results for the Atoms Helium to Radon. Atomic Data;* Academic Press, Inc.: Cambridge, MA ,1972; Vol. 4, pp 301–399.
8. Waber, J. T.; Cromer, D. T. *J. Chem. Phys.* **1965,** *42* (12), 4116–4123.
9. Clementi, E.; Raimondi, D. L. *J. Chem. Phys.* **1963,** *38* (11), 2686–2689. *J. Chem. Phys.* **1967,** *47* (4), 1300–1307.
10. Kodolov, V. I. *Polymeric Composites and Technology of Producing Aircraft Engines from Them;* Izhevsk Mechanical Institute, Izhevsk, Russia, 1992; p 200.
11. Allen, K. U. *Astrophysical Magnitudes;* Mir: Moscow, 1977; p 446.
12. Ayushev, T. Yu. *Geometric Aspects of Adaptive Technology of Producing Structures by Winding from Fibrous Composite Materials;* BNC SO RAS Publishers: Ulan-Ude, Russia, 2005; p 212.
13. Pidgainy, Yu. M.; Morozova, V. M.; Dudko, V. A. Methodology for Calculating Characteristics of Geodesic Winding of Shells of Rotational Bodies. *Mech. Polym.* **1967,** *6,* 1096–1104.
14. Nobel Lecture by Edward Moser in Physiology. Aired March 11, 2015, on TV Channel Science.
15. Korablev, G. A.; Vasiliev, Yu. G.; Zaikov, G. E. Hexagonal Structures in Nanosystems. In *Chemical Physics and Mesoscopy;* 2015; Vol. 17, pp 424–429.

CHAPTER 6

SPATIAL-ENERGY INTERACTIONS AND ENTROPY'S CURVES

G. A. KORABLEV[1,2*], V. I. KODOLOV[2,3], G. E. ZAIKOV[4], YU. G. VASIL'EV[5], and N. G. PETROVA[6]

[1]*Department of Physics, Izhevsk State Agricultural Academy, Studencheskaya St. 11, Izhevsk, Russia*

[2]*Basic Research-High Educational Centre of Chemical Physics and Mesoscopy, Ural Division of Russian Academy of Sciences, Izhevsk, Udmurt Republic, Russia*

[3]*Department of Chemistry and Chemical Technology, Kalashnikov Izhevsk State Technical University, Studencheskaya St. 7, Izhevsk, Russia*

[4]*Department of Bio-Chemistry, N. M. Emanuel Institute of Biochemical Physics, Russian Academy of Science, Kosygin St. 4, Moscow 119991, Russia*

[5]*Department of Veterinary, Izhevsk State Agricultural Academy, Studencheskaya St. 11, Izhevsk, Russia*

[6]*Department of Security and Communications, Agency of Informatization and Communication, Udmurt Republic, Russia*

Corresponding author. E-mail: korablevga@mail.ru

ABSTRACT

In the systems in which the interactions proceed along the potential gradient (positive performance), the resulting potential energy is found based on the principle of adding reciprocals of the corresponding energies of subsystems. The idea of entropy is diversified in physical and chemical, economic, engineering, and other natural processes that are confirmed by their nomograms. The application of spatial-energy parameter (SEP or P-parameter) methodology allows evaluating the possibility of conformation processes in different biosystems based on energy characteristics of free atoms.

SECTION 1
EQUILIBRIUM-EXCHANGE SPATIAL-ENERGY INTERACTIONS

ABSTRACT

The notion of spatial-energy parameter (SEP or P-parameter) is introduced based on the modified Lagrangian equation for relative motion of two interacting material points, which is a complex characteristic of important atomic values. Wave properties of P-parameter are found, its wave equation having a formal analogy with the equation of Ψ-function is given.

Some correlations of P-parameter values with Lagrangian and Hamiltonian functions are obtained. With the help of P-parameter technique, numerous calculations of exchange structural interactions (for nanosystems as well) are done, the applicability of the model for the evaluation of their intensity is given.

6.1 INTRODUCTION

To obtain the dependence between energy parameters of free atoms and degree of structural interactions in simple and complex systems is one of strategic tasks in physical chemistry. Classical physics and quantum mechanics widely use Coulomb interactions and their varieties for this.

According to Rubin,[1] Van der Waals' orientation and charge–dipole interactions are referred to electron–conformation interactions in biosystems and as a particular case—to exchange resonance transfer of energy. But biological and many cluster systems are electroneutral in structural basis. And non-Coulomb equilibrium-exchange spatial-energy interactions, that is, noncharge electrostatic processes, are mainly important for them.

The structural interactions of summed electron densities of valence orbitals of corresponding conformation centers take place—processes of equilibrium flow of electron densities due to overlapping of their wave functions.

Heisenberg and Dirac[2] proposed the exchange Hamiltonian derived in the assumption on direct overlapping of wave functions of interacting

Spatial-Energy Interactions and Entropy's Curves 59

centers: $\bar{H} = I_0 S_1 S_2$ where $\bar{H}$ is the spin operator of isotropic exchange interaction for pair of atoms, I_0 is the exchange constant, S_1 and S_2 are the overlapping integrals of wave functions.

In this model, electrostatic interactions are modeled by effective exchange Hamiltonian action in the space of spin functions.

In particular, such approach is applied to the analysis of structural interactions in cluster systems. It is demonstrated in Anderson's works[3] that in compounds of transition elements when the distance between paramagnetic ions considerably exceeds the total of their covalent radii, "superexchange" processes of overlapping cation orbitals take place through the anion between them.

In this work, similar equilibrium-exchange processes are evaluated through the notion of —P-parameter.

6.2 TWO PRINCIPLES OF ADDING ENERGY CHARACTERISTICS

The analysis of kinetics of various physical and chemical processes shows that in many cases the reciprocals of velocities, kinetic or energy characteristics of the corresponding interactions are added.

Some examples are ambipolar diffusion, resulting in velocity of topochemical reaction, change in the light velocity during the transition from vacuum into the given medium, and effective permeability of bio-membranes.

In particular, such supposition is confirmed by the formula of electron transport possibility (W_∞) due to the overlapping of wave functions 1 and 2 (in steady state) during electron–conformation interactions:

$$W_\infty = \frac{1}{2} \frac{W_1 W_2}{W_1 + W_2} \tag{6.1}$$

Equation 6.1 is used when evaluating the characteristics of diffusion processes followed by non-radiating transport of electrons in proteins.[1]

And also: "From classical mechanics it is known that the relative motion of two particles with the interaction energy $U(r)$ takes place as the motion of material point with the reduced mass μ:

$$\frac{1}{\mu} = \frac{1}{m_1} + \frac{1}{m_2} \tag{6.2}$$

in the field of central force $U(r)$, and general translational motion—as a free motion of material point with the mass:

$$m = m_1 + m_2 \qquad (6.3)$$

Such things take place in quantum mechanics as well."[4]

The task of two-particle interactions taking place along the bond line was solved in the times of Newton and Lagrange:

$$E = \frac{m_1 v_1^2}{2} + \frac{m_2 v_2^2}{2} + U\left(\bar{r}_2 - \bar{r}_1\right), \qquad (6.4)$$

where E is the total energy of the system, first and second elements are the kinetic energies of the particles, third element is the potential energy between particles 1 and 2, and vectors $\bar{r}_2$ and $\bar{r}_1$ characterize the distance between the particles in final and initial states.

For moving thermodynamic systems the first commencement of thermodynamics is as follows:

$$\delta E = d\left(U + \frac{mv^2}{2}\right) \pm \delta A, \qquad (6.5)$$

where δE is the amount of energy transferred to the system;

element $d\left(U + \dfrac{mv^2}{2}\right)$ characterize the changes in internal and kinetic energies of the system;

$+ \delta E$ is the work performed by the system and
$- \delta A$ is the work performed with the system.

As the work value numerically equals the change in the potential energy, then:

$$+ \delta A = -\Delta U \qquad (6.6)$$

and

$$- \delta A = +\Delta U. \qquad (6.7)$$

It is probable that not only in thermodynamic but in many other processes in the dynamics of moving particles interaction, not only the value of potential energy is critical, but its change as well. Therefore, similar to eq 6.4, the following should be fulfilled for two-particle interactions:

$$\delta E = d\left(\frac{m_1 v_1^2}{2} + \frac{m_2 v_2^2}{2}\right) \pm \Delta U \qquad (6.8)$$

Here

$$\Delta U = U_2 - U_1, \tag{6.9}$$

where U_2 and U_1 are the potential energies of the system in final and initial states.

At the same time, the total energy (E) and kinetic energy $\left(\dfrac{mv^2}{2}\right)$ can be calculated from their zero value, then only the last element is modified in eq 6.4.

The character of the change in the potential energy value (ΔU) was analyzed by its sign for various potential fields and the results are given in Table 6.1.

From Table 6.1 it is seen that the values $-\Delta U$ and accordingly $+\delta A$ (positive work) correspond to the interactions taking place along the potential gradient, and ΔU and $-\delta A$ (negative work) occur during the interactions against the potential gradient.

The solution of two-particle task of the interaction of two material points with masses m_1 and m_2 obtained under the condition of the absence of external forces, corresponds to the interactions flowing along the gradient, the positive work is performed by the system (similar to the attraction process in the gravitation field).

The solution of this equation via the reduced mass (μ) is the Lagrange equation for the relative motion of the isolated system of two interacting material points with masses m_1 and m_2, which in coordinate x is as follows:

$$\mu \cdot x'' = -\frac{\partial U}{\partial x}; \quad \frac{1}{\mu} = \frac{1}{m_1} + \frac{1}{m_2}.$$

Here U is the mutual potential energy of material points and μ is the reduced mass. At the same time $x'' = a$ (feature of the system acceleration). For elementary portions of the interactions Δx can be taken as follows:

$$\frac{\partial U}{\partial x} \approx \frac{\Delta U}{\Delta x}$$

That is $\mu a \Delta x = -\Delta U.$

Then

$$\frac{1}{1/(a\Delta x)}\frac{1}{(1/m_1 + m_2)} \approx -\Delta U; \quad \frac{1}{1/(m_1 a\Delta x) + 1/(m_2 a\Delta x)} \approx -\Delta U$$

TABLE 6.1 Directedness of the Interaction Processes.

No.	Systems	Type of potential field	Process	U	r_2/r_1 (x_2/x_1)	U_2/U_1	Sign ΔU	Sign δA	Process directedness in potential field
1	Opposite electrical charges	Electrostatic	Attraction	$-k\dfrac{q_1 q_2}{r}$	$r_2 < r_1$	$U_2 > U_1$	−	+	Along the gradient
			Repulsion	$-k\dfrac{q_1 q_2}{r}$	$r_2 > r_1$	$U_2 < U_1$	+	−	Against the gradient
2	Similar electrical charges	Electrostatic	Attraction	$-k\dfrac{q_1 q_2}{r}$	$r_2 < r_1$	$U_2 > U_1$	+	−	Against the gradient
			Repulsion	$-k\dfrac{q_1 q_2}{r}$	$r_2 > r_1$	$U_2 < U_1$	−	+	Along the gradient
3	Elementary masses m_1 and m_2	Gravitational	Attraction	$-\gamma\dfrac{m_1 m_2}{r}$	$r_2 < r_1$	$U_2 > U_1$	−	+	Along the gradient
			Repulsion	$-\gamma\dfrac{m_1 m_2}{r}$	$r_2 > r_1$	$U_2 < U_1$	+	−	Against the gradient
4	Spring deformation	Field of elastic forces	Compression	$k\dfrac{\Delta x^2}{2}$	$x_2 < x_1$	$U_2 > U_1$	+	−	Against the gradient
			Extension	$k\dfrac{\Delta x^2}{2}$	$x_2 > x_1$	$U_2 > U_1$	+	−	Against the gradient
5	Photo effect	Electrostatic	Repulsion	$-k\dfrac{q_1 q_2}{r}$	$r_2 > r_1$	$U_2 < U_1$	−	+	Along the gradient

Spatial-Energy Interactions and Entropy's Curves

or

$$\frac{1}{\Delta U} \approx \frac{1}{\Delta U_1} + \frac{1}{\Delta U_2} \tag{6.10}$$

where ΔU_1 and ΔU_2 are potential energies of material points on the elementary portion of interactions, ΔU is the resulting (mutual) potential energy of this interactions.

Thus:
1. In the systems in which the interactions proceed along the potential gradient (positive performance) the resulting potential energy is found based on the principle of adding reciprocals of the corresponding energies of subsystems.[5] Similarly, the reduced mass for the relative motion of two-particle system is calculated.
2. In the systems in which the interactions proceed against the potential gradient (negative performance) the algebraic addition of their masses as well as the corresponding energies of subsystems are performed (by the analogy with Hamiltonian).

6.3 SPATIAL-ENERGY PARAMETER (P-PARAMETER)

From eq 6.10, it is seen that the resulting energy characteristic of the system of two material points interaction is found based on the principle of adding reciprocals of initial energies of interacting subsystems.

"Electron with the mass m moving near the proton with the mass M is equivalent to the particle with the mass: $\mu = \dfrac{mM}{m+M}$."[6]

Therefore when modifying eq 6.10, we can assume that the energy of atom valence orbitals (responsible for interatomic interactions) can be calculated[5] by the principle of adding reciprocals of some initial energy components based on the following equations:

$$\frac{1}{q^2/r_i} + \frac{1}{W_i n_i} = \frac{1}{P_E} \tag{6.11}$$

or

$$\frac{1}{P_0} = \frac{1}{q^2} + \frac{1}{(Wrn)_i}; \tag{6.12}$$

$$P_E = P_0/r_i . \tag{6.13}$$

Here W_i is the electron orbital energy;[7] r_i is the orbital radius of I;[8] $q = Z^*/n^*$;[9] n_i is the number of electrons of the given orbital; Z^* and n^* are nucleus effective charge and effective main quantum number and r is the bond dimensional characteristics.

P_0 is called a SEP, and P_E is the effective P–parameter (effective SEP). Effective SEP has a physical sense of some average energy of valence electrons in the atom and is measured in energy units, for example, electron volts (eV).

The values of P_0-parameter are tabulated constants for the electrons of the given atom orbital.

For dimensionality SEP can be written down as follows:

$$[P_0] = [q^2] = [E]\cdot[r] = [h]\cdot[v] = \frac{kg\cdot m^3}{s^2} = J\cdot m,$$

where [E], [h], and [v] are dimensions of energy, Planck constant, and velocity, respectively. Thus P-parameter corresponds to the processes going along the potential gradient.

The introduction of P-parameter should be considered as further development of quasi-classical notions using quantum-mechanical data on atom structure to obtain the criteria of energy conditions of phase formation. At the same time, for the systems of similarly charged (e.g., orbitals in the given atom), homogeneous systems the principle of algebraic addition of such parameters is preserved:

$$\sum P_E = \sum (P/r_i); \tag{6.14}$$

$$\sum P_E = \frac{\sum P_0}{r}, \tag{6.15}$$

$$\sum P_0 = P_0' + P_0'' + P_0''' + ... ; \tag{6.16}$$

$$r\sum P_E = \sum P_0 \tag{6.17}$$

Here P-parameters are summed on all atom valence orbitals.

To calculate the values of P_E-parameter at the given distance from the nucleus depending on the bond type either atom radius (R) or ion radius (r_i) can be used instead of r.

Now, let us briefly explain the reliability of such approach. The calculations demonstrated that the values of P_E-parameters numerically equal (within 2%) total energy of valence electrons (U) by the atom statistic model. Using the known correlation between the electron density (β) and

Spatial-Energy Interactions and Entropy's Curves

interatomic potential by the atom statistic model,[10] we can obtain the direct dependence of P_E-parameter upon the electron density at the distance r_i from the nucleus.

The rationality of such approach is confirmed by the calculation of electron density using wave functions of Clementi and its comparison with the value of electron density calculated via the value of P_E-parameter.

6.4 WAVE EQUATION OF P-PARAMETER

To characterize atom spatial-energy properties two types of P-parameters are introduced. The bond between them is a simple one: $P_E = \dfrac{P_0}{R}$ where R is the atom dimensional characteristic. Taking into account additional quantum characteristics of sublevels in the atom, this equation can be written down in coordinate x as follows:

$$\Delta P_E \approx \frac{\Delta P_0}{\Delta x} \qquad \text{or} \qquad \partial P_E = \frac{\partial P_0}{\partial x}$$

where the value ΔP equals the difference between P_0-parameter of i orbital and P_{CD} is the countdown parameter (parameter of main state at the given set of quantum numbers).

According to the Korablev's[5] rule of adding P-parameters of similarly charged or homogeneous systems for two orbitals in the given atom with different quantum characteristics and according to the energy conservation rule we have:

$$\Delta P''_E - \Delta P'_E = P_{E,\lambda}$$

where $P_{E,\lambda}$ is the SEP of quantum transition.

Taking for the dimensional characteristic of the interaction $\Delta\lambda = \Delta x$, we have:

$$\frac{\Delta P''_0}{\Delta\lambda} - \frac{\Delta P'_0}{\Delta\lambda} = \frac{P_0}{\Delta\lambda} \qquad \text{or} \qquad \frac{\Delta P'_0}{\Delta\lambda} - \frac{\Delta P''_0}{\Delta\lambda} = -\frac{P_0\lambda}{\Delta\lambda}$$

Let us again divide by $\Delta\lambda$ term by term: $\left(\dfrac{\Delta P'_0}{\Delta\lambda} - \dfrac{\Delta P''_0}{\Delta\lambda}\right) \bigg/ \Delta\lambda = -\dfrac{P_0}{\Delta\lambda^2},$ where

$$\left(\frac{\Delta P'_0}{\Delta\lambda} - \frac{\Delta P''_0}{\Delta\lambda}\right) \bigg/ \Delta\lambda \sim \frac{d^2 P_0}{d\lambda^2}, \quad \text{i.e.,:} \quad \frac{d^2 P_0}{d\lambda^2} + \frac{P_0}{\Delta\lambda^2} \approx 0$$

Taking into account only those interactions when $2\pi\Delta x = \Delta\lambda$ (closed oscillator), we have the following equation:

$$\frac{d^2 P_0}{4\pi^2 dx^2} + \frac{P_0}{\Delta\lambda^2} = 0 \text{ or } \frac{d^2 P_0}{dx^2} + 4\pi^2 \frac{P_0}{\Delta\lambda^2} \approx 0$$

Since $\Delta\lambda = \dfrac{h}{mv}$,

then: $\dfrac{d^2 P_0}{dx^2} + 4\pi^2 \dfrac{P_0}{h^2} m^2 v^2 \approx 0$

or

$$\frac{d^2 P_0}{dx^2} + \frac{8\pi^2 m}{h^2} P_0 E_k = 0 \tag{6.18}$$

where $E_k = \dfrac{mV^2}{2}$ is the electron kinetic energy.

Schrodinger equation for the stationary state in coordinate x:

$$\frac{d^2\psi}{dx^2} + \frac{8\pi^2 m}{h^2} \psi E_k = 0$$

When comparing these two equations, we see that P_0-parameter numerically correlates with the value of Ψ-function: $P_0 \approx \psi$, and is generally proportional to it: $P_0 \sim \Psi$. Taking into account the broad practical opportunities of applying the P-parameter methodology, we can consider this criterion as the materialized analog of Ψ-function.

Since P_0-parameters like Ψ-function have wave properties, the superposition principles should be fulfilled for them, defining the linear character of the equations of adding and changing P-parameter.

6.5 COMPARISONS OF LAGRANGE AND HAMILTON FUNCTIONS WITH SPATIAL-ENERGY PARAMETER

Lagrange (L) and Hamilton (H) functions are the main provisions of analytical mechanics.

Lagrange function is the difference between kinetic (T) and potential (U) energies of the system:

$$L = T - U.$$

Spatial-Energy Interactions and Entropy's Curves 67

For uniform functions of the second degree, Hamilton function can be considered as the sum of potential and kinetic energies, that is, as the total mechanical energy of the system:

$$H = T + U.$$

From these equations and in accordance with energy conservation law we can see that:

$$H = L + 2T, \tag{6.19}$$

$$H - L = 2U. \tag{6.20}$$

Let us try to assess the movement of an isolated system of a free atom as a relative movement of its two subsystems: nucleus and orbital.

The atom structure is formed of oppositely charged masses of nucleus and electrons. In this system, the energy characteristics of subsystems are the orbital energy of electrons (W_i) and effective energy of atom nucleus taking screening effects into account.

In a free atom, its electrons move in Coulomb field of nucleus charge. The effective nucleus charge characterizing the potential energy of such subsystem taking screening effects into account equals q^2/r_i, where $q = Z^*/n^*$.

Here Z^* and n^* are effective nucleus charge and effective main quantum number; r_i is the orbital radius.

It can be presumed that orbital energy of electrons during their motion in Coulomb field of atom nucleus is mainly defined by the value of kinetic energy of such motion.

Thus, it is assumed that:

$$T \sim W \text{ and } U \sim q^2/r_i$$

In such approach the total of the values W and is analogs to Hamilton function (H):

$$W + q^2/r_i \sim H. \tag{6.21}$$

The analogous comparison of P-parameter with Lagrange function can be carried out when investigating Lagrange equation for relative motion of isolated system of two interacting material points with masses m_1 and m_2

in coordinate x. The principle of adding reciprocals of energy values is modeling their algebraic difference by Hamiltonian.

Thus, it is presumed that $P_E \sim L$.

Then, eq 6.19 is as follows

$$\left(W + \frac{q^2}{r_i}\right) + P_E \approx 2W. \qquad (6.22)$$

Using the values of electron bond energy as the orbital electron energy, we calculated the values of P_E-parameters of free atoms (Table 6.2) by eqs 6.11–6.13. When calculating the values of effective P_E-parameter, mainly the atom radius values by Belov–Bokiy or covalent radii (for non-metals) were applied as dimensional characteristics of atom (R).

At the same time, the averaged values of total energy, valence orbitals dividing their values by a number of valence electrons considered (N):

$$\left(\frac{q^2}{r_i} + W\right)\frac{1}{N} + P_E \approx 2W. \qquad (6.22a)$$

This energy in terms of one valence electron is the analog of Hamilton function (H).

In free atoms of 1a and 2a groups of periodic system S-orbital is the only valence orbital, and that was considered via the introduction of the coefficient $K = n/n^*$, where n is the main quantum number and n^* is the effective main quantum number by the equation:

$$\left(\frac{q^2}{r_i} + W\right)\frac{1}{\hat{E}N} + P_E \approx 2W. \qquad (6.23)$$

Thus in 1a and 2a subgroups in short periods $K = 1$, and then $K = \dfrac{4}{3.7}; \dfrac{5}{4}; \dfrac{6}{4.2}$

for 4, 5, and 6 periods of the system only for these subgroups. For all other cases $K = 1$.

Besides, for the elements only of 1a group of periodic system the value $2r_i$ (i.e., the orbital radius of i) was used as a dimensional characteristic in the first component of eq 6.23.

Spatial-Energy Interactions and Entropy's Curves

Taking into account the pointed out remarks for the initial equation the values $\left(\dfrac{q^2}{r_i} + W\right)\dfrac{1}{KN} + P_E$ and $2W$ for 65 elements were calculated and compared. Some results are given in Table 6.2.

The analysis of the data given in Table 6.2 reveals that the proximity of the values investigated is mostly within 5%. Thus there is a certain analog of eqs 6.19 and 6.23, and the value of P_E-parameter can be considered as the analog of Lagrange function and value $\left(\dfrac{q^2}{ri} + W\right)\dfrac{1}{KN}$ —analog of Hamilton function.[11]

6.6 STRUCTURAL EXCHANGE OF SPATIAL-ENERGY INTERACTIONS

In the process of solid solution formation and other structural equilibrium-exchange interactions the single electron density should be set in the points of atom-component contact. This process is accompanied by the redistribution of electron density between the valence areas of both particles and transition of the part of electrons from some external spheres into the neighboring ones. Apparently, frame atom electrons do not take part in such exchange.

Obviously, when electron densities in free atom-components are similar, the transfer processes between boundary atoms of particles are minimal; this will be favorable for the formation of a new structure. Thus the evaluation of the degree of structural interactions in many cases means the comparative assessment of the electron density of valence electrons in free atoms (on averaged orbitals) participating in the process, which can be correlated with the help of P-parameter model.

The less the difference $(P'_0/r'_i - P''_0/r''_i)$, the more favorable is the formation of a new structure or solid solution from the energy point.

In this regard, the maximum total solubility, evaluated via the coefficient of structural interaction α, is determined by the condition of minimum value α, which represents the relative difference of effective energies of external orbitals of interacting subsystems:

$$\alpha = \frac{P'_0/r_i' - P''_0/r_i''}{(P'_0/r_i' + P''_0/r_i'')/2}\,100\%, \tag{6.24}$$

$$\alpha = \frac{P_C' - P_C''}{P_C' + P_C''}\,200\%, \tag{6.25}$$

TABLE 6.2 Comparison of Basic Energy Atomic Characteristics.

Element	Valence electrons	W (eV)	r_i (Å)	q^2 (eVÅ)	P_0 (eVÅ)	R (Å)	$P_E = P_0/R$ (eV)	N	$\left(\dfrac{2}{\frac{q}{r_i}} + W\right)\left[\dfrac{1}{NK}\right] + P_E$	2W (eV)
Li	$2S^1$	5.3416	1.586·2	5.8892	3.475	1.55	2.2419	1	9.440	10.683
Be	$2S^2$	8.4157	1.040	13.159	7.512	1.13	6.6478	2	17.182	16.831
B	$2P^1$	8.4315	0.776	21.105	4.9945	0.91	5.4885	3	17.365	16.863
	$2S^2$	13.462	0.769	23.890	11.092	0.91	12.189	3	27.032	26.924
C	$2P^2$	11.792	0.596	35.395	10.061	0.86	11.699/2	4	23.645	23.584
	$2S^2$	19.201	0.620	37.240	14.524	0.77	18.862	4	38.824	38.402
N	$2P^3$	15.445	0.4875	52.912	15.830	0.71	22.296/3	5	32.228	30.890
	$2S^2$	25.724	0.521	53.283	17.833	0.71	19.788/3	5	31.392	51.448
O	$2P^1$	17.195	0.4135	71.383	6.4663	0.66	9.7979/4	6	34.087	34.390
	$2S^2$	33.859	0.450	72.620	21.466	0.66	32.524	6	65.064	67.718
F	$2P^1$	19.864	0.3595	93.625	6.6350	0.64	10.367/5	7	42.115	39.728
	$2S^2$	42.792	0.396	94.641	24.961	0.64	39.002	7	79.257	85.584
Na	$3S^1$	4.9552	1.713·2	10.058	4.6034	1.89	2.4357	1	10.327	9.9104
Mg	$3S^1$	6.8859	1.279	17.501	5.8588	1.60	3.6618	2	13.946	13.772
Al	$3P^1$	5.713	1.312	26.443	5.840	1.43	4.084	3	12.707	11.426
	$3S^2$	10.706	1.044	27.119	12.253	1.43	8.5685	3	20.796	21.412

TABLE 6.2 *(Continued)*

Element	Valence electrons	W (eV)	r_i (Å)	q^2 (eVÅ)	P_0 (eVÅ)	R (Å)	$P_E = P_0/R$ (eV)	N	$\left(\dfrac{2}{\dfrac{q}{r_i}+W}\right)\dfrac{1}{NK}+P_E$	2W (eV)
Si	3P¹	8.0848	1.068	29.377	6.6732	1.17	5.7036	4	14.600	16.170
	3S²	14.690	0.904	38.462	15.711	1.17	13.428	4	17.737	29.380
P	3P³	10.659	0.9175	38.199	16.594	1.30	12.765/3	3	21.686	21.318
	3S¹	18.951	0.803	50.922	11.716	1.10	10.651	3	38.106	37.902
S	3P¹	11.901	0.808	48.108	8.0143	1.04	7.7061	4	25.566	23.802
	3P²	11.901	0.808	48.108	13.740	1.04	13.215/2	4	24.468	23.802
	3P⁴	11.904	0.808	48.108	21.375	1.04	20.553/4	4	22.998	23.808

where P_s is the structural parameter found by the equation:

$$\frac{1}{P_C} = \frac{1}{N_1 P'_E} + \frac{1}{N_2 P''_E} + ... \tag{6.26}$$

where N_1 and N_2 are the number of homogeneous atoms in subsystems.

The isomorphism degree and mutual solubility are evaluated in many (over one thousand) simple and complex systems (including nanosystems). The calculation results are in compliance with theoretical and experimental data.

6.7 CONCLUSION

1. The introduced P-parameter can be considered as materialized analog of Ψ-function.
2. The application of such methodology allows modeling physical–chemical processes based on energy characteristics of a free atom.

SECTION 2
S-LINE AND ENTROPY IN TECHNOLOGY, ECONOMICS, AND PHYSICO-CHEMICAL STUDIES

ABSTRACT

The concept of the entropy of spatial-energy interactions is used similarly to the ideas of thermodynamics on the static entropy. The idea of entropy appeared based on the second law of thermodynamics and ideas of the adduced quantity of heat. These rules are general assertions independent of microscopic models. Therefore, their application and consideration can result in a large number of consequences which are most fruitfully used in statistic thermodynamics.

In this research, we are trying to apply the concept of entropy to assess the degree of spatial-energy interactions using their graphic dependence, and in other fields. The nomogram to assess the entropy of different processes is obtained. The variability of entropy demonstrations is discussed, in biochemical processes, economics, and engineering systems, as well.

6.8 INTRODUCTION

In statistic thermodynamics the entropy of the closed and equilibrium system equals the logarithm of the probability of its definite macrostate:

$$S = k \ln W, \qquad (6.27)$$

where W is the number of available states of the system or degree of the degradation of microstates and k is the Boltzmann's constant.

Or
$$W = e^{S/k}. \qquad (6.28)$$

These correlations are general assertions of macroscopic character, they do not contain any references to the structural elements of the systems considered and they are completely independent of microscopic models.

Therefore the application and consideration of these laws can result in a large number of consequences.

At the same time, the main characteristic of the process is the thermodynamic probability W. In actual processes in the isolated system the entropy growth is inevitable—disorder and chaos increase in the system, the quality of internal energy goes down.

The thermodynamic probability equals the number of microstates corresponding to the given macrostate.

Since the system degradation degree is not connected with the physical features of the systems, the entropy statistic concept can also have other applications and demonstrations (apart from statistic thermodynamics).

"It is clear that out of the two systems completely different by their physical content, the entropy can be the same if their number of possible microstates corresponding to one macro parameter (whatever parameter it is) coincide. Therefore the idea of entropy can be used in various fields. The increasing self-organization of human society ... leads to the increase in entropy and disorder in the environment that is demonstrated, in particular, by a large number of disposal sites all over the earth."[12]

In this research, we are trying to apply the concept of entropy to assess the degree of spatial-energy interactions using their graphic dependence and in other fields.

6.9 ENTROPIC NOMOGRAM OF THE DEGREE OF SPATIAL-ENERGY INTERACTIONS

Applying the reliable experimental data we obtain the nomogram of structural interaction degree dependence (ρ) on coefficient α, the same for a wide range of structures (Fig. 6.1).

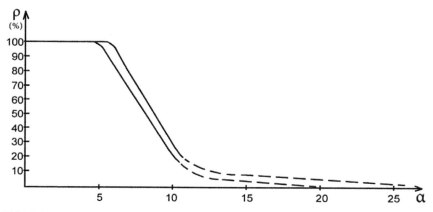

FIGURE 6.1 Nomogram of structural interaction degree dependence (ρ) on coefficient α.

This approach gives the possibility to evaluate the degree and direction of the structural interactions of phase formation, isomorphism and solubility processes in multiple systems, including molecular ones.

Such nomogram can be demonstrated[5] as a logarithmic dependence:

$$a = \beta (\ln \rho)^{-1}, \qquad (6.29)$$

where coefficient β is the constant value for the given class of structures. β can structurally change mainly within ±5% from the average value. Thus coefficient α is reversely proportional to the logarithm of the degree of structural interactions and therefore can be characterized as the entropy of spatial-energy interactions of atomic-molecular structures.

Actually, the more is ρ, the more probable is the formation of stable ordered structures (e.g., the formation of solid solutions), that is the less is the process entropy. But also the less is coefficient α.

Equation 6.29 does not have the complete analogy with Boltzmann's eq 6.27 as in this case not absolute but only relative values of the corresponding characteristics of the interacting structures are compared which can be expressed in percent. This refers not only to coefficient α but also to the comparative evaluation of structural interaction degree (ρ), for example—the percent of atom content of the given element in the solid solution relative to the total number of atoms. Therefore in eq 6.28 coefficient $k = 1$.

Thus, the relative difference of SEPs of the interacting structures can be a quantitative characteristic of the interaction entropy: $\alpha \equiv S$.

6.10 ENTROPIC NOMOGRAM OF SURFACE-DIFFUSION PROCESSES

As an example, let us consider the process of carbonization and formation of nanostructures during the interactions in polyvinyl alcohol gels and metal phase in the form of copper oxides or chlorides. At the first stage, small clusters of inorganic phase are formed surrounded by carbon-containing phase. In this period, the main character of atomic-molecular interactions needs to be assessed via the relative difference of P-parameters calculated through the radii of copper ions and covalent radii of carbon atoms.

In the next main carbonization period the metal phase is being formed on the surface of the polymeric structures.

From this point, the binary matrix of the nanosystem C→Cu is being formed.

The values of the degree of structural interactions from coefficient α are calculated, that is $\rho_2 = f\left(\frac{1}{\alpha_2} \right)$—curve 2 given in Figure 6.2. Here, the graphical dependence of the degree of nanofilm formation (ω) on the process time is presented by the data from Kodolov et al.[13]—curve 1 and previously obtained nomogram in the form $\rho_1 = f\left(\frac{1}{\alpha_1} \right)$—curve 3.

The analysis of all the graphical dependencies obtained demonstrates the practically complete graphical coincidence of all three graphs: $\omega = f(t)$, $\rho_1 = f\left(\frac{1}{\alpha_1} \right)$, $\rho_2 = f\left(\frac{1}{\alpha_2} \right)$ with slight deviations in the beginning and end of the process. Thus, the carbonization rate, as well as the functions of many other physical-chemical structural interactions, can be accessed via the values of the calculated coefficient α and entropic nomogram.

76 — Nanoscience and Nanoengineering: Novel Applications

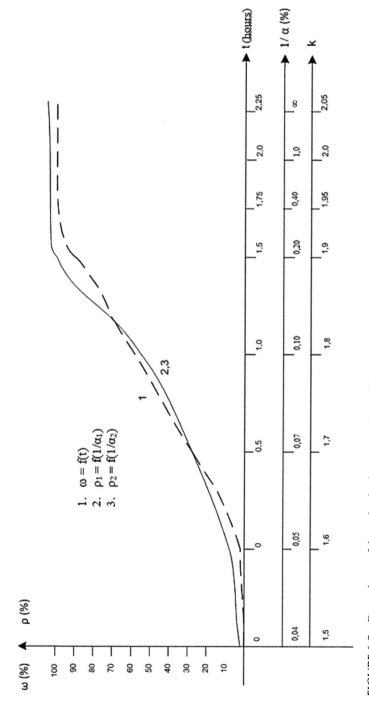

FIGURE 6.2 Dependence of the carbonization rate on the coefficient α.

6.11 NOMOGRAMS OF BIOPHYSICAL PROCESSES

1) On the kinetics of fermentative processes

"The formation of ferment-substrate complex is the necessary stage of fermentative catalysis ... At the same time, n substrate molecules can join the ferment molecule."[1, p. 58]

For ferments with stoichiometric coefficient n not equal to one, the type of graphical dependence of the reaction product performance rate (μ) depending on the substrate concentration (c) has[1] a sigmoid character with the specific bending point (Fig. 6.3).

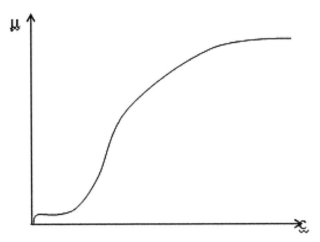

FIGURE 6.3 Dependence of the fermentative reaction rate (μ) on the substrate concentration (c).

In Figure 6.3 it is seen that this curve generally repeats the character of the entropic nomogram in Figure 6.2.

The graph of the dependence of electron transport rate in biostructures on the diffusion time period of ions is similar.[1, p. 278]

In the procedure of assessing fermentative interactions (similarly to the previously applied for surface-diffusive processes) the effective number of interacting molecules over 1 is applied.

In the methodology of P-parameter, a ferment has a limited isomorphic similarity with substrate molecules and does not form a stable compound with them, but, at the same time, such limited reconstruction of chemical bonds which "is tuned" to obtain the final product is possible.

2) Dependence of biophysical criteria on their frequency characteristics
   a) The passing of alternating current through live tissues is characterized by the dispersive curve of electrical conductivity—this is the graphical dependence of the tissue total resistance (z-impedance) on the alternating current frequency logarithm (log ω). Normally, such curve, on which the impedance is plotted on the coordinate axis, and log ω—on the abscissa axis, formally, completely corresponds to the entropic nomogram (Fig. 6.1).
   b) The fluctuations of biomembrane conductivity (conditioned by random processes) "have the form of Lorentz curve." In this graph, the fluctuation spectral density (ρ) is plotted on the coordinate axis, and the frequency logarithm function (log ω)—on the abscissa axis.

The type of such curve also corresponds to the entropic nomogram in Figure 6.1.

6.12 LORENTZ CURVE OF SPATIAL-TIME DEPENDENCE

In Lorentz curve[14] the space-time graphic dependence (Fig. 6.4) of the velocity parameter (θ) on the velocity itself (β) is given, which completely corresponds to the entropic nomogram in Figure 6.2.

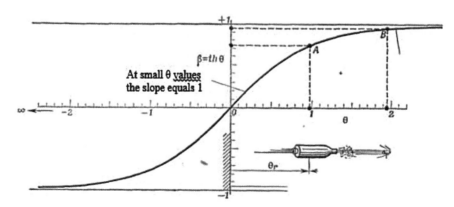

FIGURE 6.4 Concentration between the velocity parameter Θ and velocity itself $\beta = \text{th}\Theta$.

6.13 ENTROPIC CRITERIA IN BUSINESS AND NATURE

The concept of thermodynamic probability as a number of microstates corresponding to the given macrostate can be modified as applicable to the processes of economic interactions that directly depend on the parameters of business structures.

A separate business structure can be taken as the system macrostate, and as the number of microstates—number of its workers (N)—which is the number of the available most probable states of the given business structure. Thus it is supposed that such number of workers of the business structure is the analog of thermodynamic probability as applicable to the processes of economic interactions in business.

Therefore it can be accepted that the total entropy of business quality consists of two entropies characterizing: (1) decrease in the competition efficiency (S_1) and (2) decrease in the personal interest of each worker (S_2), that is: $S = S_1 + S_2$. S_1 is proportional to the number of workers in the company: $S \sim N$, and S_2 has a complex dependence not only on the number of workers in the company but also on the efficiency of its management. It is inversely proportional to the personal interest of each worker. Therefore it can be accepted that $S_2 = 1/\gamma$, where γ is the coefficient of personal interest of each worker.

By analogy with Boltzmann's equation (6.27) we have:

$$s = \ln\left(\frac{N}{\gamma}\right) \tag{6.30}$$

In Table 6.3 you can see the approximate calculations of business entropy by eq 6.30 for three main levels of business: small, medium, and large. At the same time, it is supposed that number N corresponds to some average value from the most probable values.

When calculating the coefficient of personal interest, it is considered that it can change from 1 (one self-employed worker) to zero (0), if such worker is a deprived slave, and for larger companies, it is accepted as $\gamma = 0.1$–0.01.

Despite the rather approximate accuracy of such averaged calculations, we can make quite a reliable conclusion on the fact that business entropy, with the aggregation of its structures, sharply increases during the transition from the medium to large business as the quality of business processes decreases.

Comparing the nomogram (Fig. 6.1) with the data from the Table 6.3, we can see the additivity of business entropy values (S) with the values of the coefficient of spatial-energy interactions (a), that is $S = a$.

TABLE 6.3 Entropy Growth with the Business Increase.

Structure parameters	Business		
	Small	Average	Large
N_1–N_2	10–50	100–1000	10000–100000
γ	0.9–0.8	0.6–0.4	0.1–0.01
S	2.408–4.135	5.116–7.824	11.513–16.118
(S)	3.271	6.470	13.816

It is known that the number of atoms in polymeric chain maximally acceptable for a stable system is about 100 units, which is 10^6 in the cubic volume. Then we again have log $10^6 = 6$.

6.14 S-CURVES ("LIFE LINES")

Already in the last century, some general regularities in the development of some biological systems depending on time (growth in the number of bacteria colonies, population of insects, weight of the developing fetus, etc.) were found.[15] The curves reflecting this growth were similar, first of all, by the fact that three successive stages could be rather vividly emphasized on each of them: slow increase, fast burst-type growth, and stabilization (sometimes decrease) of number (or another characteristic). Later it was demonstrated that engineering systems go through similar stages during their development. The curves drawn up in coordinate system where the numerical values of one of the most important operational characteristics (e.g., aircraft speed, electric generator power, etc.) were indicated along the vertical and the "age" of the engineering system or costs of its development along the horizontal, were called S-curves (by the curve appearance), and they are sometimes also called "life lines."

As an example, the graph of the changes in steel specific strength with time (by years) is demonstrated[15] (Fig. 6.5).

Thus, the similarity between S-curves and entropic nomogram in Figure 6.2 is observed.

And in this case, the same as before, the time dependence (t) is proportional to the entropy reverse value ($1/\alpha$). As applicable to business, such curves characterize the process intensity, for example, sale of the given products.

At the same time, entropic nomograms in accordance with Figure 6.1 assess the business quality (ordinate in such graphs).

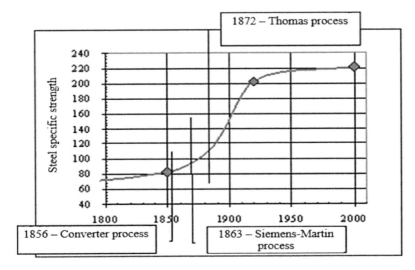

FIGURE 6.5 Dependence of steel specific strength on time.

6.15 ENTROPIC TRANSITIONS IN PHYSIOLOGICAL SYSTEMS

In actual processes, the resources often come to an end not because the system spent them, but because the new system appeared which starts more effectively to perform the similar function and attracts the resources to itself[16] (Fig. 6.6).

This characteristic is not only for short-term acting processes but also for general development of interstructural and cellular interactions.

It is much more complicated to consider and mathematically analyze heteromorphic and rather dynamic intercellular and cellular–cellular interactions.

Neurogenesis can serve as an example of such structural transformations in the organization of biological system. In particular, let us consider the neurogenesis on the example of motor nucleus of trifacial nerve. It is known that on the 10th to 12th day of embryogenesis rats are characterized by high mitotic activity. In this period, the after brain is composed of externally isomorphic populations of medulloblasts different on molecular and submolecular levels. During the indicated period, the size and shape of cells are approximately the same and variety of sizes of cellular populations in mantle layer is limited. From the 12th day, the processes of neuroblast immigration become activated that is followed by the end of their proliferative activity and activation of axon growth.

At this moment, the entropy is demonstrated in dividing the cell groups into mitotically active populations and neuroblasts with different number of arms, which stopped splitting. As a result, at the moment of birth, the variety of sizes of nerve cells reach considerable values from 7 to 8 mm in diameter in small neurons and neuroblasts and up to 25–30 mm in large neurons. The differences are also revealed in the number of arms, degree of morphological maturity of nerve cells. By the age of 9 months a pubertal animal demonstrates the stagnation processes with externally rich variety of neurons inside the nucleus. This tendency of anti-entropy growth proceeds in accordance with Figure 6.2 and is demonstrated in Table 6.4. The factors are compared with a pubertal rat. At the same time, there are vividly expressed methods inside each population with the distribution curve close to a normal one.

TABLE 6.4 Ratio of Types of Neuroblastic Cells in Motor Nucleus of Trifacial Nerve for Medium Rats (M ± m).

Development periods	Percentage of cell type ratio
Newly born	59.0 ± 2.6
1 week	67.4 ± 3.2
1 month	40.8 ± 3.2
Pubertal	26.6 ± 2.3

Thus, in a complex biological object—a mammal brain—the moments in the development of cellular populations are found when the significant dispersion is observed followed by temporary manifestations of entropy increase, which is in compliance with node points of accelerated development and transition to new qualitative change in the population composition (Fig. 6.6).

It is the increase in the structure variety, and thus, controllable transient enhancement of the system chaotic character that can be the basis initiating the transition to new states, to the development of certain cells, cell populations in general (in accordance with Fig. 6.6).

Apparently, the self-organizing processes generally follow the same principle: slow development from the structural variety, fast growth, and stabilization of the renewed biosystem. In such way, nature is struggling with entropy development in organism, maintaining it on the constant level as the main condition of stationary state.

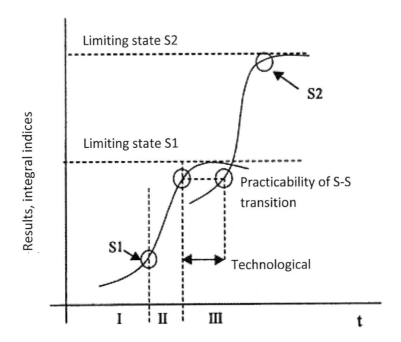

FIGURE 6.6 Entropic transitions by time *t*.

6.16 CONCLUSION

As it is known the entropy of isolated systems does not fall down. In open systems, the entropy growth is compensated by the negative entropy due to the interaction with the environment.

All the systems discussed above can be considered as open ones. This also refers to spatial-energy processes at which any change in dimensional energy characteristics is conditioned by interactions with external systems.

The same is apparently observed in connection with technical and technological systems, in whose development dynamics the additional innovations, modifications, and financial investments take place.

The entropy in thermodynamics is considered as the measure of irreversible energy dissipation. From the position of technological and economic principles, entropy is essentially the measure of irrational consumption of energy resources. With the increase in time dependence, such processes stabilize in accordance with nomograms to the most optimal values—together with anti-entropy growth, that is the value of $1/\alpha = 1/s$.

The similar growth by time of rationality of technological, economic, and physical and mathematical parameters indicates that their nomograms are universal for the majority of main processes in nature, technology, and economics.

It is known that the entropy of isolated systems decreases. The entropy growth in open systems is compensated by the negative entropy due to the interaction with the environment.

All the above systems can be considered as open ones. This also refers to spatial-energy processes, when any changes in quantitative energy characteristics are conditioned by the interaction with external systems.

It is obviously observed in engineering and technological systems, the development of which is followed by additional innovations, modifications, and financial investments.

The entropy in thermodynamics is considered as the measure of nonreversible energy dissipation. From the point of technological and economic principles, the entropy is mainly the measure of irrational energy resource utilization. With the time dependence increase, such processes stabilize in accordance with the nomogram to more optimal values—together with the growth of anti-entropy, that is the value $1/\alpha = 1/s$.

The similar growth with time of rationality of technological, economic and physical, and chemical parameters proves that such nomograms are universal for the majority of main processes in nature, technology, and economy.

General conclusion: The idea of entropy is diversified in physical and chemical, economic, engineering and other natural processes that are confirmed by their nomograms.

SECTION 3
ENERGY CRITERIA OF CONFORMATION INTERACTIONS IN BIOSYSTEMS

ABSTRACT

The notion of SEP, which is a complex characteristic of important atomic values responsible for interatomic interactions, is used to evaluate energy criteria of conformation interactions in bio-structures. The rationality of applying such methodology when investigating the conformation processes of polypeptide chains and fragments of DNA molecules is demonstrated.

Spatial-Energy Interactions and Entropy's Curves

The principles of bio-system formation and stabilization have certain analogy with the conditions of wave processes. Some principles of cluster bio-system formation are analyzed. Several examples of hexagonal formation of cellular structures, in particular, in the course of morphogenesis and formation of complex connexins are given. At the same time, hexagonality distortion of such protein structures is typical for integration problems in the areas of poorly taken transplants and signs of malignant tumor growth. The results obtained do not contradict the experimental data.

6.17 INTRODUCTION

In this work similar equilibrium-exchange processes of structures are evaluated through the notion of P-parameter.[5] Table 6.5 contains the calculation results following this technique for a number of atoms in which: W is the electron bond energy by Fisher C.F., r_i is the orbital radius by Waber J.T., q is the effective charge by Clementi E., and R is the atom dimensional characteristic.

TABLE 6.5 P-parameters of Atoms Calculated via the Electron Bond Energy.

Atom	Valence electrons	W (eV)	r_i (Å)	q_0^2 (eVÅ)	P_0 (eVÅ)	R (Å)	$P_E = P_0/R$ (eV)
1	2	3	4	5	6	7	8
H	$1S^1$	13.595	0.5292	14.394	4.7969	0.5292	9.0644
						0.375	12.792
						0.28	17.132
C	$2P^1$	11.792	0.596	35.395	5.8680	0.77	7.6208
						0.67	8.7582
						0.60	9.780
	$2P^2$	11.792	0.596	35.395	10.061	0.77	13.066
						0.67	15.016
						0.60	16.769
	$2P^3_r$				13.213	0.77	17.160
	$2S^1$	19.201	0.620	37.240	9.0209	0.77	11.715
	$2S^2$				14.524	0.77	18.862
	$2S^1+2P^3_r$				22.234	0.77	28.875
	$2S^1+2P^1_r$				13.425	0.77	17.435
	$2S^2+2P^2$				24.585	0.77	31.929
					24.585	0.67	36.694
						0.60	40.975

TABLE 6.5 *(Continued)*

Atom	Valence electrons	W (eV)	r_i (Å)	$q_0{}^2$ (eVÅ)	P_0 (eVÅ)	R (Å)	$P_E = P_0/R$ (eV)
N	$2P^1$	15.445	0.4875	52.912	6.5916	0.70	9.4166
						0.55	11.985
	$2P^2$				11.723	0.70	16.747
						0.63	18.608
	$2P^3$				15.830	0.70	22.614
						0.55	28.782
	$2S^2$	25.724	0.521	53.283	17.833	0.70	25.476
	$2S^2+2P^3$				33.663	0.70	48.090
O	$2P^1$	17.195	0.4135	71.383	6.4663	0.66	9.7979
	$2P^1$					0.55	11.757
	$2P^2$	17.195	0.4135	71.383	11.858	0.66	17.967
						0.59	20.048
	$2P^4$	17.195	0.4135	71.383	20.338	0.66	30.815
						0.59	34.471
	$2S^2$	33.859	0.450	72.620	21.466	0.66	32.524
	$2S^2+2P^4$				41.804	0.66	63.339
						0.59	70.854

6.18 FORMATION OF POLYPEPTIDE CHAIN

Main components of organic compounds constituting 98% of cell element composition: carbon, oxygen, hydrogen, and nitrogen. The polypeptide bond formed by COOH and NH_2 groups of amino acid CONH is the binding base of protein biopolymers of a cell. At the same time, carbon atoms are more frequently found in polypeptide chain nodes, and sometimes in nitrogen atoms.

Fragments of polypeptide chain nodes are formed of CH, OH, CO, NH, NH_2, COOH atom groups and some radicals. In accordance with conformation energy criteria of the initial methodology,[17] the approximate equality of P-parameters in the process of such chain conformation needs to be followed, both separately for all fragments and chain atomic nodes. The possibility of such process is estimated via the relative difference of P-parameters (coefficient α). The calculations of possible variants are given in Table 6.6. Its analysis results in the conclusion of the existence possibility of three series of such relations. Their summarized data are given in Table 6.6. Such cyclicity of functional correlations can be evaluated from the point of quantum-wave properties of P-parameter.[18]

Spatial-Energy Interactions and Entropy's Curves

TABLE 6.6 Structural P_C-parameters Calculate via the Electron Bond Energy.

Radicals, molecule fragments	P_E' (eV)	P_E'' (eV)	P_C (eV)	Orbitals
OH	9.7979	9.0644	4.7084	O $(2P^1)$
	30.815	17.132	11.011	O $(2P^4)$
	17.967	17.132	8.7710	O $(2P^2)$
H_2O	2·9.0644	17.967	9.0237	O $(2P^2)$
	2·17.132	17.967	11.786	O $(2P^2)$
CH_2	17.160	2·9.0644	8.8156	C $(2S^12P^3_r)$
	31.929	2·17.132	16.528	C $(2S^22P^2)$
	36.694	2·9.0644	12.134	C $(2S^12P^3_r)$
CH_3	31.929	3·17.132	19.694	C $(2S^22P^2)$
	15.016	3·9.0644	9.6740	C $(2P^2)$
CH	36.694	17.132	11.679	C $(2S^22P^2)$
	17.435	17.132	8.6423	C $(2S^22P^2)$
NH	16.747	17.132	8.4687	N $(2P^2)$
	48.090	17.132	12.632	N $(2S^22P^3)$
NH_2	18.608	2·9.0644	9.1827	N $(2P^2)$
	16.747	2·17.132	12.631	N $(2P^2)$
	28.782	2·17.132	18.450	N $(2P^3)$
C_2H_5	2·31.929	5·17.132	36.585	C $(2S^22P^2)$
NO	18.608	17.967	9.1410	N $(2P^2)$
	28.782	20.048	11.817	N $(2P^3)$
CH_2	31.929	2·9.0644	11.563	C $(2S^22P^2)$
CH_3	16.769	3·17.132	12.640	C $(2P^2)$
CH_3	17.160	3·17.132	12.865	C $(2P^3_r)$
COOH	8.4405	8.7710	4.3013	C $(2P^2)$
CO	31.929	20.048	12.315	C $(2S^22P^2)$
C=O	15.016	20.048	8.4405	C $(2P^2)$
C=O	31.929	34.471	16.576	O $(2P^4)$
CO=O	36.694	34.471	17.775	O $(2P^4)$
C—CH_3	31.929	19.694	12.181	C $(2S^22P^2)$
C—CH_3	17.435	19.694	9.2479	C $(2S^12P^1)$
C—NH_2	31.929	18.450	11.693	C $(2S^22P^2)$
C—NH_2	17.435	18.450	8.8844	C $(2S^12P^1)$
C—OH	8.7572	8.7710	4.3821	–

The interference minimum, oscillation weakening (in anti-phase) takes place if the difference in wave move (Δ) equals the odd number of semi-waves:

$$\Delta = (2n+1)\frac{\lambda}{2} = \lambda(n+\frac{1}{2}), \text{ where } n = 0, 1, 2, 3, \ldots \qquad (6.31)$$

It means that the minimum of interactions takes place if P-parameters of interacting structures is also "in anti-phase"—either oppositely charged systems or heterogeneous atoms are interacting (e.g. during the formation of valence-active radicals CH, CH_2, CH_3, NO_2 ..., etc.).

In this case, P-parameters are summed up based on the principle of adding the reciprocals of P-parameters (Table 6.6).

The difference in wave move (Δ) for P-parameters can be evaluated via their relative value $\left(\gamma = \dfrac{P_2}{P_1}\right)$ or via relative difference of P-parameters (coefficient a), which give an odd number at the minimum of interactions:

$$\gamma = \frac{P_2}{P_1} = \left(n+\frac{1}{2}\right) = \frac{3}{2};\frac{5}{2}\ldots \quad \text{At } n = \text{(basic state)} \quad \frac{P_2}{P_1} = \frac{1}{2} \qquad (6.32)$$

Interference maximum, oscillation enhancing (in phase) takes place if the difference in wave move equals an even number of semi-waves:

$$\Delta = 2n\frac{\lambda}{2} = \lambda n \quad \text{or} \quad \Delta = \lambda(n+1) \qquad (6.33)$$

As applicable to P-parameters, the maximum enhancement of interaction in the phase corresponds to the interactions of similarly charged systems or systems homogeneous by their properties and functions (e.g., between the fragments or blocks of complex organic structures, such as CH_2 and NNO_2).

And then:
$$\gamma = \frac{P_2}{P_1} = (n+1) \qquad (6.34)$$

By this model, the maximum of interactions corresponds to the principle of algebraic addition of P-parameters. When $n = 0$ (basic state), we have $P_2 = P_1$, or: the maximum of structure interaction occurs if their P-parameters of subsystems are equal. This postulate and eq 6.34 are used as basic conditions for the formation of stable structures and conformation processes of the fragments of complex systems.

Hydrogen atom, element no. 1 with orbital $1S^1$ defines the main energy criteria of structural interactions (their "ancestor"). Table 6.5 shows its three P_E-parameters corresponding to three different characteristics of the atom.

Spatial-Energy Interactions and Entropy's Curves

$R_1 = 0.5292$ A^0; orbital radius (quantum-mechanical characteristic) gives the initial main value of P_E-parameter equaled to 9.0644 eV and

$R_2 = 0.375$ A^0; distance equaled to the half of the bond energy in H_2 molecule. But if hydrogen atom is bound with other atoms, its covalent radius is ≈ 0.28 A^0.

In accordance with eq 6.34 $P_2 = P_1 (n+1)$, therefore

$$P_1 \approx 9.0644 \text{ eV, } P_2 \approx 18.129 \text{ eV.}$$

These are the values of possible energy criteria of stable (stationary) structures. The dimensional characteristic 0.375 A^0 does not satisfy them, therefore, there is a transition onto the covalence radius $R_3 \approx 0.28$ A^0, which provides the value of P-parameter approximately equaled to P_2.

Three series with approximately equal values of P-parameters of atoms or radicals at $\alpha < 7.5\%$ are given in Table 6.7.

First series for $P_E = 9.0644$ eV; main, initial, where H, C, O, and N atoms have P_E-parameters only of the first electron and interactions proceed in the phase.

Second series for $P''_E = 12.792$ eV is the non-rational, pathological as it more corresponds to the interactions in anti-phase: by eq 6.32 $P''_E = 13.596$ eV.

Third series for $P''''_E = 17.132$ eV; stationary as the interactions are in the phase: by eq 6.33 $P''''_E = 18.129$ eV ($\alpha = 5.5\%$).

With specific local energy actions (electromagnetic fields, radiation, etc.), the structural formation processes can grow along the pathological series II that can result in the destruction of the main conformation chain.[18]

Therefore during the transplantation and use of stem cells, the condition of approximate equality of P-parameters of the corresponding structures should be observed (not by the series II).

In particular, this similarity is expressed in analogs organization of glycoprotein receptor-membrane complexes of glycocalyxes of transplanted cell populations and in structural proximity of active and signal surface areas of proteins forming them.[19] This allows donor cells to most probably form adhesive intercellular communicative interactions both between themselves and cell populations of the recipient, which is the basis for the recommendation to introduce recipient tissues and cells into the areas similar to the localization of these cells that are well known in practice.

This fact is also important for the transplantation of neuroblastic cells into a recipient's brain. It was thought before that one of the most important factors of neuron survival during the transplantation is the lack of neuron ability to form major histocompatibility *complex* (MHC) I-complexes on

TABLE 6.7 Bio-structural Spatial-energy Parameters (eV).

Series	H	C	N	O	CH	CO	NH	NH$_2$	OH	C—NH$_2$	C—CH$_3$	$\langle P_E \rangle$	$\langle \alpha \rangle$
I	9.0644 (1S^1)	8.7582 (2P^1) 9.780 (2P^1)	9.4166 (2P^1)	9.7979 (2P^1)	8.6423 (2S^{2}2P^2–1S^1)	8.4405 (2P^2–2P^2)	8.4687 (2P^2–1S^1)	9.1827	8.7710	8.8844 2S^{1}2P^1_r– (2P^3–1S^1)	9.2479 2S^{1}2P^1_r– (2S^{2}2 P^2–1S^1)	8.9905	0.82
II	12.792 (1S^1) 11.715 (1S^1)	13.066 (2P^2)	11.985 (2P^1)	11.757 (2P^1)	11.679 (2S^{2}2 P^2–1S^1) 12.081 (2S^{2}2 P^2–1S^1)	12.315 (2S^{2}2 P^2–2P^2)	12.632 (2S^{2}2 P^3–1S^1)	12.631	11.011	11.693 2S^{2}2P^2– (2P^3–1S^1)	12.181 2S^{2}2P^2– (2S^{2}2 P^2–1S^1)	12.138	5.25
III	17.132 (1S^1)	16.769 (2P^2) 17.435 (2S^{1}2P^1)	16.747 (2P^2)	17.967 (2P^2)	Blocks C and H	16.576 (2S^{2}2 P^2–2P^4)	Blocks N and H	18.450	Blocks CO and OH	Blocks C and NH$_2$	Blocks C and NH$_2$	17.104	0.16–4.92

Note: Designations of interacting orbitals are indicated in brackets.

glycocalyx. Nevertheless, the ability to form MHC I-complexes of neurons, even under the conditions of possible development of immune reaction, serves as the basis for their synthesis taking into account the important role of affinity of these factors to enhance the interactions with the nearest cell medium and to direct the further differentiation.[20]

6.19 CONFORMATIONS OF HEXAGONAL STRUCTURES IN CELLULAR SYSTEMS

In the Nobel lecture by Edward Moser[21] we can, in particular, point out the following problematic results:

1. Cluster structures of cells form geometrically symmetrical hexagonal systems.
2. Cells themselves statistically concentrate along coordinate axes of symmetry with deviations not exceeding 7.5% (Fig. 6.7).

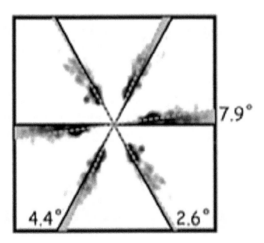

FIGURE 6.7 Statistic distribution of cells along coordinate axes.[7]

It is demonstrated by Korablev et al.[17] that the approximate equality of effective interaction energies along all the bond directions which are evaluated via the values of P-parameter is the main condition for forming stable biosystems. In this regard, hexagonal cycles will be the prevailing structures, and pentagonal—to lesser extent.

A number of facts of hexagonal formation of biological systems can be also given as examples. For instance, the collocation of thin and thick myofilaments in skeletal muscle fibers and cardiomyocytes. At the same time, six thin myofilaments are revealed around each thick one. This system of functionally linked macromolecular complexes is constituted of calcium-dependent transient bonds between myosins and actins.

Also, mechanotoropic interactions in surface layers of multilayer flat epitheliums conjugated with ample desmosomal contacts creating force fields on the background of the available hydrostatic pressure in epithelial cells are naturally followed by the formation of ordered epidermal columns with flat cells in the surface mainly of hexagonal shape and more rarely of pentagonal one.

Apical structures of intermediate contacts in the course of morphogenesis are another example of forming hexagonally organized tension lines in inter-cellular communications. In particular, this phenomenon is revealed in the area of primary stripe at the second stage of gastrulation of vertebrata. The peculiarities in distributing E-cadherins and possible directions of tension lines from regular hexagonal to irregular pentagonal ones depending on the functional state of epithelial cells are an important factor in this process.[22]

The regular structure of connexin proteins in the course of forming connexin complexes in fissure contacts is known. It is demonstrated that the possible option of changing their organization with significant hexagonality distortion during mutations and pathologies with long-term distortion of intercellular ionic homeostasis blocks possible direct informational inter-cellular communications, and, thus, disorganizes synchronizing influence in related cell populations.[23] Such phenomenon is typical for integration problems in the areas of poorly taken transplants and signs of malignant tumor growth.

6.20 CONFORMATION OF FRAGMENTS OF DNA MOLECULES

Pyrimidines—cytosine (C) and thymine (T), as well as purines—adenine (A) and guanine (G) are canonic foundations in DNA molecule. They comprise nitrogen and carbon atoms, as well as molecular groups CH, OH, NH, NH_2, and CH_3. The main property of pyrimidines and purines is nitrogen atoms can attach protons.[24] The values of P-parameters of all these fragment components are given in Tables 6.5—6.7.

Such data are also systemized in operation series with numerical values to be quantized from the initial value of hydrogen atom (9.0644 eV)

Spatial-Energy Interactions and Entropy's Curves 93

approximately two times more. Obviously, the series of structures with the parameter value two times less from the initial one (about 4.533 eV) are also possible. If we take 9.0644 eV as the unit of energy content ($X = 1$), then in the first series $X = 1$ with deviations of coefficient α to ±0.82% for practically all structures. And for radical OH the value of X can be both 0.5 and 1. In Table 6.8 you can see the numerical composition of purines and pyrimidines, as well as the values of their energy contents. Out of two values of X for OH group, one ($X = 1$) corresponds to the possibility of DNA fragment formation. Another one ($X = 0.5$), obtained taking into account the interactions of the most valent-active orbitals of hydrogen and oxygen atoms, apparently characterizes the possibility of structural interactions in the formations already formed that is considered in Table 6.8.

Despite a rather simplified approach in this model, the results obtained to match the experimental data. Thus, from Table 6.8 it follows that $X_A + X_T$ equals the sum of $X_C + X_G$, which indicates the vertical stability of DNA branches. It is known that the horizontal stability of such structure is defined, in particular, by the fact that molecules A and T are connected with two hydrogen bonds, and G and C are connected with three hydrogen bonds.

TABLE 6.8 Numerical and Energy Composition of DNA Fragments.

	x	A	T	C	G	$X_A + X_T$	$X_C + X_G$
C	1	3	3	2	4		
N	1	3	2	2	3		
CH	1	2	1	2	1		
NH	1	1	–	–	1		
NH_2	1	1	–	1	1		
CH_3	1	–	1	–	–		
OH	0.5	–	2	1	1		
$\Sigma <x>$		10	8	7.5	10.5	18	18

Such correlation defines the equality of their energy contents. From Table 6.8 we have $X_T + 2 = X_A$ (two hydrogen bonds are added), and $X_C + 3 = X_G$ (three hydrogen bonds are added).

According to Chargaff's rule,[24] for DNA molecule the number of fragments A equals T and number of fragments C equals G that also corresponds to the equality principle of summed up P-parameters for all fragments of DNA molecule.

6.21 CONCLUSION

The application of SEP methodology allows evaluating the possibility of conformation processes in different biosystems based on energy characteristics of free atoms.

KEYWORDS

- **spatial-energy parameter**
- **wave functions**
- **electron density**
- **structural interactions**
- **entropy**
- **nomogram**
- **biophysical processes**

REFERENCES

1. Rubin, A. B. Biophysics. *Theoretical Biophysics;* Vysshaya Shkola: Moscow, 1987; p 319.
2. Dirac, P. A. *Quantum Mechanics;* Oxford University Press: London, 1935.
3. Anderson. P. W. In *Magnetism;* Academic Press: Cambridge, MA, 1963; p 25.
4. Blokhintsev, D. I. Basics of Quantum Mechanics. Vysshaya Shkola: Moscow, 1961; p 512.
5. Korablev, G. A. *Spatial-Energy Principles of Complex Structures Formation;* Brill Academic Publishers and VSP: Leiden, Netherlands, 2005; p 426.
6. Eyring, G.; Walter, J.; Kimball, G. *Quantum Chemistry;* F. L.: Moscow, 1948; p 528.
7. Fischer, C. F. *Average-energy of Configuration Hartree-Fock Results for the Atoms Helium to Radon. Atomic Data;* Academic Press, Inc.: Cambridge, MA, 1972; Vol. 4, pp 301–399.
8. Waber, J. T.; Cromer, D. T. *J. Chem. Phys.* **1965,** *42* (12), 4116–4123.
9. Clementi, E.; Raimondi, D. L. Atomic Screening Constants from S.C.F. Functions, 1. *J. Chem. Phys.* **1963,** *38* (11), 2686–2689.
10. Gombash, P. *Statistic Theory of an Atom and Its Applications;* I. L.: Moscow, 1951; p 398.
11. Korablev, G. A.; Kodolov, V. I.; Lipanov, A. M. Analog Comparisons of Lagrange and Hamilton Functions with Spatial-energy Parameter. *Chem Phys Mesoscopy* **2004,** *6* (1), 5–8.

12. Gribov, L. A.; Prokofyeva, N. I. *Basics of Physics;* Vysshaya Shkola: Moscow, 1992; p 430.
13. Kodolov, V. I.; Khokhriakov, N. V.; Trineeva, V. V.; Blagodatskikh, I. I. *Chem. Phys. Mesoscopy* **2008,** *10* (4), pp 448–460.
14. Taylor, E.; Wheeler, J. Spacetime Physics. Mir Publishers: Moscow, 1987; p 320.
15. Kynin, A. T.; Lenyashin, V. A. Assessment of the Parameters of Engineering Systems Using the Growth Curves. http://www.metodolog.ru/01428/01428.html.
16. Lyubomirsky, A.; Litvin, S. Laws of Engineering System Development. http://www.metodolog.ru/00767/00767.html.
17. Korablev, G. A.; Vasiliev, Yu. G.; Zaikov, G. E. Hexagonal Structures in Nanosystems. *Chem. Phys. Mesoscopy* **2015,** *17* (3), 424–429.
18. Korablev, G. A.; Zaikov, G. E. In Bio-structural Energy Criteria of Functional States in Normal and Pathological Conditions. *Kinetic Catalysis and Mechanism;* Nova Science Publishers Inc.: Hauppauge, NY, 2012; pp 351–354.
19. Forrest, L. R.; Woolf, T. B. Discrimination of Native Loop Conformations in Membrane Proteins: Decoy Library Design and Evaluation of Effective Energy Scoring Functions. *Proteins* **2003,** *52* (4), 492–509.
20. Joseph, M. S.; Bilousova, T., et al. Transgenic Mice With Enhanced Neuronal Major Histocompatibility Complex Class I Expression Recover Locomotor Function Better after Spinal Cord Injury. *J. Neurosci. Res.* **2011,** *89* (3), 365–372.
21. Nobel Lecture by Edward Moser in Physiology. Aired March 11, 2015, on TV Channel Science.
22. Lecuit, T.; Yap, A. S. E-cadherin Junctions as Active Mechanical Integrators in Tissue Dynamics. *Nat Cell Biol.* **2015,** *17,* 533–539.
23. Zonta, F.; Buratto, D., et al. Molecular Dynamics Simulations Highlight Structural and Functional Alterations in Deafness–related M34T Mutation of Connexin. *Front. Physiol.* **2014,** *26,* 5–85.
24. Volkenshtein, M. V. *Biophysics;* Nauka: Moscow, 1988; p 592.

CHAPTER 7

FUNCTIONAL DEPENDENCIES IN THE EQUATIONS OF MOTION OF THE PLANETS

G. A. KORABLEV[1,2*]

[1]*Department of Physics, Izhevsk State Agricultural Academy, Studencheskaya St. 11, Izhevsk , Russia*

[2]*Basic Research-High Educational Centre of Chemical Physics and Mesoscopy, Ural Division of Russian Academy of Sciences, Izhevsk, Udmurt Republic, Russia*

*E-mail: korablevga@mail.ru

ABSTRACT

Equations of dependence of rotational and orbital motions of planets are given, their rotation angles are calculated. Wave principles of direct and reverse rotation of planets are established. The established dependencies are demonstrated at different scale levels of structural interactions in biosystems as well. The accuracy of calculations corresponds to the accuracy of experimental data.

7.1 INTRODUCTION

The world around us is in constant motion. Main types of functionally interconnected mechanical motions (translational, rotational, and oscillatory) determine the dynamic stability of systems. Vast theoretical and experimental experience in physical and mathematical properties of simple and complex compounds and principles of their self-organization at different scale levels of such conformation has been gained till now.

But the problem of establishing the most general regularities of these processes is topical. "However, the science is still far from making it happen in the general form."[1] Thus, applying the entire set of analytical and qualitative methods, the celestial mechanics provides the solution for many problems on the motion of solids.[2,3] But some other issues of celestial mechanics require further discussion, for instance, the functional dependence of rotational and orbital motion of planets, as well as the initial principles of forming the direct and reverse motion of planets. Therefore, in this chapter, we attempt to investigate such problematic issues with the help of conception of corpuscular–wave dualism proposed earlier.[4]

7.2 INITIAL CRITERIA

1. In the systems in which the interaction proceeds along the potential gradient (positive work), the resultant potential energy is found based on the principle of adding reciprocals of corresponding energies of subsystems. Similarly, the reduced mass for the relative motion of the isolated systems of two particles is calculated.
2. In the systems in which the interaction proceeds against the potential gradient (negative performance), the algebraic addition of their masses, as well as the corresponding energies of subsystems is performed (similar to Hamiltonian).
3. Two principles of adding energy characteristics of structural interactions can be transformed into the corpuscular-wave dualism processes. Corpuscular processes flow in all interactions along the potential gradient, and wave dualism corresponds to the interactions against the potential gradient.
4. Act of quantum action expressed via Plank's constant is narrowed to the energy equilibrium-exchange redistribution between the corpuscular and wave processes.
5. Phase difference between electric and magnetic oscillations in electromagnetic wave is $\pi/2$. Applying $(2/\pi)^2$ as the proportionality coefficient, we have the equation for Plank's constant with the accuracy close to the accuracy of the initial data:

$$h = \left(\frac{4}{\pi^2} + a_0 \right) P_e \frac{\varepsilon}{\mu},$$

where $a_0 = 0.0023293$—experimental quantum correction to spin g_s is the factor, ε is the electric constant, and μ is the magnetic constant. Here $P_e = wr$, where w is the energy of a free electron, and r is its classic radius.

6. It is assumed that during the rotational-translation motion of an electron, the energies are redistributed in the system "particle-wave" that is demonstrated via the angular vector of such motion (winding angle)—Θ.

This angular vector of electron motion is quantized by an integer number through the tangent square of this angle: $tg^2\varphi_r = 2$; $tg^260° = 3$; $tg^245° = 1$, where $\varphi_r = 54.73°$—a so-called "geodesic angle," which is widely spread in engineering, for example, in spaceship production.

The quantum functions of square tangent $k = 1, 2$, and 3 numerically determine the ratios of two triangle legs, whose values characterize energy dependencies via axial and circumferential stresses in the system with corpuscular and wave processes.

7. In quantum mechanics, the ratio between the particle magnetic moment and its mechanical moment is the magnetomechanic ratio—g. At the same time, $g_s = 2$, if the electron magnetic moment is conditioned only by the spin component and $g = 1$, if it is produced by the electron orbital motion. Their ratio $g_s/g = 2$ that, the same way as $tg^2\varphi_r = 2$, characterizes the corresponding corpuscular-wave dependencies in this approach.

7.3 EQUATION OF DEPENDENCE OF ROTATIONAL AND ORBITAL MOTION OF PLANETS

The foregoing principles of corpuscular–wave mechanism give the possibility to consider the unified positions of many structural and dynamic processes, which are different in nature and scale. For example, the characteristic of spin-orbital interaction, fine structure constant $\alpha = r/\lambda$, where r is the electron classic radius and λ is its Compton wavelength.

Formally, but similarly, interaction force of two long conductors with the current proportion to the ratio $l/2\pi r$, where l is the length of conductors and r is the distance between them.

And the number 2π widely used in physical regularities, equals the circumference ratio to its radius: $2\pi = l/R$.

In these examples, as in many others, this approach allows evaluating structural interactions based on the ratios of corpuscular and wave

100 Nanoscience and Nanoengineering: Novel Applications

spatial-energy parameters in each action. Obviously, such principles are also demonstrated in Kepler's third law, which can be given as follows:[3]

$$Gm = 4\pi^2 \frac{a^3}{T^2} \qquad (7.1)$$

where G is the gravitational constant, m is the planet mass, a is the distance to barycenter (system mass center), and T is the planet revolution period. Entering the relative mass $m_0 = m/m_3$ (where m_3 is the Earth mass), we can have:

$$\frac{a^3}{T^2 m_0} = \text{const.}$$

or

$$\frac{1}{(T^2)^{\frac{1}{2}} m_0^{\frac{1}{3}}} = \text{const.} \qquad (7.2)$$

Since the masses of planets are rather small in comparison with their distance to the sun, then at the first approximation they can be considered as mathematical points and the equation of mathematical pendulum period can be applied to them:

$$T^2 = 4\pi^2 \frac{l}{g},$$

where g is the free fall acceleration. The average radius of the planet orbital motion R can be taken as the pendulum length l. Then, similarly to the foregoing dependencies, we can introduce the planet radius r into the numerator of eq 7.2 and then we have:

$$\frac{r}{\left(\frac{4\pi^2}{g} Rm_0\right)^{\frac{1}{3}}} = \gamma \qquad (7.3)$$

This expression in the units $m/s^{2/3}$ satisfies the principle of corpuscular-wave mechanism for space macrostructures. But in Kepler's third law only the orbital motion was considered, but in this case two motions, each of which has its own wave part. Therefore, the interference of coherent waves occurs.

Similar to the foregoing examples, the coherence can be considered as the ratio of the difference of the travel path to the length of the coherent wave (Δ/λ). The interference principle is most easily performed for liquid–gaseous planets (planets in Jupiter system) as shown in Table 7.1.

TABLE 7.1 Characteristics of Rotational and Orbital Motion of Planets.

Planet	$R \times 10^6$, m	$R \times 10^9$, m	$\dfrac{m}{\gamma}$, $sec^{2/3}$	Direction of rotation	Calculation formula for β	n	$\dfrac{m}{\beta}$, $sec^{2/3}$	$\delta = \dfrac{\beta}{\beta_0}$	Θ°_1	Θ°_2	$\Theta^{\circ\ 3,5,6}$
Mercury	2.4397	57.9	1039.7	+	$n^{1/2}\dfrac{\gamma}{2}$	2	735.2	0.9735		86.6	87.0
Venus	6.0515	108.2	855.17	−		1	740.6	0.9810		87.2	87.0
Earth	6.3780	149.6	755.2	+	$2n^{1/2}\dfrac{\gamma}{2}$	1	755.2	1	66.56		66.556
Mars	3.3970	227.9	734.7	+	$2n^{1/2}\dfrac{\gamma}{2}$	1	734.7	0.9728	64.75		64.8
Jupiter	71.492	778.6	715.8	+		1	715.8	$0.9478^{1/2}$		86.6	86.9
Saturn	60.268	1433.7	735.9	+		1	735.9	0.9744	64.85		64.3
Uranus	25.596	2870.4	463.6	−		1	696.8	0.9227		82.07	82.0
Neptune	24.764	4491.1	365.49	+		2	730.98	0.96793^2	62.5		61.68
Pluto	1.1510	5868.9	295.55	−		2	738.75	0.9782			
							$\langle\beta\rangle=732.5$				

$$\beta_+ = 2n\frac{\gamma}{2},\tag{7.4}$$

boosting of oscillations during the direct rotation of planets.

$$\beta_- = \frac{(2n+1)\gamma}{2},\tag{7.5}$$

damping of oscillations during the reverse rotation of planets.

Here n is the integer number.

The intensity of wave propagation depends on the medium density and its distribution in the planet volume.

The value characterizing the planet density increase toward its center is called "dimensionless moment of inertia" (I^*). The ratio of average values of I^* for solid and liquid–gaseous planets based on different data[3,5,6] is within 1.4 and 1.45 as demonstrated in Table 7.2.

TABLE 7.2 Ratio of Dimensionless Moments of Inertia of Solid and Liquid–gaseous Planets.

Planet	Mercury	Venus	Earth	Mars	$\langle I^*_S \rangle$
I^*_S	0.324	0.333	0.33076	0.377	0.341
Planet	Jupiter	Saturn	Uranus	Neptune	$\langle I^*_{LG} \rangle$
I^*_{LG}	0.20	0.22	0.23	0.29	0.235

The average value $I^*_S / I^*_{LG} = 1.451$.
Source: Adapted from Refs. [3,5].

Such property for solid planets is taken into account in Table 7.1 and eqs 7.4 and 7.5 by introducing the values $n^{1/2}$ and $(2n + 1)^{1/2}$. Such approach also refers to Mercury as it is the nearest to the sun, to its liquid–gaseous structure.

In general, the application of corpuscular-wave mechanism to space macro-systems explains the specifics of formation of direct or reverse rotation of the planets.

7.4 ROTATION ANGLES OF PLANETS

In physical sense, the parameter β characterizes the motion difference of interfering waves and γ is the wavelength. The average value of $\beta = 732.5$ m/s$^{2/3}$ with the deviation of most of the planets under 2% (apart from Uranus).

Functional Dependencies in the Equations of Motion

The equation $tg^2\, \Theta = k$ was used to evaluate quantum transitions in atoms.[4] The squares and cubes of initial parameters are applied in Kepler's equation and other regularities of space macro-systems. In this approach, as the calculations demonstrated, the semi-empirical equation is performed:

$$\beta = (tg^2\theta)^4 \tag{7.6}$$

where Θ is the rotation angle of planets.

For Earth $\beta = 755.2$ and based on eq 7.6 $\Theta_0 = 66.455°$. For more accurate calculation, taking into account some analogy of macro- and micro-processes, we use, as before, the experimental quantum correction in the form of $a_0 = 1.0023293$ by the following equation:

$$\theta_1 = a_0\theta_0. \tag{7.7}$$

The calculation by equations for Earth (7.6 and 7.7) gives the value $\Theta_0 = 66.455°$ with the deviation from the experimental value by 0.007%. The sun has the same value of the rotation angle.

As applicable to the rest of the planets, the value $\delta = \beta/\beta_0$ (where $\beta_0 - \beta$ value for Earth) is introduced in eq 7.7 based on the following equations:

$$\theta_1 = a_0\,\delta\theta_0 \tag{7.8}$$

$$\text{or} \qquad \theta_1 = a_0\,\frac{\beta}{\beta_0}\theta_0 \tag{7.8a}$$

$$\theta_1 = \frac{4}{3}a_0\delta\theta_0 \tag{7.9}$$

$$\text{or} \qquad \theta_2 = \frac{4}{3}a_0\,\frac{\beta}{\beta_0}\theta_0 \tag{7.9a}$$

Equations 7.8 and 7.8a are performed for the planets, whose rotation angle is less than Earth's one. For the other planets, eqs 7.9 and 7.9a are performed. Those are the planets, which are in the beginning of the planet subsystem by the value of the dimensionless moment of inertia (Mercury and Jupiter), as well as the planets with the reverse rotation (Venus and Uranus). The calculation results are given in Table 7.1.

The coefficient 4/3 has been applied before for the comparative evaluation of quantum transitions with different complexity types.[4] In this research, the average ratio of the angles by the experimental data[3,5,6] given in Table 7.3 also has the value $1.336 \approx 4/3$.

TABLE 7.3 Ratio of Rotation Angles of Planets.

Planet	Mercury	Venus	Earth	Mars	$\langle \Theta_2 \rangle$
Θ_2	87.00	87.00	86.90	82.00	85.73
Planet	Jupiter	Saturn	Uranus	Neptune	$\langle \Theta_1 \rangle$
Θ_1	66.556	64.80	64.30	61.00	64.16

The average value $\Theta_2/\Theta_1 = 1.336 \approx 4/3$.
Source: Adapted from Refs. [3, 5, and 6].

For the value δ, the influence of the medium density distribution is taken into account via the transition factor from one distribution level to another. Since for Jupiter in Table 7.2 the value Γ^* is less in comparison with Earth's Γ^* in 1.45 times, therefore, in the calculations $\delta = 0.9478^{1/2}$. On the contrary with Jupiter, for Neptune Γ^* increases in 1.45 times, therefore, in the calculations $\delta = 0.96793^2$.

All the calculated results are good in accordance with the experimental data.

7.5 CONCLUSION

1. Semi-empirical equations of the dependence of rotational and orbital motion of planets are obtained.
2. Many structural-dynamic processes in macro- and micro-systems, including the specifics of formation of direct or reverse rotation of the planets are explained based on the previously proposed method to evaluate the corpuscular-wave mechanism.
3. The given calculations of rotation angles of the planets are within the accuracy of the experimental data.

KEYWORDS

- orbital and rotational motion of planets
- direct and reverse rotation of planets
- rotation angles
- complex compounds
- scale levels

REFERENCES

1. Erden-Gruz, T. *Basics of Matter Composition;* Mir: Moscow, 1976; 438.
2. Duboshin, G. N. *Celestial mechanics;* Nauka: Moscow, 1978; p 456.
3. Panteleev, V. L. Physics of Earth and Planets. *Lectures;* MSU: Moscow. http://www.astronet.ru/db/msg/1169697/node17.html.
4. Korablev, G. A. *Spatial-energy Parameter and Its Application in Research;* LAP LAMBERT Academic Publishing: Germany, 2016; pp 1–65.
5. *Encyclopedia in Physics*; Moscow: Soviet Encyclopedia, **1988,** *1,* 704; Moscow: Soviet Encyclopedia, **1990,** *2,* 704; Moscow: Russian Encyclopedia, **1992,** *3,* 672.
6. Allen, K. U. *Astrophysical Magnitudes;* Mir: Moscow, 1977; p 446.

CHAPTER 8

EXCHANGE SPATIAL-ENERGY INTERACTIONS

G. A. KORABLEV[1,2] and G. E. ZAIKOV[3*]

[1]*Department of Physics, Izhevsk State Agricultural Academy, Studencheskaya St. 11, Izhevsk, Russia*

[2]*Basic Research-High Educational Center of Chemical Physics and Mesoscopy, Ural Division of Russian Academy of Sciences, Izhevsk, Udmurt Republic, Russia*

[3]*Department of Bio-Chemistry, N. M. Emanuel Institute of Biochemical Physics, Russian Academy of Science, Kosygina St. 4, Moscow 119991, Russia*

[*]*Corresponding author. E-mail: chembio@sky.chph.ras.ru*

ABSTRACT

The notion of spatial-energy parameter (SEP or P-parameter) is introduced based on the modified Lagrangian equation for relative motion of two interacting material points, and is a complex characteristic of important atomic values responsible for interatomic interactions and having the direct connection with electron density inside an atom. Wave properties of P-parameter are found, its wave equation having a formal analogy with the equation of Ψ-function is given.

With the help of P-parameter technique, numerous calculations of exchange structural interactions are done, the applicability of the model for the evaluation of intensity of fundamental interactions is demonstrated, initial theses of quark screw model are given.

8.1 SPATIAL-ENERGY PARAMETER

When oppositely charged heterogeneous systems interact, there is a certain compensation of the volume energy, which results in decrease in the resultant energy (e.g., during the hybridization of atom orbitals), takes place.[1] But this is not the direct algebraic deduction of corresponding energies. The comparison of numerous regularities in physical and chemical processes lets us assume that in such and similar cases the principle of adding reverse values of volume energies or kinetic parameters of interacting structures are observed. For instance:

1. During the ambipolar diffusion: when joint motion of oppositely charged particles is observed in the given medium (in plasma or electrolyte), the diffusion coefficient (D) is found as follows:

$$\frac{\eta}{D} = \frac{1}{a_+} + \frac{1}{a_-}$$

where a_+ and a_- are charge mobility of both atoms and η is the constant coefficient.

2. Total velocity of topochemical reaction (v) between the solid and gas is found as follows:

$$\frac{1}{v} = \frac{1}{v_1} + \frac{1}{v_2}$$

where v_1 is the reagent diffusion velocity and v_2 is the velocity of reaction between the gaseous reagent and solid.

3. Change in the light velocity (Δv) when moving from vacuum to given medium is calculated by the principle of algebraic deduction of reverse values of the corresponding velocities:

$$\frac{1}{\Delta v} = \frac{1}{v} - \frac{1}{c},$$

where c is the light velocity in vacuum.

4. Lagrangian equation for relative motion of the system of two interacting material points with masses m_1 and m_2 in coordinate x is as follows:

$$m_{np} x'' = -\frac{\partial U}{\partial x} \tag{8.1}$$

Exchange Spatial-Energy Interactions

where

$$\frac{1}{m_r} = \frac{1}{m_1} + \frac{1}{m_2} \qquad (8.1a)$$

where U is the mutual potential energy of material points and m_r is the reduced mass. At the same time $x'' = a$ (characteristic of system acceleration).

For elementary interaction areas Δx: $\dfrac{\partial U}{\partial x} \approx \dfrac{\Delta U}{\Delta x}$

Then, $m_r a\Delta x = -\Delta U$; $\quad \dfrac{1}{1/(a\Delta x)} \cdot \dfrac{1}{(1/m_1 + 1/m_2)} \approx -\Delta U \qquad$ or:

$$\frac{1}{1/(m_1 a\Delta x) + 1/(m_2 a\Delta x)} \approx -\Delta U$$

Since in its physical sense, the product $m_i a\Delta x$ equals the potential energy of each material point $(-\Delta U_i)$, then:

$$\frac{1}{\Delta U} \approx \frac{1}{\Delta U_1} + \frac{1}{\Delta U_2} \qquad (8.2)$$

Thus, the resultant energy characteristic of the interaction system of two material points is found by the principle of adding the reverse values of initial energies of interacting subsystems.

Therefore, assuming that the energy of atom valent orbitals (responsible for interatomic interactions) can be calculated by the principle of adding the reverse values of some initial energy components and the introduction of spatial-energy parameter (SEP or P-parameter) as the averaged energy characteristic of valent orbitals is postulated based on the following equations:

$$\frac{1}{q^2/r_i} + \frac{1}{W_i n_i} = \frac{1}{P_E} \qquad (8.3)$$

or

$$\frac{1}{P_0} = \frac{1}{q^2} + \frac{1}{(Wrn)_i}; \qquad (8.4)$$

$$P_E = P_0/r_i \qquad (8.5)$$

where W_i is the orbital energy of electrons,[2] r_i is the orbital radius of i orbital,[3] $q = Z^*/n^*$,[4,5] n_i is the number of electrons of the given orbital, Z^* and n^* is the effective charge of the nucleus and effective main quantum number, and r is the bond dimensional characteristics.

The value P_0 will be called SEP, and value P_E is the effective P-parameter. Effective SEP has a physical sense of some averaged energy of valent electrons in the atom and is measured in the energy units, for example, in electron volts (eV).

Values of P_0-parameter are tabulated constant values for the electrons of the atom given orbital.

For the dimensionality SEP can be written down as follows:

$$[Ð_0] = [q^2] = [E] \cdot [r] = [h] \cdot [\upsilon] = \frac{kgm^3}{s^2} = Jm$$

where $[E]$, $[h]$, and $[\upsilon]$ are the dimensionalities of energy, Plank's constant, and velocity.

The introduction of P-parameter should be considered as further development of quasi-classic notions using quantum-mechanical data on the atom structure to obtain the criteria of phase-formation energy conditions. At the same time, for similarly charged systems (e.g., orbitals in the given atom) and homogeneous systems the principle of algebraic addition of these parameters will be preserved:

$$\sum P_E = \sum (P_0/r_i); \tag{8.6}$$

$$\sum P_E = \frac{\sum P_0}{r} \tag{8.7}$$

or:
$$\sum P_0 = P_0' + P_0'' + P_0''' + \dots; \tag{8.8}$$

$$r\sum P_E = \sum P_0 \tag{8.9}$$

Here, P-parameters are summed by all atom valent orbitals.

To calculate the values of P_E-parameter at the given distance from the nucleus either atomic radius (R) or ionic radius (r_i) can be used instead of r depending on the bond type.

Applying eq 8.8 to hydrogen atom we can write down the following:

$$K\left(\frac{e}{n_1}\right)^2 = K\left(\frac{e}{n_2}\right)^2 + mc^2\lambda \tag{8.10}$$

where e is the elementary charge, n_1 and n_2 are main quantum numbers, m is the electron mass, c is the electromagnetic wave velocity, λ is the wavelength, and K is a constant.

Exchange Spatial-Energy Interactions

Using the known correlations $v = C/\lambda$ and $\lambda = h/mc$ (where h is the Plank's constant and v is the wave frequency) from eq 8.10, the equation of spectral regularities in hydrogen atom can be obtained, in which $2\pi^2 e^2/hC = K$.

8.2 EFFECTIVE ENERGY OF VALENT ELECTRONS IN ATOM AND ITS COMPARISON WITH STATISTIC MODEL

The modified Thomas–Fermi equation converted to a simple form by introducing dimensionless variables[6] is as follows:

$$U = e\left(V_i - V_0 + \tau_0^2\right) \tag{8.11}$$

where: V_0 is the countdown potential, e is the elementary charge, τ_0 is the exchange and correlation corrections, V_i is the interatomic potential at the distance r_i from the nucleus, and U is the total energy of valent electrons.

For the 21st element, the comparisons of the given value U with the values of P_E-parameter are partially given in Table 8.1.

As it is seen from Table 8.1, the parameter values of U and P_E are practically the same (in most cases with the deviation not exceeding 1–2%) without any transition coefficients. Multiple corrections introduced into the statistic model are compensated with the application of simple rules of adding reverse values of energy parameters, and SEP quite precisely conveys the known solutions of Thomas–Fermi equation for interatomic potential of atoms at the distance r_i from the nucleus. Namely, the following equality takes place:

$$U = P_E = e\ (V_i - V_0 + \tau^2_0) \tag{8.12}$$

Using the known correlation[6] between the electron density (β_i) and interatomic potential (V_i) we have:

$$\beta^{2/3}_i \approx (3e/5) \times (V_i - V_0);\ \beta^{2/3}_i \approx Ae \times (V_i - V_0 + \tau^2_0)$$
$$= [Ae \times r_i \times (V_i - V_0 + \tau^2_0)]/r_i \tag{8.13}$$

where A is a constant. According to eqs 8.12 and 8.13, we have the following correlation:

$$\beta^{2/3}_i = A\ P_0/r_i \tag{8.14}$$

setting the connection between P_0-parameter and electron density in the atom at the distance r_i from the nucleus.

TABLE 8.1 Comparison of Total Energy of Valent Electrons in Atom Calculated in Thomas–Fermi Statistic Atom Model (U) And With the Help of Approximation.

Atom	Valent electrons	r_i (Å)	X	φ (X)	U (eV)	W_i (eV)	n	q^2 (eV Å)	P_E (eV)
1	2	3	4	5	6	7	8	9	10
Ar	$3P^4$	0.639	3.548	0.09–0.084	35.36–33.02	12+	4	73.196	33.45
	$3S^2$	0.607	3.268	0.122–0.105	47.81	34.8 (t)	2	96.107	48.44
					44.81	29.0	2	96.107	42.45
	$2P^4$	0.146	0.785	0.47	834.25	246	4	706.3	817.12
V	$4S^2$	1.401	8.508	0.0325	7.680	7.5	2	22.33	7.730
			8.23	0.0345	8.151				
Cr	$4S^2$	1.453	8.95	0.0295	7.013	7	2	23.712	7.754
			8.70	0.0313	7.440				
Mn	$4S^2$	1.278	7.76	0.0256	10.89	6.6 (t)	2	25.12	7.895
						7.5	2	25.12	10.87
Fe	$4S^2$	1.227	7.562	0.0282	8.598	8.00	2	26.57	9.201
						7.20 (t)	2	26.57	8.647
Co	$4S^2$	1.181	7.565	0.02813	9.255	8	2	27.98	10.062
			7.378	0.03075	10.127	7.5 (t)	2	27.98	9.187
Ni	$4S^2$	1.139	7.2102	0.02596	9.183	9	2	29.348	10.60
						7.7 (t)	2	29.348	9.640
Cu	$4S^2$	1.191	7.633	0.0272	9.530	7.7	2	30.717	9.639
Jn	$5S^2$	1.093	8.424	0.033	21.30	11.7	2	238.3	21.8
			8.309	0.03415	22.03*				
	$4d^{10}$	0.4805	3.704	0.106	155.6	20	10	258.23	145.8
Te	$5p^4$	1.063	8.654	0.0335	23.59	9.8	4	67.28	24.54
			8.256	0.0346	24.37*				
	$5S^2$	0.920	7.239	0.0326	26.54	19	2	90.577	27.41
			7.146	0.0341	27.72*	17	2	90.537	25.24

Note: (1) Bond energies of electrons W_i are obtained: "t"—theoretically (by Hartree–Fock method), "+"—by XPS method, all the rest—by the results of optic measurements and (2) "*"—energy of valent electrons (U) calculated without Fermi-Amaldi amendment.

Exchange Spatial-Energy Interactions

From the value $e(V_i - V_0 + \tau_0^2)$ in Thomas–Fermi model, there is a function of charge density, P_0-parameter is a direct characteristic of electron charge density in atom.

This is confirmed by an additional check of equality correctness (8.14) using Clementi function.[7] A good correspondence between the values β_i, calculated via the value P_0 and obtained from atomic functions (Fig. 8.1) is observed.

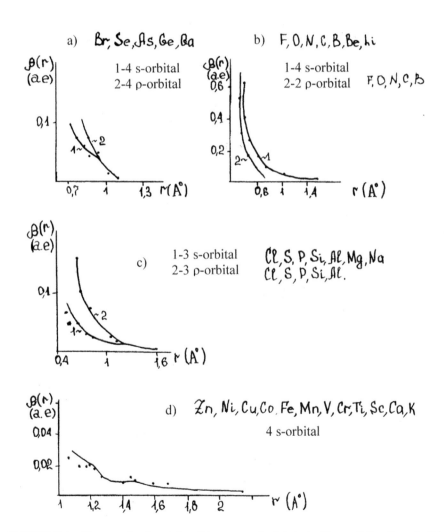

FIGURE 8.1 Electron density at the distance r_i, calculated via Clementi functions (solid lines) and with the help of P-parameter (dots).

8.3 WAVE EQUATION OF P-PARAMETER

For the characteristic of atom spatial-energy properties two types of P-parameters with simple correlation between them are introduced: where R is atom dimension characteristic. Taking into account additional quantum characteristics of sublevels in the atom, this equation in coordinate x can be written down as follows:

$$\Delta P_E \approx \frac{\Delta P_0}{\Delta x} \quad \text{or} \quad \partial P_E = \frac{\partial P_0}{\partial x}$$

where the value ΔP equals the difference between P_0-parameter of i-orbital and P_{CD} countdown parameter (parameter of basic state at the given set of quantum numbers).

According to the established rule[8] of adding P-parameters of similarly charged or homogeneous systems for two orbitals in the given atom with different quantum characteristics and in accordance with the law of energy conservation we have:

$$\Delta P_E'' - \Delta P_E' = P_{E,\lambda}$$

where $P_{E,\lambda}$ is the P-parameter of quantum transition.

Taking as the dimension characteristic of the interaction $\Delta \lambda = \Delta x$, we have:

$$\frac{\Delta P_0''}{\Delta \lambda} - \frac{\Delta P_0'}{\Delta \lambda} = \frac{P_0}{\Delta \lambda} \quad \text{or} \quad \frac{\Delta P_0'}{\Delta \lambda} - \frac{\Delta P_0''}{\Delta \lambda} = -\frac{P_{0\lambda}}{\Delta \lambda}$$

We divide term wise by $\Delta \lambda$: $\quad \dfrac{\dfrac{\Delta P_0'}{\Delta \lambda} - \dfrac{\Delta P_0''}{\Delta \lambda}}{\Delta \lambda} = -\dfrac{P_0}{\Delta \lambda^2}$, where

$$\frac{\dfrac{\Delta P_0'}{\Delta \lambda} - \dfrac{\Delta P_0''}{\Delta \lambda}}{\Delta \lambda} \sim \frac{d^2 P_0}{d\lambda^2}, \quad \text{that is,} \quad \frac{d^2 P_0}{d\lambda^2} + \frac{P_0}{\Delta \lambda^2} \approx 0$$

Taking into account the interactions where $2\pi\Delta x = \Delta \lambda$ (closed oscillator), we have the following equation:

$$\frac{d^2 P_0}{dx^2} + 4\pi^2 \frac{P_0}{\Delta \lambda^2} \approx 0$$

Exchange Spatial-Energy Interactions   115

As $\Delta\lambda = \dfrac{h}{mv}$, then: $\dfrac{d^2 P_0}{dx^2} + 4\pi^2 \dfrac{P_0}{h^2} m^2 v^2 \approx 0$

or $\dfrac{d^2 P_0}{dx^2} + \dfrac{8\pi^2 m}{h^2} P_0 E_k = 0$ (8.15)

where $E_k = \dfrac{mV^2}{2}$ is the electron kinetic energy.

Schrödinger equation for stationary state in coordinate x:

$$\frac{d^2 \psi}{dx^2} + \frac{8\pi^2 m}{h^2} \psi E_k = 0 .$$ (8.16)

Comparing eqs 8.15 and 8.16, we can see that P_0-parameter correlates numerically with the value of Ψ-function: $P_0 \approx \Psi$ and in general case is proportional to it: $P_0 \sim \Psi$. Taking into account wide practical application of P-parameter methodology, we can consider this criterion the materialized analog of Ψ-function.

Since P_0-parameters like Ψ-function possess wave properties, the principles of superposition should be executed for them, thus determining the linear character of equations of adding and changing P-parameters.

8.4 WAVE PROPERTIES OF P-PARAMETERS AND PRINCIPLES OF THEIR ADDITION

Since P-parameter possesses wave properties (by the analogy with Ψ'-function), then the regularities of the interference of corresponding waves should be mainly executed with structural interactions.

Interference minimum, oscillation attenuation (in anti-phase) takes place if the difference in wave motion (Δ) equals the odd number of semi-waves:

$$\Delta = (2n+1)\frac{\lambda}{2} = \lambda(n + \frac{1}{2}), \qquad \text{where } n = 0, 1, 2, 3, \dots \quad (8.17)$$

As applied to P-parameters, this rules means that interaction minimum occurs if P-parameters of interacting structures are also "in anti-phase"—there is an interaction either between oppositely charged systems or heterogeneous atoms (e.g., during the formation of valent-active radicals CH, CH_2, CH_3, NO_2, etc.).

In this case, the summation of P-parameters takes place by the principle of adding the reverse values of P-parameters—eqs 8.3 and 8.4).

The difference in wave motion (Δ) for P-parameters can be evaluated via their relative value $\left(\gamma = \dfrac{P_2}{P_1} \right)$ or via the relative difference in P-parameters (coefficient α), which with the minimum of interactions produce an odd number:

$$\gamma = \frac{P_2}{P_1} = \left(n + \frac{1}{2} \right) = \frac{3}{2}; \frac{5}{2} \dots$$

When $n = 0$ (main state) $\dfrac{P_2}{P_1} = \dfrac{1}{2}$.

$$(8.18)$$

Let us mention that for stationary levels of one-dimensional harmonic oscillator, the energy of these levels $\varepsilon = hv\ (n + \dfrac{1}{2})$, therefore, in quantum oscillator, in contrast to a classical one, the minimum possible energy value does not equal to zero.

In this model, the interaction minimum does not produce the zero energy, corresponding to the principle of adding the reverse values of P-parameters—eqs 8.3 and 8.4. Interference maximum and oscillation amplification (in phase) take place if the difference in wave motion equals the even number of semi-waves:

$$\Delta = 2n\ \frac{\lambda}{2} = \lambda n \quad \text{or } \Delta = \lambda(n+1).$$

As applied to P-parameters the maximum amplification of interactions in the phase corresponds to the interactions of similarly charged systems or systems homogeneous in their properties and functions (e.g., between the fragments and blocks of complex organic structures, such as CH_2 and NNO_2 in octogen).

Then,
$$\gamma = \frac{P_2}{P_1} = (n+1).$$

$$(8.19)$$

By the analogy, for "degenerated" systems (with similar values of functions) of two-dimensional harmonic oscillator the energy of stationary states:

$$\varepsilon = hv\ (n + 1).$$

In this model, the interaction maximum corresponds to the principle of algebraic addition of P-parameters—eqs 8.6–8.8. When $n = 0$ (basic state), we have $P_2 = P_1$, or: interaction maximum of structures takes place when

Exchange Spatial-Energy Interactions 117

their P-parameters equal. This postulate can be used as[8] the main condition of isomorphic replacements.

8.5 STRUCTURAL EXCHANGE SPATIAL-ENERGY INTERACTIONS

In the process of solution formation and other structural interactions, the single electron density should be set in the points of atom-component contact. This process is accompanied by the redistribution of electron density between the valent areas of both particles and transition of the part of electrons from some external spheres into the neighboring ones. Apparently, frame atom electrons do not take part in such exchange.

Obviously, when electron densities in free atom-components are similar, the transfer processes between boundary atoms of particles are minimal; this will be favorable for the formation of a new structure. Thus, the evaluation of the degree of structural interactions in many cases means the comparative assessment of the electron density of valent electrons in free atoms (on averaged orbitals) participating in the process.

The less the difference $(P'_0/r'_i - P''_0/r''_i)$, the more favorable is the formation of a new structure or solid solution from the energy point.

In this regard, the maximum total solubility, evaluated via the coefficient of structural interaction α is determined by the condition of minimum value α which represents the relative difference of effective energies of external orbitals of interacting subsystems:

$$\alpha = \frac{P'_o / r'_o - P''_o / r''_i}{(P'_o / r'_i + P''_o / r''_i)/2}100\% \tag{8.20}$$

$$\alpha = \frac{P'_S - P''_S}{P'_S + P''_S}200\% \tag{8.20a},$$

where P_S is structural parameter is found by the equation:

$$\frac{1}{P_S} = \frac{1}{N_1 P'_E} + \frac{1}{N_1 P''_E} + ... \tag{8.20b},$$

where N_1 and N_2 are numbers of homogeneous atoms in subsystems.

The nomogram of the dependence of structural interaction degree (ρ) on coefficient α, unified for the wide range of structures, was prepared based on all the data obtained. In Figure 8.2, you can see such nomogram obtained using P_E-parameters calculated via the bond energy of electrons (w_i) for structural interactions of isomorphic type.

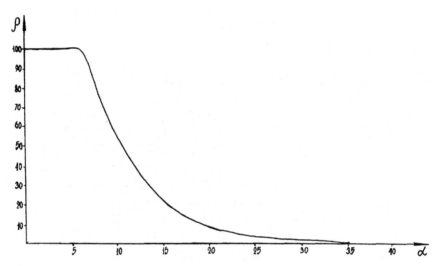

FIGURE 8.2 Dependence of the structural interaction degree (ρ) on the coefficient α.

The mutual solubility of atom-components in many (over a thousand) simple and complex systems was evaluated using this technique. The calculation results are in compliance with theoretical and experimental data.[8]

Isomorphism as a phenomenon is used to be considered as applicable to crystalline structures. But similar processes can obviously take place between molecular compounds, where their role and importance are not less than those of purely coulomb interactions.

In complex organic structures during the interactions the main role can be played by separate "blocks" or fragments. Therefore, it is necessary to identify these fragments and evaluate their SEP. Based on the wave properties of P-parameter, the overall P-parameter of each fragment should be found by the principle of adding the reverse values of initial P-parameters of all atoms. The resultant P-parameter of the fragment block or all the structure is calculated by the rule of algebraic addition of P-parameters of the fragments constituting them.

The role of the fragments can be played by valent-active radicals, for example, CH, CH_2, $(OH)^-$, NO, NO_2, $(SO_4)^{2-}$, etc. In complex structures, the given carbon atom usually has two or three side bonds. During the calculations by the principle of adding the reverse values of P-parameters the priority belongs to those bonds, for which the condition of interference minimum is better performed. Therefore, the fragments of the bond C—H (for CH, CH_2, CH_3 ...) are calculated first, then separately the fragments N—R, where R is the binding radical (e.g., for the bond C—N).

Exchange Spatial-Energy Interactions

Apparently spatial-energy exchange interactions (SEI) based on equalizing electron densities of valent orbitals of atom-components have in nature the same universal value as purely electrostatic coulomb interactions, but they supplement each other. Isomorphism known from the time of E. Mitscherlich (1820) and D. I. Mendeleev (1856) is only the particular manifestation of this general natural phenomenon. The numerical side of the evaluation of isomorphic replacements of components both in complex and simple systems rationally fit in the frameworks of P-parameter methodology. More complicated is to evaluate the degree of structural SEI for molecular, including organic structures. The technique for calculating P-parameters of molecules, structures, and their fragments is successfully implemented. But such structures and their fragments are frequently not completely isomorphic with respect to each other. Nevertheless, there is SEI between them, the degree of which in this case can be evaluated only semi-quantitatively or qualitatively. By the degree of isomorphic similarity all the systems can be divided into three types:

I Systems mainly isomorphic to each other—systems with approximately similar number of dissimilar atoms and summarily similar geometrical shapes of interacting orbitals.

II Systems with the limited isomorphic similarity—such systems, which:
1. either differ by the number of dissimilar atoms but have summarily similar geometrical shapes of interacting orbitals;
2. or have definite differences in geometrical shapes of orbitals but similar number of interacting dissimilar atoms.

III Systems do not have isomorphic similarity—such systems, which considerably differ by both the number of dissimilar atoms and geometric shapes of their orbitals.

Then, taking into account the experimental data, all types of SEI can be approximately classified as follows:

Systems I

1. $\alpha < (0–6)\%$; $\rho = 100\%$.
   Complete isomorphism, there is complete isomorphic replacement of atom-components.
2. $6\% < \alpha < (25–30)\%$; $\rho = 98 – (0–3)\%$.
   There is either wide or limited isomorphism according to nomogram 1.
3. $\alpha > (25–30)\%$.
   no SEI.

Systems II

1. $\alpha < (0\text{--}6)\%$.
   a. There is the reconstruction of chemical bonds, can be accompanied by the formation of a new compound.
   b. Breakage of chemical bonds can be accompanied by separating a fragment from the initial structure but without attachments or replacements.
2. $6\% < \alpha < (25\text{--}30)\%$.
   Limited internal reconstruction of chemical bonds without the formation of a new compound or replacements is possible.
3. $\alpha > (20\text{--}30)\%$.
   no SEI.

Systems III

1. $\alpha < (0\text{--}6)\%$.
   a. Limited change in the type of chemical bonds of the given fragment, internal regrouping of atoms without the breakage from the main part of the molecule and without replacements.
   b. Change in some dimensional characteristics of the bond is possible.
2. $6\% < \alpha < (25\text{--}30)\%$.
   Very limited internal regrouping of atoms is possible.
3. $\alpha > (25\text{--}30)\%$.
   no SEI.

Nomogram (Fig. 8.2) is obtained for isomorphic interactions (systems of types I and II).

In all other cases, the calculated values α and ρ refer only to the given interaction type, the nomogram of which can be clarified by reference points of etalon systems. If we take into account the universality of spatial-energy interactions in nature, this evaluation can be significant for the analysis of structural rearrangements in complex biophysical–chemical processes.

Fermentative systems contribute a lot to the correlation of structural interaction degree. In this system, the ferment structure active parts (fragments, atoms, and ions) have the value of P_E-parameter that is equal to P_E-parameter of the reaction final product. This means the ferment is structurally "tuned" via SEI to obtain the reaction final product, but it will not include into it due to the imperfect isomorphism of its structure (in accordance with III).

Exchange Spatial-Energy Interactions

The most important characteristics of atomic-structural interactions (mutual solubility of components, energy of chemical bond, energetics of free radicals, etc.) were evaluated in many systems using this technique.[8-15]

8.6 TYPES OF FUNDAMENTAL INTERACTIONS

According to modern theories, the main types of interactions of elementary particles, their properties, and specifics are mainly explained by the availability of special complex "currents"—electromagnetic, proton, lepton, etc. Based on the foregoing model of SEP, the exchange structural interactions finally come flowing and equalizing the electron densities of corresponding atomic-molecular components. The similar process is obviously appropriate for elementary particles as well. It can be assumed that in general case interparticle exchange interactions come to the redistribution of their energy masses M.

The elementary electrostatic charge with the electron as a carrier is the constant of electromagnetic interaction.

Therefore, for electromagnetic interaction, we will calculate the system proton–electron.

For strong inter nucleon interaction that comes to the exchange of π-mesons, let us consider the systems nuclides-π-mesons. Since the interactions can take place with all three mesons (π^-, π^0, and π^+), we take the average mass in the calculations ($<M> = 136.497$ MeV/s^2).

Rated systems for strong interaction:

1. P—(π^-, π^0, π^+);
2. $(P–n)$—(π^-, π^0, π^+);
3. $(n–P–n)$—(π^-, π^0, π).

Neutrino (electron and muonic) and its antiparticles were considered as the main representatives of weak interaction.

Dimensional characteristics of elementary particles (r) were evaluated in femtometer units (1 fm = 10^{-15} m) according to Murodyan and PEChAYa.[16]

At the same time, the classic radius: $r_e = e^2/m_e s^2$ was used for electron, where e is the elementary charge, m_e is the electron mass, and s is the light speed in vacuum.

The fundamental Heisenberg length (6.690×10^{-4} fm) was used as the dimensional characteristic of weak interaction for neutrino.[16]

The gravitational interaction was evaluated via the proton P-parameter at the distance of gravitational radius (1.242×10^{-39} fm).

In the initial eq 8.3 for free atom, P_0-parameter is found by the principle of adding the reverse values q^2 and wr, where q is the nucleus electric charge and w is the bond energy of valent electron.

Modifying eq 8.3, as applied to the interaction of free particles, we receive the addition of reverse values of parameters $P = Mr$ for each particle by the equation:

$$1/P_0 = 1/(Mr)_1 + 1/(Mr)_2 + \ldots \tag{8.21}$$

where M is the energy mass of the particle (MeV/s^2).

By eq 8.21 using the initial data[16] P_0-parameters of coupled strong and electromagnetic interactions were calculated in the following systems:

1. nuclides-π-mesons—(P_n-parameters);
2. proton–electron—(P_E-parameter).

For weak and gravitational interactions only the parameters $P_v = Mr$ and $P_r = Mr$ were calculated, as in accordance with eq 8.21 the similar nuclide parameter with greater value does not influence the calculation results.

The relative intensity of interactions (Table 8.2) was found by the equations for the following interactions:

1. strong $\qquad \alpha_B = <P_n>/<P_n> = P_n/P_n = 1 \qquad (8.22a)$

2. electromagnetic $\qquad \alpha_B = P_e/<P_n> = 1/136.983 \qquad (8.22b)$

3. weak $\qquad \alpha_B = P_g/<P_n>, \alpha_B = 2.04 \times 10^{-10}; 4.2 \times 10^{-6} \qquad (8.22c)$

4. gravitational $\qquad \alpha_B = P_r/<P_n> = 5.9 \times 10^{-39}. \qquad (8.22d)$

In the calculations of α_B the value of P_n-parameter was multiplied by the value-equaled $2\pi/3$, that is, $<P> = (2\pi/3)P_n$. Number 3 for nuclides consisting of three different quarks is "a magic" number (see the next section for details). As it is known, number 2π has a special value in quantum mechanics and physics of elementary particles. In particular, only the value of 2π correlates theoretical and experimental data when evaluating the sections of nuclide interaction with each other.[17]

As it is known[18], "very strong," "strong," and "moderately strong" nuclear interactions are distinguished. For all particles in the large group with relatively similar mass values of—unitary multiplets or supermultiplets—very strong interactions are the same.[18] In the frames of the given model, a very strong interaction between the particles corresponds to the

Exchange Spatial-Energy Interactions

TABLE 8.2 Types of Fundamental Interactions.

Interaction type		M, $<M>$ (MeV/s²)	r (fm)	Elementary particles	M, $<M>$ (MeV/s²)	r (fm)	P_n, P_e, P_v, P_z (MeV fm/s²)	$\frac{2\pi}{3}P_n$ $= <P_n>$	$a_e, <a_e>$ (calculation) —by eq 8.22	a_e (experiment)
Electromagnetic	P	938.28	0.856	e^-	0.5110	2.8179	$P_e = 1.4374$	—	1/136.983	1/137.04
Strong	P	938.28	0.856	π^-, π^0, π^+	136.497	0.78	$P_n = 94.0071$	196.89	1	1
	P–n	938.92	0.856	π^-, π^0, π^+	136.497	0.78	$P_n = 94.015$	196.90	1	1
	n–P–n	939.14	0.856	π^-, π^0, π^+	136.497	0.78	$P_n = 94.018$	196.91	1	1
Weak				$\upsilon_e, \overline{\upsilon}_e$	$<6\times10^{-5}$	6.69×10^{-4}	$P_v = 4.014\times10^{-8}$		$<2.04\times10^{-10}$	$10^{-10}-10^{-14}$
				$\upsilon_\mu, \overline{\upsilon}_\mu$	<1.2	6.69×10^{-4}	$P_v = 8.028\times10^{-4}$		$<4.2\times10^{-6}$	$10^{-5}-10^{-6}$
Gravitational	P	938.28	1.242×10^{-39}				$P_r = 1.17\times10^{-36}$		5.9×10^{-39}	$10^{-38}-10^{-39}$

maximum value of P-parameter $P = Mr$ (coupled interaction of nuclides). Taking into consideration the equality of dimensional characteristics of proton and neutron, by eq 8.21 we obtain the values of P_n-parameter equaled to 401.61, 401.88, and 402.16 (MeVfm/s²) for coupled interactions p–p, p–n, and n–n, respectively, thus obtaining the average value $\alpha_g = 4.25$. It is a very strong interaction. For eight interacting nuclides, $\alpha_g \approx 1.06$—a strong interaction.

When the number of interacting nuclides increases, α_g decreases—moderately strong interaction. Since the nuclear forces act only between neighboring nucleons, the value α_g cannot be very small.

The expression of the most intensive coupled interaction of nuclides is indirectly confirmed by the fact that the life period of double nuclear system appears to be much longer than the characteristic nuclear time.[19]

Thus, it is established that the intensity of fundamental interactions is evaluated via P_n-parameter calculated by the principle of adding the reverse values in the system nuclides-π-mesons. Therefore, it has the direct connection with Plank's constants:

$$(2\pi/3)P_n \approx Er = 197.3 \text{ MeVfm/s}^2 \qquad (8.23)$$

$$(2\pi/3)P_\tau \approx M_n \lambda_k = \text{MeVfm/s}^2 \qquad (8.23a)$$

where E and r are Plank's energy and Plank's radius calculated via the gravitational constant; M_n and λ_k are energy mass and nuclide Compton wavelength.

In eq 8.21, the exchange interactions are evaluated via the initial P-parameters of particles equaled to the product of mass by the dimensional characteristic: $P = Mr$.

Since these P-parameters can refer to the particles characterizing fundamental interactions, their direct correlation defines the process intensity degree (α_g):

$$\alpha_a = \frac{P_i}{P_n} = \frac{(Mr)_i}{(Mr)_n} \qquad (8.24)$$

The calculations by eq 8.24 using the known Plank's values and techniques are given in Table 8.3. As before, the energy and dimensional characteristics are taken by Murodyan and PEChAYa.[16]

The results obtained are in accordance with theoretical and experimental data, for example, Bukhbinder[20] and Okun.[21]

TABLE 8.3 Evaluation of the Intensity of Fundamental Interactions Using Plank's Constants and Parameter $P = Mr$.

Interaction type	Particles, constants	M (MeV/s^2)	r (fm)	Mr (MeVfm/s^2)	$\alpha_e = Mr/(Mr)_p$ (calculation)	α_e (experiment)
Strong	Proton	938.28	$\Lambda = 0.2103$	197.3	1	1
	Plank's values	1.221×10^{22}	1.616×10^{-20}	197.3		
Electromagnetic	Electron	0.5110	2.8179	1.43995	$1/137 \times 02$	1/137.036
Weak	$\upsilon_e, \bar{\upsilon}_e$	$<6 \times 10^{-5}$	6.69×10^{-4}	$<4.014 \times 10^{-8}$	$<2.03 \times 10^{-10}$	10^{-10} to 10^{-14}
	$\upsilon_\mu \bar{\upsilon}_\mu$	<1.2	6.69×10^{-4}	$<8.028 \times 10^{-4}$	$<4.07 \times 10^{-6}$	10^{-5} to 10^{-6}
Gravitational	Proton	938.28	1.242×10^{-39}	1.165×10^{-36}	5.91×10^{-39}	10^{-38} to 10^{-39}
			Gravitational radius			

8.7 ON QUARK SCREW MODEL

Let us proceed from the following theses and assumptions:

1. By their structural composition, macro- and micro-world resemble a complex matreshka. One part has some similarity with the other: solar system–atom–atom nucleus–quarks.
2. All parts of this "matreshka" are structural formations.
3. Main property of all systems—motion: translatory, rotary, and oscillatory.
4. Description of these motions can be done in Euclid three-dimensional space with coordinates x, y, and z.
5. Exchange energy interactions of elementary particles are carried out by the redistribution of their energy mass M (MeV/s^2).

Based on these theses we suggest discussing the following screw model of the quark.

1. Quark structure is represented in certain case as a spherical one, but in general, quark is a flattened (or elongated) ellipsoid of revolution. The revolution takes place around the axis (x) coinciding with the direction of angular speed vector, perpendicular to the direction of ellipsoid deformation.
2. Quark electric charge (q) is not fractional but integer, but redistributed in three-dimensional space with its virtual concentration in the directions of three coordinate axes: $q/3$.
3. Quark spherical or deformed structure has all three types of motion. Two of them—rotary and translatory are in accordance with the screw model, which besides these two motions, also performs an oscillatory motion in one of three mutually perpendicular planes: xoy, xoz, and yoz (Fig. 8.3).
4. Each of these oscillation planes corresponds to the symbol of quark color (e.g., xoyred, xozblue, and yozgreen).
5. Screw can be "right" or "left." This directedness of screw rotation defines the sign of quark electric charge. Let us assume that the left screw corresponds to positive and right screw corresponds to negative quark electric charge.
6. Total number of quarks is determined by the following scheme: for each axis (x, y, and z) of translator motion two screws (right and left) with three possible oscillation planes.

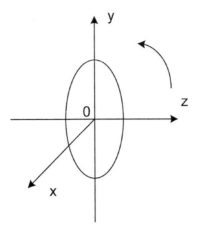

FIGURE 8.3 Structural scheme of quark in section yoz.

We have 3 × 2 × 3 = 18. Besides, there are 18 antiquarks with opposite characteristics of screw motions. Total: 36 types of quarks.

7. These quark numbers can be considered as realized degrees of freedom of all three motions (three translatory + two rotary + three oscillatory).
8. Translatory motion is preferable by its direction, coinciding with the direction of angular speed vector. Such elementary particles constitute our World. The reverse direction is less preferable—this is "Antiworld."
9. Motion along axis x in the direction of the angular speed vector, perpendicular to the direction of ellipsoid deformation, is apparently less energy consumable and corresponds to the quarks U and d, forming nuclides. Such assumption is in accordance with the values of energy masses of quarks in the composition of andirons: 0.33; 0.33; 0.51; 1.8; 5; (?) in GeV/s² for d, u, s, c, b, and t—types of quarks, respectively.

The quark screw model can be proved by other calculations and comparisons.

8.7.1 CALCULATION OF ENERGY MASS OF FREE NUCLIDE (ON THE EXAMPLE OF NEUTRON)

Neutron has three quarks d_1-u-d_2 with electric charges −1, +2, and −1, distributed in three spatial directions, respectively. Quark u cements the system electrostatically. Translatory motions of the screws d_1-u-d_2 proceed along axis x, but oscillatory ones—in three different mutually perpendicular planes (Pauli principle is realized).

Apparently, in the first half of oscillation period u-quark oscillates in the phase with d_1-quark, but in the opposite phase with d_2-quark. In the second half of the period, everything is vice versa. In general, such interactions define the geometrical equality of directed spatial-energy vectors, thus providing the so-called quark discoloration.

The previously formulated rules of adding P-parameters spread to both types of P-parameters (P_0 and P_E). In this case, there is an addition of energy P_E-parameters, since the subsystems of interactions possess similar dimensional characteristics. As both interactions are realized inside the overall system, P_E-parameters are added algebraically and more accurately in this case—geometrically by the following equation:

$$\frac{M}{2} = \sqrt{m_1^2 + m_2^2}$$

where M is the energy mass of free neutron, $m_1 = m_2 = 330$ MeV/s^2 masses of quarks u and d (in the composition of andirons).

The calculation gives $M = 9\,33.38$ MeV/s^2. This is for strong interactions. Taking into account the role of quarks in electromagnetic interactions,[21] we get the total energy mass of a free neutron: $M = 933.38 + 933.38/137 = 940.19$ MeV/s^2. With the experimental value, $M = 939.57$ MeV/s^2 the relative error in calculations is 0.06%.

8.7.2 CALCULATION OF BOND ENERGY OF DEUTERON VIA THE MASSES OF FREE QUARKS

The particle deuteron is formed during the interaction of a free proton and neutron. The bond energy is usually calculated as the difference of mass of free nucleons and mass of free deuteron. Let us demonstrate the dependence of deuteron bond energy on the masses of free quarks. The quark masses are added algebraically in the system already formed: in proton $m_1 = 5+5+7 = 17$ MeV/s^2, in neutron $m_2 = 7+7+5 = 19$ MeV/s^2. As a dimensional characteristic of deuteron bond, we take the distance corresponding to the maximum value of nonrectangular potential pit of nucleon interaction. By the graphs experimentally obtained we know that such distance approximately equals 1.65 fm. Exchange energy interactions between proton and neutron heterogeneous systems are evaluated based on eq 8.21. Then we have:

Exchange Spatial-Energy Interactions   129

$$1/(M_C 1.65K) = 1/(17 \cdot 0.856) + 1/(19 \cdot 0.856),$$

where $K = 2\pi/3$. Based on the calculations we have $M_C = 2.228$ MeV/s², this practically coincides with reference data ($M_C = 2.225$ MeV/s²).

Modifying the basic theses of quark screw model, it can also be applied to other elementary particles (proton, electron, neutron, etc.). For instance, an electrically neutral particle neutron can be considered as a mini-atom, the analog of hydrogen atom.

8.8 CONCLUSIONS

1. The notion of P-parameter is introduced based on the simultaneous accounting of important atomic characteristics and modified Lagrangian equation.
2. Wave properties of P-parameter are found, its wave equation formally similar to the equation of ψ-function is obtained.
3. Applying the methodology of P-parameter:
   a) most important characteristics of exchange energy interactions in different systems are calculated;
   b) intensities of fundamental interactions are calculated;
   c) initial theses of quark screw model are given.

KEYWORDS

- **Lagrangian equations**
- **wave functions**
- **spatial-energy parameter**
- **electron density**
- **elementary particles**
- **quarks**

REFERENCES

1. Batsanov, S. S.; Zvyagina, R. A. Overlap Integrals and Challenge of Effective Charges. Nauka: Novosibirsk, Russia, 1966; p 386.

2. Fischer, C. F. In Average-Energy of Configuration Hartree-Fock Results for the Atoms Helium to Radon. *Atomic Data;* Academic Press, Inc.: Cambridge, MA, 1972; Vol. 4, p 301–399.
3. Waber, J. T.; Cromer, D. T. Orbital Radii of Atoms and Ions. *J. Chem. Phys.* **1965,** *42* (12), 4116–4123.
4. Clementi, E.; Raimondi, D. L. Atomic Screening Constants from S. C. F. Functions. *J. Chem. Phys.* **1963,** *38* (11), 2686–2689.
5. Clementi, E.; Raimondi, D. L. Atomik Screening Constants from S.C.F. Functions, II. *J. Chem. Phys.* **1967,** *47* (4), 1300–1307.
6. Gombash, P. *Atom Statistic Theory and Its Application;* I. L.: Moscow, 1951; p 398.
7. Clementi, E. Tables of Atomic Functions J.B.M. S. *Res. Develop. Suppl.* **1965,** *9* (2), 76.
8. Korablev, G. A. *Spatial Energy Principles of Complex Structures Formation;* Brill Academic Publishers and VSP: Leiden, Netherlands, 2005; p 426.
9. Korablev, G. A.; Kodolov, V. I.; Lipanov, A. M. *Analog Comparisons of Lagrangian and Hamiltonian Functions with Spatial-energy Parameter. Chemical Physics and Mesoscopy;* URC RAS: Moscow, 2004; Vol. 6 (1), pp 5–18.
10. Korablev, G. A.; Zaikov, G. E. Energy of Chemical Bond and Spatial-energy Principles of Hybridization of Atom Orbitals. *J. Appl. Polym. Sci.USA.* **2006,** *101* (3), 2101–2107.
11. Korablev, G. A.; Zaikov, G. E. P-*Parameter as and-Objective Characteristics of Electronegativity. Reactions and Properties of Monomers Polymers;* Nova Science Publishers Inc.: Hauppauge, NY, 2007; pp 203–213.
12. Korablev, G. A.; Zaikov, G. E. Spatial-energy Interactions of Free Radicals. *Success in Gerontology;* Eskulap: St. Petersburg, Russia, 2008; Vol. 21 (4), pp 535–563.
13. Korablev, G. A.; Zaikov, G. E. *Formation of Carbon Nanostructures and Spatial-energy Criterion of Stabilization. Mechanics of Composition Materials and Constructions;* RAS: Moscow, 2009; Vol. 15 (1), pp 106–118.
14. Korablev, G. A.; Zaikov, G. E. *Energy of Chemical Bond and Spatial-energy Principles of Hybridization of Atom Orbitals. Chemical Physics;* RAS: Moscow, 2006; Vol. 25 (7), pp 24–28.
15. Korablev, G. A.; Zaikov, G. E. *Calculations of Activation Energy of Diffusion and Self-diffusion. Monomers, Oligomers, Polymers, Composites and Nanocomposites Research;* Nova Science Publishers: Hauppauge, NY, 2008; pp 441–448.
16. Murodyan, R. M.; PEChAYa, M. *Physical and Astrophysical Constants and Their Dimensional and Dimensionless Combinations;* Atomizdat: Moscow, 1977; Vol. 8 (1), pp 175–192.
17. Barashenkov, V. S. *Sections of Interactions of Elementary Particles;* Nauka: Moscow, 1966; p 532.
18. Yavorsky, B. M.; Detlav, A. A. *Reference-book in Physics;* Nauka: Moscow, 1968; p 940.
19. Volkov, V. V.; PEChAYa, M. *Exchange Reactions with Heavy Ions;* Atomizdat: Moscow, 1975; Vol. 6 (4), pp 1040–1104.
20. Bukhbinder, I. L. Fundamental Interactions. *Sorov Educ. J.* **1997,** *5.* http://nuclphys.sinp.msu.ru/mirrors/fi.htm.
21. Okun, L. B. Weak Interactions. http://www.booksite.ru/fulltext/1/001/008/103/116.htm.

PART II
Nanosystems Formation Processes

CHAPTER 9

INVESTIGATION OF THE FORMATION OF NANOFILMS ON THE SUBSTRATES OF POROUS ALUMINUM OXIDE

A. V. VAKHRUSHEV[1,2*], A. YU. FEDOTOV[1], A. V. SEVERYUKHIN[1], and R. G. VALEEV[3]

[1]Department of Mechanics of Nanostructure, Institute of Mechanics, Ural Division, Russian Academy of Sciences, T. Baramzinoy 34, Izhevsk, Russia

[2]Department of Nanotechnology and Microsystems, Kalashnikov Izhevsk State Technical University Studencheskaya St. 7, Izhevsk, Udmurt Republic, Russia

[3]Department of Physics and Chemistry of Surface, Physical-Technical Institute, Ural Branch of the Russian Academy of Science, Kirova St. 132, Izhevsk 426000, Udmurt Republic, Russia

*Corresponding author. E-mail: Vakhrushev-a@yandex.ru

ABSTRACT

Different processes of nanostructure interactions and mechanisms of template and pore siltation were registered for different types of precipitated atoms. Single atoms, which reached the pore bottom, were observed for all types of precipitated atoms. The most complete and dense pore siltation was registered in the process of gallium epitaxy. The pore-filled with atoms can be considered as a quantum point and used to obtain optic and electric effects. When studying the siltation of coatings with pores of different dimensions with gallium atoms, it was found out that the number of atoms actively grows in the pore in the time interval of 20–120 pcs. The pore siltation after 120 pcs of condensation time is accompanied by the reconstruction of atomic structure that contributes to the stabilization of

dependencies and slight decrease in the percentage of gallium atoms, which penetrated the pore.

9.1 INTRODUCTION

Owing to its hexagonal-ordered arrangement of pores vertically aligned to the film surface, porous anodic aluminum oxide (AAO) is quite often used as a template to synthesize different nanostructures: nanowires, nanopoints, nanorings, nanotubes, etc.[1,2] AAO can also be successfully used as a carrier of catalytically active nanoparticles,[3,4] as well as nanostructures of semiconductors.[5,6] This gives the possibility to form the ordered aggregates of nanostructures of semiconductor fluorescent material of the same size and shape that allows representing each nanostructure as a separate light emitter. The coherent addition of radiation from each light source results in significant light intensity increase.[7]

The work objective is to develop the algorithms and methods for modeling the processes of precipitation of nanosized films onto the templates of porous aluminum oxide and to study the kinetics of the foregoing processes and structure of nanosized films with the help of mathematical modeling methods. In the course of the work, we also investigated the influence of dimensional parameters of the pores in aluminum oxide matrix on the siltation processes and formation of nanofilm coatings. Gold, silver, chromium, copper, iron, gallium, germanium, titanium, platinum, and vanadium were used as precipitated materials while modeling.

9.2 PROBLEM SETTING AND MATHEMATICAL MODEL

The problem of precipitating nanofilms onto porous templates of aluminum oxide was solved by molecular dynamics method (MD method).[8] MD method has been widely used when modeling the behavior of nanosystems due to the simplicity of implementation, satisfactory accuracy, and low costs of computational resources. The solution of Newton's differential equation of motion for each particle forms the basis of this method. The potential was used the potential modified method of the immersed atom (MEAM).

In MEAM, the full energy of the system is recorded as the total of energies of single atoms[9]:

$$E = \sum_i E_i = \sum_i \left(F_i(\bar{\rho}_i) + \frac{1}{2}\sum_{j \neq i} \varphi_{ij}(r_{ij}) \right),$$

where E is the full energy of the system; E_i is energy of i atom used to calculate the interaction forces of the atoms in motion equations; F_i is embedding function for i atom located in the medium with background electron density $\overline{\rho}_i$; $\varphi_{ij}(r_{ij})$ is pair potential between i and j atoms located at the distance r_{ij}. More details of MEAM model can be found in the works of the authors of this chapter on similar topic.[10–12]

Software package for parallel computational processes large-scale atomic/molecular massively parallel simulator (LAMMPS) is applied to conduct theoretical research. LAMMPS has been developed by the group of Sandia National Laboratories and is free software for mathematical models of different levels, including classical MD.[13]

The problem of modeling the formation of nanofilm coatings was solved in several steps. At the first stage, the template from amorphous aluminum oxide was formed. Aluminum and oxygen atoms are put into the computational cell in the required proportion (2:3) with periodic boundary conditions on each side (Fig. 9.1a). The template stabilizes and comes to rest under the action of potential forces under normal thermodynamic conditions (Fig. 9.1b). The template stabilization is conditioned by potential forces, in particular, as it is formed due to the self-organization of aluminum and oxygen atoms. At the same time, heat fluctuations and diffusion are present in the range of the set temperature in the template formed, but there is no essential reconstruction of its structure, the atoms slightly oscillate near the positions they occupy. The hole is cut in the template at the second stage—the pore with the required radius and depth (the cutting of the template with the pore is demonstrated in Fig. 9.1c). Later, this pore will be silted with atoms of different types (Fig. 9.1d). The general pattern of the problem of forming heterogeneous electro-optical coatings is given in Figure 9.1.

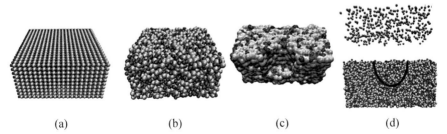

FIGURE 9.1 Steps of solving the problem of forming nanofilm coatings based on porous aluminum oxide: (a) initial state of the system; (b) template relaxation without the pore; (c) template relaxation with the pore; (d) precipitation.

136 Nanoscience and Nanoengineering: Novel Applications

The porous template was silted by homogeneous precipitation of atoms along the normal against the template. The atoms being precipitated were added in the region above the template during the siltation stage. Their position above the template was determined by the uniform random distribution law. The number of atoms added to a time unit and their total number was the process control parameters. The initial velocity of the precipitated atoms was constant. The velocity parameters were changed only under the interaction of the precipitated atoms with the template. To conduct the test calculations, the single nanostructure was considered in the air-free atmosphere and its dynamics during the relaxation self-organization of atoms.

9.3 MODELING RESULTS AND THEIR ANALYSIS

The amorphous aluminum oxide templates with the following dimensions length 12.4 nm, width 12.4 nm, and height 6.2 nm were used in the modeling process. The total number of atoms in the template after the pore formation was about 60.5 thousand. Before the precipitation process, the template was at rest, at the beginning its temperature was 300 K and it was further maintained at the same level.

The pore with the radius of 2 and 4 nm deep was cut in aluminum oxide template. The pore siltation with aluminum oxide was not observed at rest without the atom precipitation. The lower template layer was fixed to avoid its vertical movement at the precipitation stage. The rest of the atoms was not fixed and could freely move in any direction.

Different types of atoms were precipitated onto aluminum oxide template in this work. The number of precipitated atoms was 20,000. The precipitation was uniform along the whole template surface and with the same intensity in time. The atom velocity at epitaxy was 0.05 nm/ps. The translation results of the periodic computational cell relative to the perpendicular of the precipitation plane during the epitaxy of iron atoms are shown in Figure 9.2.

The graphs of penetration of considered substance atoms into the pore on aluminum oxide template are shown in Figure 9.3. The most penetration was demonstrated by gallium atoms and the least by palladium ones. Gold demonstrated the best penetration among noble metals. After about 120 pcs the intensity of all atom penetration goes down and further it changes insignificantly that is seen in Figure 9.3. Iron and gallium atoms demonstrated the highest intensity at the initial modeling stage (approximately up to 90 pcs).

Investigation of the Formation of Nanofilms

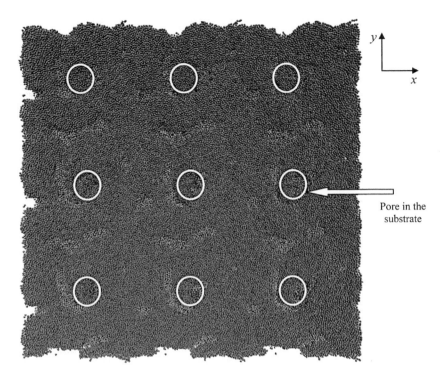

FIGURE 9.2 Siltation of the template of porous aluminum oxide with iron atoms: top view, precipitation time—0.05 ns.

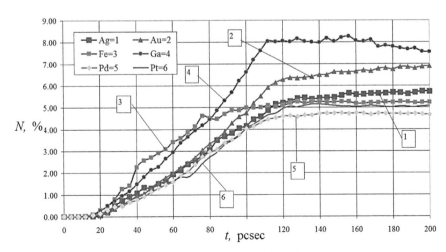

FIGURE 9.3 **(See color insert.)** Percentage of precipitated atoms that got into the pore to the total number of precipitated atoms.

The dynamics of the pore filling with precipitated atoms is given in Figure 9.4, where you can see the graphs of the depth of mass center of atoms, which penetrated the pore. The depth of mass center is calculated only relatively to the atoms, which filled the pore. Therefore, at the initial time moments (up to 20 pcs) the greater shifting of the mass center is observed. Further, the number of atoms penetrating the pore grows the dependencies in Figure 9.4 shift in the direction of the middle of the pore depth. The shift of the mass center reaches the most significant depth when iron and gallium atoms are precipitated.

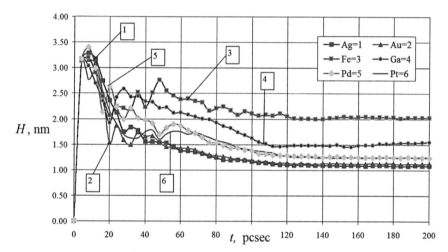

FIGURE 9.4 (See color insert.) Depth of mass center of atoms, which penetrated the pore.

For the following series of computational experiments, the pore radius in the template varied, the depth being the same (4 nm). Gallium atoms were used for precipitation as one of the most suitable to form nanostructured objects on the template. The graphs of gallium atoms, which got inside the pore, in percent relative to the total number of precipitated atoms, are demonstrated in Figure 9.5. The number of atoms actively grows in the time interval of 20–120 pcs. The pore siltation after 120 pcs of condensation is followed by the reconstruction of atomic structure that corresponds to the stabilization of dependencies and slight decrease in the percentage of atoms, which penetrated the pore.

Besides, during the siltation of pores of different radii with gallium atoms, the mass center of precipitated atoms stabilizes at different depths of the pore. For the pores with the radii of 2 and 3 nm, the mass center is formed above the middle of the pore depth. With the size growth, the mass center starts to be formed in one place—near the middle of the pore depth. This fact allows saying that the further radius growth (over 5 nm) will not significantly influence the mass center; the pore is already rather compactly filled with the precipitated atoms.

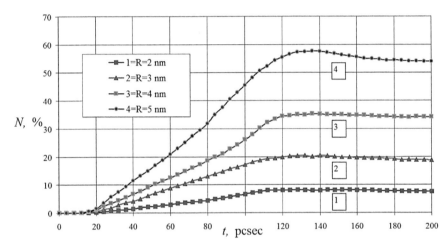

FIGURE 9.5 (See color insert.) Percentage of gallium atoms, which got into the pore in relation to the total number of precipitated atoms.

9.4 CONCLUSION

The precipitation methods of nanosized films are used for certain technological processes and applied to forecast and design nanofilm materials. Different processes of nanostructure interaction and siltation mechanisms of templates and pores are registered for different types of precipitated atoms. Single atoms, which reached the pore bottom, were observed for all types of precipitated atoms. The most complete and dense pore siltation was registered in the process of gallium epitaxy. The pore-filled with atoms can be considered as a quantum point and used to obtain optic and electric effects.

When studying the siltation with gallium atoms of the coatings with pores of different sizes, it was found out that the number of atoms actively grows in the pore in the time interval of 20–120 pcs. The pore siltation after 120 pcs of condensation time is accompanied by the reconstruction of atomic structure that contributes to the stabilization of dependencies and slight decrease in the percentage of gallium atoms, which penetrate the pore. With the pore growth, the mass center starts to be formed in one place—near the middle of the pore depth. The further growth of the pore radius (over 5 nm) does not significantly influence the mass center; the pore is already quite compactly packed with the precipitated atoms.

The modeling results can be used when developing and optimizing technological processes of optic coating formation and analysis of physical properties of nanofilms and nanostructures formed in pores and on surfaces of aluminum oxide templates.

ACKNOWLEDGMENTS

The investigation was carried out with financial support of Russian Science Foundation (project No 15-19-10002).

KEYWORDS

- **epitaxy**
- **nanofilms**
- **modeling**
- **LAMMPS**
- **crystalline structure**
- **molecular dynamics**
- **MEAM**

REFERENCES

1. Ying, J. Y. Nanoporous Systems and Templates the Unique Self-assembly and Synthesis of Nanostructures. *Science Spectra.* **1999,** *18,* 56–63.
2. Li, A. P.; Muller, F.; Birner, A.; Nielsch, K.; Gosele, U. Hexagonal Pore Arrays with a 50–420 nm Interpore Distance Formed by Self-organization in Anodic Alumina. *J. Appl. Phys.* **1998,** *84* (11), 6023–6026.
3. Doroshenko, M. N.; Gerasimchuk, A. I.; Mazurenko, E. A. Kataliticheskoe Vliyanie Poverkhnosti na Formirovanie Nanotrubok Germaniya PE MOCVD-metodom [The Catalytic Effect of the Surface on the Formation of Germanium Nanotubes PE

Investigation of the Formation of Nanofilms   141

CVD-method]. *Khimiya, Fizika i Tekhnologiya Poverkhnosti [Chem. Phys. Surf. Technol.].* **2013,** *4* (4), 366–372.

4. Mu, C.; Yu, Y.; Liao, W.; Zhao, X.; Xu, D.; Chen, X.; Yu, D. Controlling Growth and Field Emission Properties of Silicon Nanotube Arrays by Multistep Template Replication and Chemical Vapour Deposition. *Appl. Phys. Lett.* **2005,** *87* (11), 1–3.

5. Melnik, Yu. V.; Nikolaev, A. E.; Stepanov, S. I.; Zubrilov, A. S.; Nikitina, I. P.; Vassilevski, K. V.; Tsvetkov, D. V.; Babanin, A. I.; Musikhin, Yu. G.; Tretyakov, V. V.; Dmitriev, V. A. AlN/GaN and AlGaN/GaN Heterostructures Grown by HVPE on SiC Substrates. *Mat. Res. Soc. Symp. Proc.* **1998,** *482,* 245–249.

6. Nikolaev, A. E.; Melnik, Yu. V.; Kuznetsov, N. I.; Strelchuk, A. M.; Kovarsky, A. P.; Vassilevski, K. V.; Dmitriev, V. A. GaN pn-Structures Grown by Hydride Vapor Phase Epitaxy. *Mat. Res. Soc. Symp. Proc.* **1998,** *482,* 251–256.

7. Xu, H. J.; Li, X. J. Structure and Photoluminescent Properties of a ZnS/Si Nanoheterostructure Based on a Silicon Nanoporous Pillar Array. *Semicond. Sci. Technol.* **2009,** *24* (7), 075008.

8. Lennard-Jones, J. E. On the Determination of Molecular Fields. II. From the Equation of State of a Gas. *Proc. Roy. Soc. A.* **1924,** *106,* 463–477.

9. Jelinek, B.; Houze, J.; Kim, S.; Horstemeyer, M. F.; Baskes, M. I.; Kim S. G. Modified Embedded-Atom Method Interatomic Potentials for the Mg-Al Alloy System. *Phys. Rev. B.* **2007,** *75* (5), 054106.

10. Vakhrushev, A. V.; Fedotov, A. Yu.; Severyukhin, A. V.; Suvorov, S. V. Modelirovanie Protsessov Polucheniya Spetsial'nykh Nanostrukturnykh Sloev v Epitaksial'nykh Strukturakh Dlya Utonchennykh Fotoelektricheskikh Preobrazovateley [Modelling of Processes of Special Nanostructured Layers of Epitaxial Structures for Sophisticated Photovoltaic Cells]. *Himicheskaya fizika i mezoskopiya [Chem. Phys. Mesoscopics].* **2014,** *16* (3), 364–380.

11. Vakhrushev, A. V.; Severyukhin, A. V.; Fedotov, A. Yu.; Valeev, R. G. Issledovanie Protsessov Osazhdeniya Nanoplenok Na Podlozhku iz Poristogo Oksida Alyuminiya Metodami Matematicheskogo Modelirovaniya [Research Nanofilms Deposition Processes on a Substrate of Porous Aluminum Oxide by Means of Mathematical Simulation]. *Vychislitelnaya mehanika sploshnyh sred [Comput. Continuum. Mech].* **2016,** *9* (1), 59–72.

12. Vakhrushev, A. V.; Fedotov, A. Yu.; Severyukhin, A. V.; Valeev, R. G. Modelirovanie Protsessov Osazhdeniya Nanoplenok na Podlozhku Poristogo Oksida Alyuminiya [Simulation of Deposition Films on Nano Porous Alumina Substrate]. *Himicheskaya fizika i mezoskopiya [Chem. Phys. Mesoscopics].* **2015,** *17* (4), 511–522.

13. LAMMPS Molecular Dynamics Simulator [Electronic resource]. http://lammps.sandia. gov (accessed May 25, 2016).

CHAPTER 10

METAL/CARBON NANOCOMPOSITES AND THEIR MODIFIED ANALOGUES: THEORY AND PRACTICE

V. I. KODOLOV[1,2] and V. V. TRINEEVA[2,3*]

[1]*Basic Research-High Educational Centre of Chemical Physics and Mesoscopy, Ural Division of Russian Academy of Sciences, Izhevsk, Udmurt Republic, Russia*

[2]*Department of Chemistry and Chemical Technology, Kalashnikov Izhevsk State Technical University, Studencheskaya St. 7, Izhevsk, Russia*

[3]*Institute of Mechanics, Ural Division, Russian Academy of Sciences, T. Baramzinoy 34, Izhevsk, Russia*

[*]*Corresponding author. E-mail: Vera_Kodolova@mail.ru*

ABSTRACT

The metal/carbon nanocomposites (NCs) properties are considered in this chapter. The theoretical and experimental methods give possibility to prognosticate and to lead the experimental estimation of influence of NCs on changes in the polymeric media electron (submolecular) structures are discussed. The NC electromagnetic field action on the medium molecules leads to the medium molecule self-organization and to the formation of nanostructured fragments (fractals) in the material composition. These changes can be accompanied by the processes of active particles appearance. The changes of media electron structure are possible under the influence of minute quantities of NCs, which is confirmed by the X-ray photoelectron spectroscopic investigations. The mechanism of polymeric materials modification processes by supersmall (minute) quantities of metal/carbon

NCs is proposed. The estimation methods of self-organization in polymeric compositions and materials, modified by minute quantities of metal/carbon NC, are also proposed. The orientation processes, in which mesoscopic particles (NCs) participate, lead to the changes of submolecular structures and properties of materials. Usually in the process of polymeric matrix, self-organization stimulates the increase of material heat capacity, the material density, its thermal stability, and also the adhesive strength. The improvement of other properties of materials, modified by the supersmall quantities of metal/carbon NCs, is also possible.

10.1 INTRODUCTION

Now, we observe the creation of new synthetic science on the basis of synergetic and fractal theory, chemistry within nanoreactors, and mesoscopic physics. Everyone from these scientific trends considers the objects of nanosize level (0.1–1000 nm) and investigates near phenomena such as self-organization, self-similarity, interferential, spectrum quantization, and charge quantization. These phenomena take place at definite conditions (size of phase coherence must be less than 1000 nm).

If we classify the surrounding medium on sizes of objects, we can define the following worlds: pico (sizes: from 10^{-12} to 10^{-9}m); nano (sizes: from 10^{-9} to 10^{-6}m); micro (sizes: from 10^{-6} to 10^{-3} m). For the first case, the unstable state takes place and the orientation of the particles is possible only at field actions. When the microworld is present, the stable state characterizes and the nucleation occurs. In the case concerning to nanoworld, the metastable state, and the particles self-organization are observed. In other words, meso world is found between pico- and microworlds (intermediate state). Therefore, the "meso" prefix is more correct for the definition of notions such as mesophysics or mesoscopic physics and analogical notions, and "nano" prefix may be used for dimensions, for instance, nanoseconds, nanograms, nanometer, nanoliter, etc.

It should be noted that in the nanotechnology as well as in the nano-chemistry, the nanoparticle upper limit is taken as 100 nm. However, in the mesoscopic physics, this limit for mesoparticle (nanoparticle) equals to 1000 nm. According to mesoscopic physics, the particles size in 1000 nm corresponds to the length of phase coherence, and the interval 100–0.1 nm for these particles is determined as the electron wavelengths in different materials (on Fermi energy levels).

From the comparison of the notions and fundamentals of abovementioned scientific trends, it is possible to speak about the creation of new science—the chemical mesoscopic.

Some fundamental theses of this scientific trend on nanostructures new class investigations are considered below—synthesis and investigations of properties of metal/carbon nanocomposites (NCs).

10.2 METAL/CARBON NANOCOMPOSITES SYNTHESIS IN NANOREACTORS OF POLYMERIC MATRICES

The metal/carbon NCs were obtained within the nanoreactor of polymeric matrices. What are nanoreactors? Let us give the definition of this notion.

Nanoreactors are defined as specific nanostructures and can be nanosized cavity in polymeric matrices or space bounded part, in which reactant chemical particles are orientated with the creation of transitional state before predetermined product formation.

According to Buchachenko,[1] the following classification of nanoreactors is known:

- One-Dimensional Nanoreactors
- Two-Dimensional Nanoreactors
- Cluster Nanoreactors: Metal Clusters (Fig. 10.1).

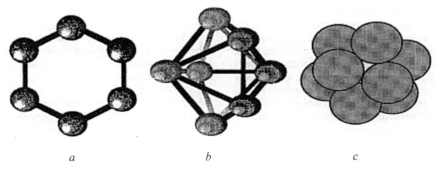

FIGURE 10.1 Elements nanoclusters, dispositional on metal electronegativity decreasing (a) Aurum, (b) silicon, and (c) aluminum.

The nanostructures formation within nanoreactors is determined by the nature of reactants, which participate in synthesis, and also by the nanoreactors energetic and geometric characteristics.

The analysis of modern scientific data and our experiments show the following peculiarities of nanostructures formation within nanoreactors:

1. The principal peculiarity—the decreasing of collateral parallel processes and the process direction to special product side.
2. The low energetic expenses and the high rates of processes.
3. The dependence of obtained nanostructures properties from the energetic and geometric characteristics of nanoreactors.

These peculiarities characterize for metal/carbon NC RedOx synthesis.

The application of mesoscopic physics principles may be useful for the most reduction–oxidation processes. The following scheme of metal/carbon NCs RedOx synthesis by use of mesoscopic physics principles is possible (Fig. 10.2).

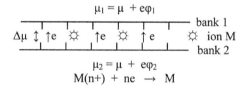

FIGURE 10.2 The scheme of RedOx synthesis of metal-containing nanostructures.

The formation of metal/carbon NCs within polymeric matrices is realized during two stages.

The first stage (preparatory) includes the formation of metal-containing phase clusters in coordination field of polymeric phase functional groups.

The second stage of metal/carbon NCs formation is presented as RedOx process on interface bound of metal-containing and polymeric phases (within nanoreactor) with one-dimensional, two-dimensional, or space growth NC according to energy expenses.

10.3 DECIPHER OF EXPERIMENTAL DATA ON STRUCTURE AND CHARACTERISTICS OF METAL/CARBON NANOCOMPOSITES AND RESULTS DISCUSSION

Earlier in the study of Kodolov et al. [2-6], the methods of metal/carbon NCs obtaining into polymeric matrices nanoreactors as well as the investigations of their characteristics were described. It is established that the metal/carbon nanocomposites are produced as metal clusters or metal nanocrystals into carbon or carbon–polymer shells (Fig. 10.3).

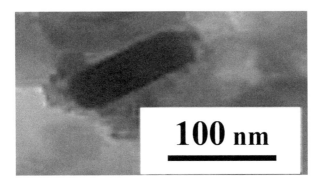

FIGURE 10.3 The nickel nanocrystal obtained from polyvinyl alcohol nanoreactor.

The shell or nanofilm image, in which metal cluster is found, changes depending on the metal nature (Fig. 10.4).

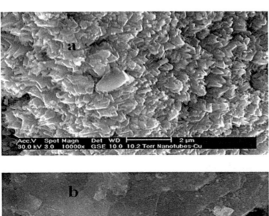
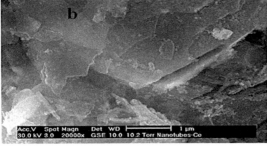

FIGURE 10.4 The nanofilms images in dependence on the metal nature which participated in process of RedOx synthesis.

According to the investigations using of transmission electron microscopy (TEM), carbon shell itself consists of carbon fiber layers, which contain acetylene and carbine fragments associated with metal clusters or metal crystals (Fig. 10.5).

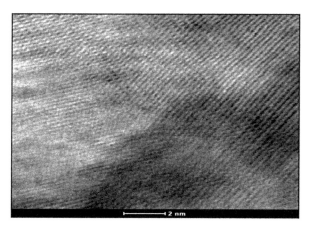

FIGURE 10.5 The microphotography of copper/carbon nanocomposite obtained by TEM.

These results in the comparison with the experimental results of other investigations are deciphered and discussed below.

At the decipher of microphotography's of metal/carbon NCs, the sizes (diameters) and the directions angles of fibers were determined and correspond to d for fiber ~0.2–0.3 nm; $\varphi = 54°44'$ (concerning to axis of cluster). The decipher of X-ray photoelectron spectra C1s has evidence about the presence of C—H bonds (285 eV) and polyacetylene fragments, and also two satellites, corresponding to sp^2 (284 and 306 eV) and sp^3 (286.2 and 312 eV) hybridization (Fig. 10.6).

According to C1s spectrum analysis, C—H groups are contained in fiber approximately 20%, C—C fragments in 2D projection (sp^2 hybridization) ~50% and corresponding C—C in 3D projection (sp^3 hybridization) ~30%.

The spectrum analysis testifies about the presence in NC 35% C—H groups, and also C—C bonds (sp-sp^2 hybridization and sp^3 hybridization) in equal quantities.

Thus, the metal/carbon NC has the shell from reversed carbon fibers, which are associated with metal clusters. Atomic magnetic moments are studied by using X-ray photoelectron spectroscopy and EPR spectroscopy (Tables 10.1 and 10.2).

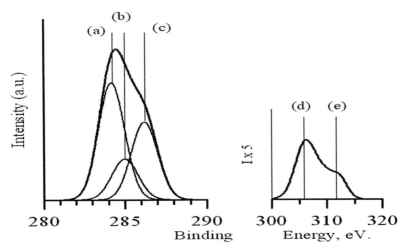

FIGURE 10.6 C1s spectrum of copper/carbon nanocomposite and spectrum of satellite structure. C1s spectrum of copper/carbon nanocomposite contains three forming parts: (a) C—C (sp^2)— 284 eV; (b) C—H—285eV; (c) C—C (sp^3)—286.2 eV and satellite structure; and (d) satellite (sp^2); (e) satellite (sp^3).

TABLE 10.1 Atomic Magnetic Moments (μ_B) of Copper and Nickel in Metal/Carbon Nanocomposites.

Metal/carbon nanocomposite/massive sample	Cu$_{NC}$/Cu$_{ms}$	Ni$_{NC}$/Ni$_{ms}$
Atomic magnetic moment, μ_B	1.3/ –	1.8/0.5

TABLE 10.2 Experimental EPR Data for Copper and Nickel/Carbon Nanocomposites.

Type of metal/carbon nanocomposite	g-factor	Number of unpaired electrons, spin/g
Copper/carbon nanocomposite	2.0036	1.2 × 10^{14}
Nickel/carbon nanocomposite	2.0036	10^{22}

10.3.1 WHAT IS METAL/CARBON NANOCOMPOSITE?

The answer may be the following:

Metal/carbon nanocomposite is the metal nanoparticles stabled by carbon nanofilm structure.

These carbon nanofilms consist of carbon fibers which contain acetylene and carbine fragments. The latter is associated with metal clusters. The metal atomic magnetic moments are increased, and there are unpaired electrons on the surface of carbon shells.

150   Nanoscience and Nanoengineering: Novel Applications

Thus, the stable complex from metal clusters and carbon shell is formed. These nanoparticles are characterized rather well. The example of copper/carbon NCs energetic characteristics is adduced below:

Relative content, Cu/C, %	50/50
Density, g/cm^3	1.71
Summary mass, au	36.75
Linear size, nm	25
Specific surface, m^2/g	160
Frequency of skeleton vibration, Hz	4×10^{11}
Middle energy of vibration, erg	1.6×10^{13}

The presence of double bonds and unpaired electrons is confirmed by the interactions of phosphorus, silicon, and sulfur substances with NCs. As the result of this interaction, the reduction of correspondent elements takes place accompanied by the growth of metal atomic magnetic moments for NCs and the increasing of their polarization action on media. The estimation of modified NCs activity is carried out with the use of IR spectroscopy, X-ray photoelectron spectroscopy, and volt–ampere-metrology (coulombmetrology). For instance, after the silicon introduction into copper/carbon NC, its properties are changed:

- Copper atomic magnetic moment grows to 3 μ_B.
- Antioxidant activity increases in 11 times.
- Dynamic viscosity of liquid glass, which contains 0.001% of Cu/C NC with silicon, is increased by 21%.

1.4 THEORETICAL AND EXPERIMENTAL METHODS OF ESTIMATION OF METAL/CARBON NANOCOMPOSITES INFLUENCE ON MEDIA AND POLYMERIC COMPOSITIONS

The modification process for media and compositions is concluded in the change of composition (medium) structure and properties through self-organization of its chemical particles in regular order. The corresponding effect may be achieved owing to the chemical interactions of a certain quantity of active substances which are introduced into medium and pretend to be the crystallization center or the chemical particles organization center.

According to Bulgakov et al.[7], the equation for the part organized particles determination is proposed as follows:

Metal/Carbon Nanocomposites and Their Modified Analogues

$$W = L \times c \times K_c \qquad (10.1)$$

where W is the organized chemical particles medium part, L is the active additive interactions number with medium particles, c is the quantity of active additive introduced into medium, and K_c is the quantity of medium particles adsorbed layers on the active additive surface.

According to Lipatov[8], the percent of active additive introduced in modified medium should not exceed the limit of 5% or else, the modification effect could be decreased on account of the aggregation of active additive particles which have micron sizes.

When the additive sizes are decreased to nanometer sizes, the phenomena of mesoscopic particles appear. According to Imri[9] and Mosklets,[10] the phenomena such as interference, spectrum quantization, and charge quantization occur when the mesoparticles have the limitations in motions or in the energetic possibilities realization. These limitations are expressed in the absence of progressive motion and rotation. In this case, the mesoparticle can only vibrate and also the electron transport is possible.

The media (or polymeric compositions) properties changes under the mesoparticles influence can be achieved at equal distribution of these particles in composition volume and at its coagulation absence. Latter is possible at the following conditions:

- certain polarity and dielectric constant of medium;
- minute concentration of mesoparticles; and
- ultrasound action on the correspondent suspension for the proportional distribution of mesoparticles.

The assignment of active nanostructures (mesoparticles) during the compositions modification is concluded in the activation of matrices self-organization in the needful direction. For the realization of this goal, the determination of organized phase part is necessary. The Johnson–Mehl–Avrami–Kolmogorov equation is applied for the organized phase determination as follows:

$$W = 1 - \exp(-k\tau^n), \qquad (10.2)$$

where W is the part of organized phase, k is the parameter defined the rate of organized phase growth, τ is the duration of organized phase growth, and n is the fractal dimension.

10.5 MECHANISM OF FUNCTIONAL MATERIALS MODIFICATION PROCESSES BY METAL/CARBON NANOCOMPOSITE MINUTE QUANTITIES

According to Kodolov et al.,[11-15] the positive results on materials properties improvement are presented when the minute quantities of metal/carbon NCs are introduced in these materials. According to Khokhriakov[16], the hypothesis about nanostructures influence transmission on macromolecules of polymeric matrices is proposed. This hypothesis complies with mesoscopic physics principles which consider quantum effects at the certain conditions of mesoparticle existence.

The composition polarization is possible because there is the charge quantization with the wave expansion on polar (functional) groups of media (e.g., polymer macromolecule) (Fig. 10.7).

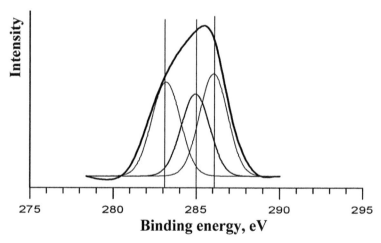

FIGURE 10.7 C1s spectrum of nickel/carbon Nnnocomposite. C1s spectrum of nickel/carbon nanocomposite contains three forming parts: (Me) C—C (sp, sp^2)—283–284 eV; C—H—285 eV; and C—C (sp^3)—286.2 eV.

The designation: MP (NC)—mesoparticle (nanocomposite); δe—charge (electron) quant; →—the polarization direction; ♀——♀—the fragment of macromolecule with functional groups; P_s—the summary polarization system; Σp_{fg}—total of functional groups polarization effect; $p_{MP(NC)}$—the polarization (or dipole moment) of MP or NC.

The quantum charge wave expansion leads to the functional group's polarization (dipole moments) change as well as the extinction increasing.

Metal/Carbon Nanocomposites and Their Modified Analogues 153

Last, bring growth of peaks intensities in IR spectra. The individual peaks growth effects in IR spectra are observed at the introduction of minute quantities of NCs (Table 10.3). Let us note that the growth of peaks intensity in IR spectra is observed when the quantity of introduced NC is decreased. This fact complies with fundamental principles of chemical mesoscopics. In illustrated case, the instance of finely dispersed suspension of Cu/C NC (hardener for epoxy resin) takes place. According to data of Table 10.3, the decreasing of NC quantity to 0.001% leads to the growth of some peaks intensity in IR spectra.

TABLE 10.3 The Peak Intensity Change in Dependence of Cu/C Nanocomposite Concentration.

No.	$V(cm^{-1})$	I_1/I_0	$I_{0.01}/I_0$	$I_{0.001}/I_0$	Atomic groups
1	1050	1.235	1.411	1.686	C—O—C st
2	1450	1.179	1.590	1.744	C—H
3	1776	1.458	1.347	1.691	C=O st as
4	1844	1.463	1.412	1.678	C=O st sy
5	2860–3090	1.182	1.545	1.750	C—H

At the second day of that suspension existence, the floccules are formed and peaks intensity sharply drops (Fig. 10.8).

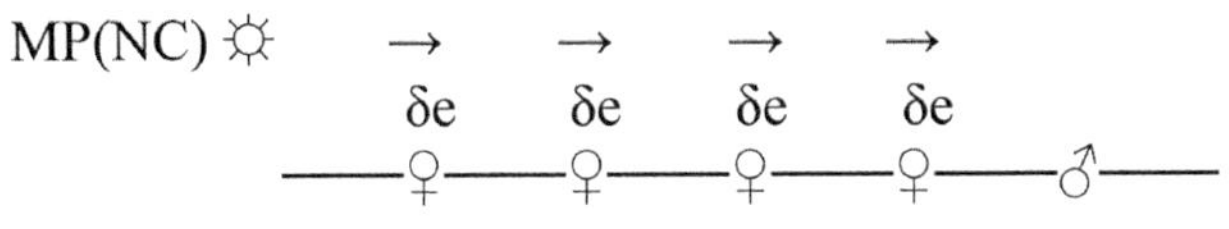

FIGURE 10.8 The medium polarization scheme (example of polymeric composition macromolecule fragment).

However, the suspension activity can be increased with the using of ultrasound treatment. The treatment optimal duration determined as 7 min. In this case, the peak intensity in IR spectra is increased in 2–4 times (Table 10.4).

TABLE 10.4 The Peaks Intensity Changes in IR Spectrum of Cu/C Nanocomposite in Dependence of the Ultrasound Treatment Duration.

No.	$v (cm^{-1})$	I_7/I_0	I_{10}/I_0	Atomic groups
1	1776.6	3.7932	0.7574	C=O st as
2	1844.1	2.5065	0.9115	C=O st sy
3	3039.1	2.3849	0.9589	C—H st

The charge (electron) quantization should lead to the macromolecule electron structure change and, as corollary, to change submolecular structures of polymeric substances. Therefore, the special films of nanostructured materials on the basis of functional polymers such as polyvinyl alcohol, polymethyl methacrylate, and polycarbonate, which contain metal/carbon NC in the minute quantities (10^{-1} to 10^{-5}%) are prepared. The films obtained are investigated by X-ray photoelectron spectroscopy (X-ray PES) and by atomic force microscopy (AFM). The investigations by means of X-ray PES show that the films based on polycarbonate have more changes of electron structure at the introduction of minute quantities of copper/carbon NC in comparison with other polymeric films because these films are more polarized. The C1s spectra for Cu/C NC and for nanostructured polycarbonate are presented below (Fig. 10.9). The expansion of C1s spectra for nanostructured polycarbonate is possible owing to the determination of the energetic states for sp, sp^2, and sp^3 satellites. From the results of C1s

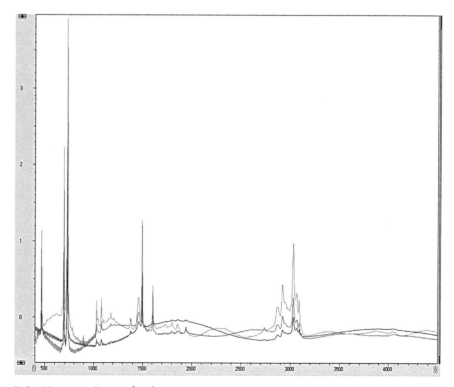

FIGURE 10.9 (See color insert.) The peaks intensity changes in IR spectra of Cu/C nanocomposite in dependence of the duration of suspension existence. Red—in the beginning; green—the first day; and blue—the third day.

spectra for polycarbonate, containing the different minute quantities of Cu/C NC, we can note that after concentration equaled to 10^{-2}% of Cu/C NC, the peaks correspondent to sp^2 and sp^3 peaks appears in these spectra. In other words, the "stamp" of NC which is used during modification is appeared. That "stamp" is observed also at the NC containing, equaled to 10^{-5} %, in polycarbonate film. It is noted that the relation between sp^2 and sp^3 peaks changes. For instance, the intensity of sp^2 hybridization carbon peak is higher to the intensity of sp^3 peak in the concentration interval from 0.01 to 0.001% of NC. The change of concentration to 10^{-4}% brings the proximity of intensities of sp^2 and sp^3 peaks.

However, when the concentration of NC is decreased in polycarbonate to 10^{-5}%, the sp^3 peak intensity becomes greater than sp^2 peak. It is possible to suppose that the self-organization in polymeric structure gets the spatial character when concentration of NC is decreased to the minute (supersmall) values. In this case, the question appears, how the submolecular structure as well as surface structure change when the electronic structure exchange (variable). For the decision of this question, the atomic force microscopy method is applied. Some images of polycarbonate nanostructured films surface are presented below. Polycarbonate is modified by minute quantities of Cu/C NC (from 10^{-1} to 10^{-4}%) (Fig. 10.10).

As follow from AFM images (Fig. 10.11), the surface layers structure strongly changes at the concentration of Cu/C NC equal to 10^{-4}% the transition from two-dimensional level to three-dimensional level of submolecular structures orientation. This fact is confirmed by the growth of sp^3 in comparison with sp^2 peak from X-ray photoelectron C1s spectra. It is interesting to observe the direction of carbon fibers in comparison with the direction of submolecular structures orientation in the nanostructured polycarbonate surface layers.

Thus, it is possible that the wave which initiates the self-organization process in polymeric composition is expensed from these fibers associated with metal cluster. The latter leads to the correspondent orientation of submolecular structures in nanostructured composite surface layers.

The self-organization mechanism for polymeric compositions modified by the minute quantities of metal/carbon NC is concluded in the conditions created for composition polarization, which brings the great change of electron and submolecular structures of materials. Certainly, these changes influence the modified materials properties.

The example of results for that modification with using of above-considered suspensions will be presented below. In this example, the epoxy compositions with different additives including NCs were investigated.

If cross-linking agent is used, the finely dispersed suspension based on iso methyl tetra hydrophthalates anhydrate and copper/carbon NC. In Figure 10.11, the results of modification epoxy compounds (materials 1 and 2) are given.

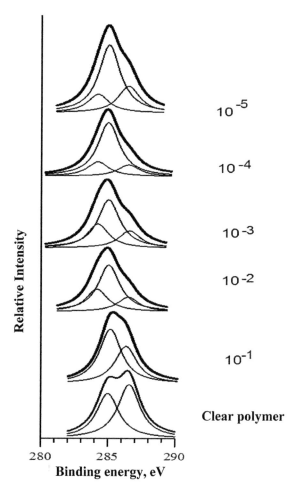

FIGURE 10.10 X-ray photoelectron C1s spectra of nanostructured polycarbonate modified by Cu/C nanocomposite minute quantities (10^{-1} to 10^{-5}%).

The modification by copper/carbon NC quantity equal to 0.005% improves the adhesion characteristics (Fig. 10.12) for the material 1 (green) on 59.77% and for the material 2 (dark blue) on 47.17%. At the same time, the following applications of metal/carbon NCs are possible:

Metal/Carbon Nanocomposites and Their Modified Analogues

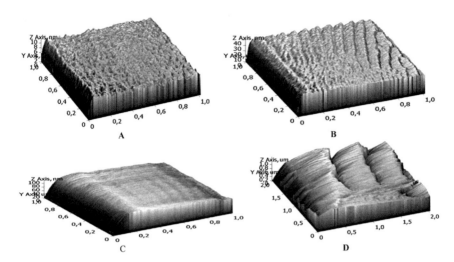

FIGURE 10.11 AFM images of polycarbonate nanostructured films surface: A—0.1% Cu/C NC; B—0.01% Cu/C NC; C—0.001% Cu/C NC; and D—0.0001% Cu/C NC.

- sorbents of toxic substances in gaseous and liquid phases and also rare metal ions in the electrochemical production;
- additives for modification of composites;
- stimulators of plant and animal growth;
- quantum nanogenerators;
- nanostuff for nanometallurgy;
- catalysts of chemical processes;
- synergists for fire resistant or thermostable systems.

Applications of nanostructures are possible owing to its uncial structure and properties.

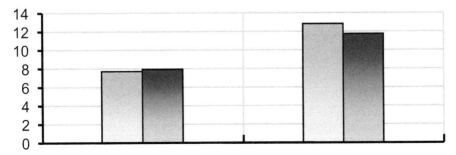

FIGURE 10.12 (See color insert.) Comparison of adhesion strength for materials. 1 (green) and 2 (dark blue) before (A) and after (B) the modification by copper/carbon nanocomposites.

10.6 CONCLUSION

The unification of four scientific trends such as synergetic, fractal theory, chemistry within nanoreactors, and mesoscopic physics are proposed on the basis of its comparison on investigated objects and observed phenomena. In this case, the new science could be named as chemical mesoscopics. Some peculiarities and regularities of this new scientific trend on the instance of creation, characteristics investigations, and application of nanostructure new class such as metal/carbon NCs are discussed. The RedOx synthesis of metal/carbon nanocomposites is considered with the mesoscopic physics principles application. The results of metal/carbon NC structure and characteristics are given. The presence of unpaired electrons on metal/carbon NC carbon shell and also the metal atomic magnetic moment growth are determined. The high activity of nanoparticles obtained in the polymeric compositions modification is shown; it is necessary to obtain positive effect introduced from 0.01 to 0.0001% of metal/carbon NC. The possible application of proposed nanostructure new class is determined.

KEYWORDS

- chemical mesoscopics
- synergetic
- fractal theory
- chemistry within nanoreactors
- mesoscopic physics
- charge quantization
- metal/carbon nanocomposites

REFERENCES

1. Buchachenko, A. L. Chemistry on the Border Centuries – Achievement and Prospects. *Russ. Chem. Rev.* **1999,** *68* (2), 85–102.
2. Kodolov, V. I.; Didik, A. A.; Volkov, A. Yu.; Volkova, E. G. The Metal Containing Carbon Nanostructures Obtaining Method from Organic Compound with Unorganic Salts Additives. Patent RF N 2221744, 2004.

3. Kodolov, V. I.; Didik, A. A.; Shayakhmetova, Ye. Sh.; Kuznetsov, A. P.; Volkov, A. Yu.; Volkova, E. G. The Carbon Metal Containing Nanostructures Obtaining Method from Aromatic Hydrocarbons. Patent RF N 2223218, 2004.
4. Kodolov, V. I.; Nikolaeva, O. A.; Zakharova, G. S.; Shayakhmetova, Ye. Sh.; Volkova, E. G.; Volkov, A. Yu.; Makarova, L. G. The Carbon Metal Containing Nanostructures Obtaining Method. Patent RF N 2225835, 2004.
5. Kodolov, V. I.; Trineeva, V. V.; Blagodatskikh, I. I.; Vasil'chenko, Yu. M.; Vakhrushina, M. A.; Bondar, A. Yu. The Nanostructures Obtaining and the Synthesis of Metal/Carbon Nanocomposites in Nanoreactors. In *Nanostructure, Nanosystems, and Nanostructured Materials Theory, Production and Development;* Apple Academic Press: Toronto/Point Pleasant, NJ, 2013; Vol. 558, pp 101–145.
6. Kodolov, V. I.; Trineeva, V. V. How Mesoscopic Physics Explains RedOx Synthesis of Metal/carbon Nanocomposites Within Nanoreactors of Functional Polymers. *Chem. Phys. Mesoscopy* **2015,** *17* (4), 580–587.
7. Bulgakov, V. K.; Kodolov, V. I.; Lipanov, A. M. *Modeling of Polymeric Materials Combustion;* Chemistry: Moscow, 1990; p 240.
8. Lipatov, Yu. S. *Physical Chemistry of Filled Polymers;* Chemistry: Moscow, 1977; p 304.
9. Imri, Yi. *Introduction in Mesoscopic Physics;* Fizmatlit: Moscow, 2004; p 301.
10. Mosklets, M. V. Fundamentals of Mesoscopic Physics. Khar'kov: NTU KhPI: Kharkiv Oblast, Ukraine, 2010; p 180.
11. Kodolov, V. I.; Trineeva, V. V.; Vasil'chenko, Yu. M. The Calculating Experiments for Metal/Carbon Nanocomposites Synthesis in Polymeric Matrices with the Application of Avrami Equations. In *Nanostructures, Nanomaterials and Nanotechnologies to Nanoindustry;* Apple Academic Press: Toronto/Point Pleasant, NJ, 2015; Vol. 400, pp 105–118.
12. Kodolov, V. I.; Trineeva, V. V.; Khokhriakov, N. V. Synthesis and Application of Metal/Carbon and Metal/Polymeric Nanocomposites: Theory, Experiment and Production. In *The Problems of Nanochemistry for the Creation of New Materials;* IEPMD: Torun, Poland, 2012; Vol. 250, pp 7–15.
13. Shabanova, I. N.; Kodolov, V. I.; Terebova, N. S.; Trineeva, V. V. *X-ray Photoelectron Spectroscopy in the Investigation Metal/Carbon Nanosystems and Nanostructured Materials;* Udmurt University: Izhevsk, Russia, 2012; p 252.
14. Kodolov, V. I.; Khokhriakov, N. V.; Trineeva, V. V.; Blagodatskikh, I. I. Problems of Nanostructures Activity Estimation, Nanostructures Directed Production and Application. In *Nanomaterials Yearbook Nanostructures, Nanomaterials and Nanotechnologies to Nanoindustry;* Nova Science Publishers, Inc.: Hauppauge, NY, 2009; Vol. 383, pp 1–17.
15. Kodolov, V. I.; Trineeva, V. V. Fundamental Definitions for Domain Nanostructures and Metal/Carbon Nanocomposites. In *Nanostructure, Nanosystems, and Nanostructured Materials. Theory, Production and Development;* Apple Academic Press: Toronto/Point Pleasant NJ, 2013; Vol. 558, pp 1–42.
16. Khokhriakov, N. V.; Kodolov, V. I.; Korablev, G. A.; Trineeva, V. V.; Zaikov, G. E. Prognostic Investigations of Metal or Carbon Nanocomposites and Nanostructures Synthesis Processes Characterization. In *Nanostructure, Nanosystems, and Nanostructured Materials. Theory, Production and Development;* Apple Academic Press: Toronto/Point Pleasant, NJ, 2013; Vol. 558, pp 43–99.

CHAPTER 11

THE METAL/CARBON NANOCOMPOSITES MODIFICATION WITH USE OF AMMONIUM POLYPHOSPHATE FOR THE APPLICATION AS NANOMODIFIER OF EPOXY RESINS

R. V. MUSTAKIMOV[1,2*], V. I. KODOLOV[1,2], I. N. SHABANOVA[2,3], and N. S. TEREBOVA[2,3]

[1]*Basic Research-High Educational Centre of Chemical Physics and Mesoscopy, Ural Branch of the Russian Academy of Science, Izhevsk, Udmurt Republic, Russia*

[2]*Department of Chemistry and Chemical Technology, Kalashnikov Izhevsk State Technical University, Studencheskaya St. 7, Izhevsk, Russia*

[3]*Laboratory of Nanostructures, Physico-Technical Institute, Ural Branch of the Russian Academy of Science, Kirov St. 132, Izhevsk 426000, Russia*

Corresponding author. E-mail: rostmust@mail.ru

ABSTRACT

The different methods of the phosphorus-containing copper/carbon nanocomposites formation are considered in this chapter. The difference of methods is concluded as follows: (1) The mixing-grinding of reagents in relation 1:1; (2) The mixing-grinding of reagents with addition of water small qualities; and (3) the mixing-grinding of reagents with last vacuum using. The estimation of investigation results is carried out by the means of IR-spectroscopy and X-ray photoelectron spectroscopy. Better results were obtained when the second variant was utilized. In this case, the atomic magnetic moment is increased from 1.3 to 3 Bohr Magneton, and the oxidation state of phosphorus atom is decreased from +5 to 0, according to X-ray photoelectron spectra. These data correspond to IR spectra.

11.1   INTRODUCTION

At present, polymer materials are widely used. Due to their unique properties, they are used as electrical insulating and structural materials in engineering and construction. Epoxy resin is one of the most used polymers. Epoxy compositions have low water permeability, high thermal stability, and stability against aggressive media which allow using them as protective coating. However, as a protective coating, epoxy resin has some disadvantages, such as combustibility and insufficient strength. In this connection, at present there is the necessity of improving performance characteristics of epoxy systems.

One of the promising trends in the modification of epoxy systems is the introduction of carbon nanostructures in their composition. However, it is necessary to select an optimal type and quantity of modifiers or to modify the nanostructures themselves. For this purpose, modification of nanostructures is used, which is intercalation of additional functional groups for improving interaction with material.[1]

Since the surface of nanostructures has low reactivity, for improving binding between the surface of nanostructures and the molecules of a surrounding medium, the modification of the nanostructure surface is used, that is, the attachment of certain chemical groups of sp-elements to the nanostructure surface, which form a covalent bond with the atoms on the nanostructure surface. In this case, the dispersancy and solubility of nanostructures improve and their coagulation in bunches is prevented due to the repulsion of the atoms of sp-elements attached to their side areas. The variation of the synthesis parameters allows obtaining nanostructures of the required dimension and imperfection. Modification provides additional conditions for improving the properties of nanostructures and effective interaction of nanostructures with a material.

For grafting additional functional groups providing a stronger interaction of carbon metal-containing (C/M) nanostructures and the matrix and,

The Metal/Carbon Nanocomposites Modification   163

thus, the improvement of the material mechanical properties, the functional groups containing phosphorus are often introduced in C/M-nanostructures.[1] As a phosphorus-containing component, ammonium phosphates including ammonium polyphosphate (APP) are selected. APP is an environmentally friendly substance and is widely used. APP is widely used for the preparation of flame-retardant coats. Thus, the modification with APP will allow using nanostructures for improving material properties and phosphorus will decrease combustibility due to its catalytic ability in the reactions of carbonization.

The goal of this work is the investigation of the functionalization of carbon copper nanostructures (C/Cu) by mechanochemical intercalation with APP for using as a modifier of epoxy resin.

11.2   INITIAL REAGENTS

11.2.1   CARBON COPPER-CONTAINING NANOSTRUCTURES

C/M-containing nanostructures are nanoparticles of copper stabilized in carbon nanofilm structures formed by carbon amorphous nanofibers associated with a metal-containing phase. Amorphous carbon fibers contain polyacetylene and carboxylic fragments. As the result of stabilization and association of chemically active nanoparticles of metal and the matrix of carbon material, a complex is formed which is stable in the air and at heating.

C/Cu nanostructures are synthesized in nanoreactors of the polymer matrices of polyvinyl alcohol. The process includes two main stages: mechanochemical mixing of reagents resulting in the formation of gels and xerogels and drying. In the result, copper/carbon nanocomposite is obtained. (Table 11.1).[2]

TABLE 11.1   Characteristics of Carbon Copper-containing Nanocomposite.

Composition Met(%):C(%)	50/50
Density, g/cm^3	1.71
Average size, nm	20
Form of metal nanoparticles	Spheres, single crystals in the form of cubes, and dodecahedra
Atomic magnetic moment, μB	1.3

11.2.2 AMMONIUM POLYPHOSPHATE

APP is a nonpolluting substance which can be widely used. Modification with APP will allow to increase the activity of nanocomposites, and the modification of polymers with such structures will improve their properties.

The chemical formula: $(NH_4PO_3)_n$ $(n > 1000)$.

11.3 EXPERIMENTAL TECHNIQUE

The process of modification APP was introduced into the studied nanostructures in the following concentration: APP:C/Cu = 1:1. In the study of Akhmetshina,[1] this ratio was determined as optimal for functionalizing C/M nanostructures.

The modification conducted had five stages. The first stage was the mechanical activation of C/Cu nanostructures in a triturating machine. The APP powder was added to the activated nanostructures and the mixture was mechanically treated (stage 2). After that, for the activation of the process and the effectiveness of triturating, a small amount of distilled water was added into the mixture and the mechanical treatment was continued (stage 3). After that, the prepared powder was dried at 75°C (stage 4). As the result of drying, the particles were agglomerated; the particle agglomerates were triturated in the triturating machine (stage 5).

For determining optimal conditions for the fictionalization, three different variants of the above method were studied. X-ray photoelectron spectroscopy was used for studying the samples prepared by the three variants of the method.

I variant, sample 1: mechanical activation of C/Cu nanostructures in the triturating machine. The APP powder was added to the activated nanostructures and the mixture was mechanically treated.

II variant, sample 2: the mechanical activation of C/Cu nanostructures in the triturating machine. The APP powder was added to the activated nanostructures and the mixture was mechanically treated. After that, for the activation of the process and the effectiveness of triturating, a small amount of distilled water was added into the mixture, and the mechanical treatment continued. After that, the prepared powder was dried at 75°C. The drying was conducted in the air. As the result of drying, the particles were agglomerated; the particle agglomerates were triturated in the triturating machine.

III variant, sample 3: the mechanical activation of C/Cu nanostructures in the triturating machine. The APP powder was added to the activated

The Metal/Carbon Nanocomposites Modification 165

nanostructures and the mixture was mechanically treated. After that, for the activation of the process and the effectiveness of triturating, a small amount of water was added into the mixture, and the mechanical treatment was continued. After that, the prepared powder was dried at 75°C; the drying was conducted under vacuum with the residual pressure of 10 mm Hg. As the result of the drying, the particles were agglomerated; the particle agglomerates were triturated in the triturating machine.

11.4 INVESTIGATION METHODS

11.4.1 IR-SPECTROSCOPY

The IR-spectroscopic studies were conducted on IR-Fourier spectrometer FCM 1201. With the use of the device, the spectra of the initial components and nanostructures (including functionalized nanostructures) were obtained.

The spectra taking was conducted on an adaptor MNPVO36 in the wave number range of 650–2000 cm^{-1}. An empty cell was used as a sample for comparison.

11.4.2 XPS STUDY

The XPS studies were conducted on an X-ray electron magnetic spectrometer with the resolution 10^{-4}, the device luminosity was 0.085% at the excitation by AlKα line 1486.5 eV in the vacuum of 10^{-8} to 10^{-10}. The use of X-ray electron magnetic spectrometers is practical because their magnetic energy-analyzer is separated from the spectrometer vacuum chamber.

11.5 RESULTS AND DISCUSSION

11.5.1 IR-SPECTROSCOPY

For determining, the necessity of the addition of a polar medium (H_2O) for improving the modification process, the IR transmission spectra of the initial C/Cu-nanocomposites and APP are compared with those of C/Cu+APP after the second stage of modification (without H_2O) (Fig. 11.1).

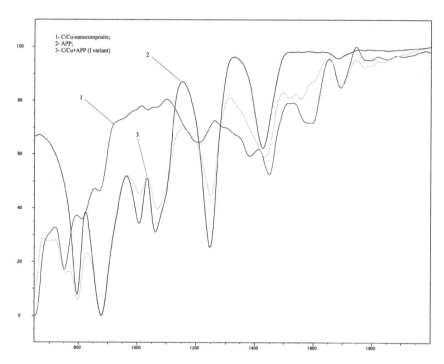

FIGURE 11.1 (See color insert.) IR spectra of C/Cu-nanocomposites, APP and C/Cu+APP (after the second stage of modification).

It is noted that after mechanochemical intercalation, the IR spectra change which indicates the decrease in the number of phosphate groups. This is indicated by the peaks at 1006, 1062, and 1246 cm^{-1} corresponding to the P—O variations. The peaks of the functionalized nanocomposite are similar to the peaks of the initial substances without any noticeable shift, which indicates that the chemical reactions between the substances are incomplete. Based on the above spectra, it is concluded that for the reaction between the mixture components a certain medium is required.

Figure 11.2 shows the transmission spectra of C/Cu-P (I variant), initial C/Cu-nanocomposites, and APP.

The peak at 877 cm^{-1} observed in the APP-spectrum relating to the bond P—O—P shifts by 11 cm^{-1} to the far-infrared region in the spectrum of C/Cu-P (I variant). Instead of the peaks at 1006 and 1062 cm^{-1} relating to the bond P—O, in the spectrum of the nanocomposite there is a maximum at 1074 cm^{-1}, which is characteristic of the bond P—O—H. The intensities of the peaks corresponding to the valence vibrations of the bond P—O

(1246 cm⁻¹) and NH_2 (795 cm⁻¹) are significantly decreased in the spectrum of C/Cu-nanocomposites, which indicates a significant decrease in the number of the mentioned bonds.

Based on the above spectra, it can be concluded that the reaction with the addition of distilled water is more complete. The content of ammonia groups decreases.

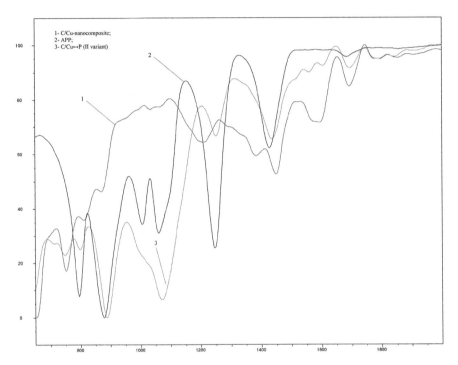

FIGURE 11.2 (See color insert.) The IR spectra of C/Cu-nanocomposites, APP, and C/Cu-P (I variant).

It is suggested that after the mechanochemical treatment of the powders with the addition of water (3 stage), the drying of the prepared mixture should be carried out under vacuum. For comparing, the variants of the method for the preparation of functionalized nanocomposites, the spectra of C/Cu-P-nanocomposites of variants I, and II are compared (Fig. 11.3).

According to the analysis of the above spectra, there are no significant changes in the nanocomposite. The drying under vacuum has not influenced the process of modification.

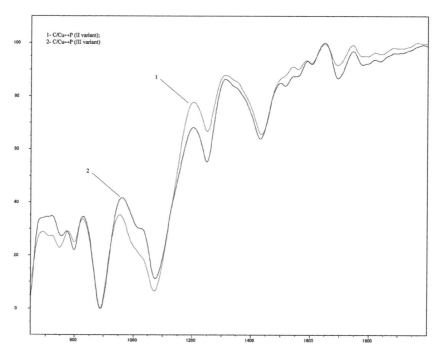

FIGURE 11.3 (See color insert.) The IR-spectra of C/Cu-P (I variant) and C/Cu-P (II variant).

11.5.2 XPS STUDY

For the determination of the chemical bond between the d-metal atoms and the sp-element atoms in the studied samples, the XPS-spectra of the core levels C1s, O1s, Cu3s, P2p, and N1s have been obtained.

Sample 1: C/Cu nanostructures modification with APP without the addition of water.

The C1s, O1s, and P2p spectra are taken at room temperature. The spectra N1s and Cu3s are weakly manifested.

The spectrum C1s (Fig. 11.4a) consists of two components C—H (285 eV) and C—O (287 eV).

In the O1s spectrum, there are two components: Oxygen bound to carbon and adsorbed oxygen.

The P2p spectrum (Fig. 11.5a) consists of one component with binding energy of 130 eV which corresponds to the bond of phosphorus and copper.[3] The spectrum Cu3s is not observed which can be explained by high looseness of the surface layer of the nanostructures.

The Metal/Carbon Nanocomposites Modification

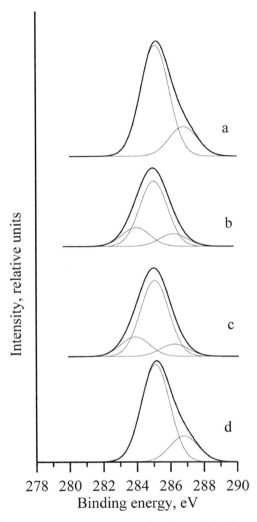

FIGURE 11.4 The XPS C1s-spectra: (a) sample 1; (b) sample 2; (c) sample 3, the spectrum is obtained at $T = 100°C$; and (d) sample 3, the spectrum is obtained at room temperature.

Sample 2: C/Cu nanostructures modification with APP with addition of water; drying in the air.

The C1s, O1s, Cu3s, and P2p spectra are obtained at room temperature and at heating up to 100°C. The N1s spectrum is weakly manifested.

At room temperature, in the Cu3s spectrum the parameters of the multiplet splitting are correlated with the number of uncompensated d-electrons of copper atoms and with the spin magnetic moment of copper.

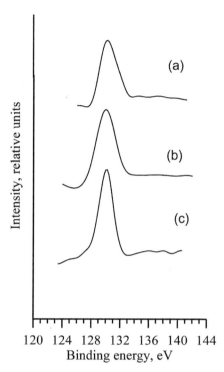

FIGURE 11.5 The XPS P2p spectra: (a) sample 1; (b) sample 2; and (c) sample 3.

The parameters of the multiplet splitting of the Cu3s spectra are given in Table 11.2, where I_2/I_1 is the ratio of the intensities of the maxima of the multiplet splitting lines; Δ is the energy distance between the maxima of the multiplet splitting in the 3s-spectra of copper, C/Cu-containing nanostructures, and modification C/Cu-containing nanostructures; μ is the magnetic moments in Bohr magnetons.

TABLE 11.2 Characteristics of X-ray Photoelectro Spectra and Atomic Magnetic Moments.

Sample	I_2/I_1	Δ, eV	μ_{Cu}, μ_B
Cu3s$_{massive}$	0	0	0
Cu3s$_{nano}$	0.2	3.6	1.3
Cu3s$_{nano\ function\ with\ P}$	0.6	3.6	3.0

The results obtained to indicate the increase of the number of uncompensated d-electrons on copper atoms, which evidences the participation of

The Metal/Carbon Nanocomposites Modification 171

d-electrons of copper atoms in the hybridized chemical bond with p-electrons of phosphorus atoms. In comparison with the unmodified C/Cu nanostructures, the copper atomic magnetic moment increases from 1.3 to 3 μ_B and the distance between the maxima of the multiplet of the Cu3s spectrum is similar to that of the unmodified C/Cu nanostructures, which indicates the similar localization of 3d-electrons in the surrounding of copper atoms of unmodified C/Cu nanostructures.

At the temperature in the range of room temperature −50°C, the P2p spectrum (Fig. 11.5b) consists of one component characteristic of the bond of phosphorus and copper (130 eV); at heating to higher than 100°C with the second component with the binding energy 135 eV appears, which is characteristic of the bond of phosphorus and oxygen.[4,5]

The C1s spectrum consists of the components sp^2, sp^3, and C—H (Fig. 11.4b).

The O1s spectrum consists of two components: oxygen bound to phosphorus and adsorbed oxygen.

Sample 3: The C/Cu nanostructures modification d with APP with addition of water; drying in vacuum.

The C1s, O1s, and P2p spectra are obtained at room temperature and at heating to 100°C. At room temperature, the Cu3s spectrum is not found. On the sample surface, the traces of nitrogen remain both at room temperature and at heating.

At room temperature, the C1s spectrum has two maxima—C—H (285 eV) and C—O (287 eV) (Fig. 11.4d).

The O1s spectrum consists of two components: oxygen bound to phosphorus and adsorbed oxygen.

At room temperature, the P2p spectrum consists of one component (132.0 eV) (Fig. 11.5c).

At heating to 50°C, the Cu3s spectrum appears.

At heating, in the C1s spectrum (Fig. 11.4c), in addition to the C—H bond, the components with sp^2- and sp^3-hybridization of the valence electrons of carbon atoms appear. Obviously, even small heating in vacuum leads to the decrease of the looseness of the surface layer and to the manifestation of the properties characteristic of the modified C/Cu nanostructures.

11.6 CONCLUSION

Thus, the investigations conducted show that modification takes place in all the three cases.

In the case, when drying takes place in the air, the magnetic moment on copper atoms changes due to the participation of d-electrons of copper atoms in the hybridized chemical bond with p-electrons of phosphorus atoms. In comparison with non-functionalized C/Cu-containing nanostructures, the atomic magnetic moment on copper atoms increases from 1.3 to 3 μ_B.

In the case, when drying takes place in vacuum, the Cu3s spectrum is not observed, which is likely to be connected with high looseness of the surface layer of the modification C/Cu nanostructure. At heating to higher than 50°C, the Cu3s spectrum appears, and in the C1s spectrum the components with sp^2- and sp^3-hybridization of the valence electrons appear which are characteristic of nanostructures.

In the case, when modification takes place without addition of water, the Cu3s spectrum is not observed even at heating. In the C1s spectrum, the components with sp^2- and sp^3-hybridization of the valence electrons are not observed either which is due to high degree of looseness of the nanostructure surface.

KEYWORDS

- **copper/carbon nanocomposite**
- **modification**
- **ammonium**
- **polyphosphate**
- **intercalation**
- **atomic magnetic moment**
- **IR spectra**
- **X-ray photoelectron spectra**

REFERENCES

1. Akhmetshina, L. F. The Development of the Method for the Functionalization of Carbon Metal-containing Nanocomposites and Methods for the Preparation of Suspensions Based of the Nanocomposites for Modifying Composite Materials. Avtoreferat Dissertatsii na Soiskanie Uchenoy Stepeni K. T. N. (Thesis of PhD), Perm, 2011.
2. Kodolov, V. I.; Vasilchenko, Yu. M.; Akhmetshina, L. F.; Shklyaev, D. A.; Sharipova, A. G.; Volkova, E. G.; Ulyanov, A. L.; Kovyazina, O. A. Patent RF 393,110, Method for

Preparation of Carbon Metal-containing Nanostructures, Applicant and Patent Holder AO "IEMZ" "KUPOL", 2012.

3. Wagner, C. D.; Riggs, W. M.; Davis, L. E.; Moulder, J. F. Handbook of X-ray Photoelectron Spectroscopy. Perkin-Elmer Corporation (Physical Electronics): Waltham, MA, 1979; p 192.

4. Beamson, G.; Briggs, D. High Resolution XPS of Organic Polymers – The Scienta ESCA 300 Database. Wiley Interscience: Hoboken, NJ, 1992; p 293.

5. Nefyodov, V. I. X-ray Photoelectron Spectroscopy of Chemical Compounds. *Reference Book;* Khimia: Moscow, 1984; p 256.

PART III
New Insights and Developments

CHAPTER 12

EFFICIENCY OF METAL/CARBON NANOCOMPOSITE APPLICATION IN LILY GROWING UNDER PROTECTED SOIL CONDITIONS

V. M. MERZLYAKOVA[1], A. A. LAPIN[2*], and V. I. KODOLOV[3,4]

[1]*Izhevskaya State Agricultural Academy, Izhevsk, Russia*

[2]*Department of Water Resource, Kazan State Power Engineering University, Krasnoselskaya Str. 51, Kazan, The Republic of Tatarstan, Russia*

[3]*Deparment of Chemistry and Chemical Technology, Kalashnikov Izhevsk State Technical University, Studencheskaya St. 7, Izhevsk, Russia*

[4]*Basic Research-High Educational Centre of Chemical Physics and Mesoscopy, Ural Division of Russian Academy of Sciences, Izhevsk, Udmurt Republic, Russia*

[*]*Corresponding author. E-mail: lapinanatol@mail.ru*

ABSTRACT

The scientific investigations and industrial experiments on the copper/carbon nanocomposites influence in the processes of bulb lily promotion carried out. Obtained results show the increase in the height of the flower sprout, the height of the stem during the cutting of plants, the number of buds, and the diameter of the open flowers. It is established that the use of 0.01% copper/ carbon nanocomposite in the cultivation of lilies led to an increase in the cost of sales and, accordingly, profit and profitability of production. Treatment of lily bulbs with a 0.01% of copper/carbon nanocomposite is energetically effective since allowed to get products of higher quality.

12.1 INTRODUCTION

Lilies are very ancient high-decorative perennial bulbous plants. A wide variety of species, varieties, and hybrid forms makes it possible to widely use them in cutting and gardening. Many varieties can serve as good material for forcing in protected ground and for cutting in open ground.

Lilies are one of the most important industrial flower crops. In terms of world production, the lily slices firmly hold the third-fourth place, behind only roses, carnations, and tulips.

Modern technologies would increase economic efficiency in the agro-industrial complex. For the effective cultivation of flower crops in protected soil, it is necessary to introduce innovative technologies that ensure high yields, which allow reducing material costs and increasing profitability. In modern industrial floriculture, methods of nanotechnology make it possible to influence the process of growing flower crops, their productivity, and the quality of flowers. A promising direction is the development and application of nanoelements for crop production with optimal particle sizes for maximum assimilation of macro and microelements.

Changes in the main characteristics of substances and elements are due not only to small dimensions but also to the manifestation of quantum mechanical effects with the dominant role of interfaces. These effects occur at a critical size that is commensurable with the so-called correlation radius of a physical phenomenon (e.g., with the mean free path of electrons, the size of the magnetic domain, etc.). An important feature of metallic nanoelements, which plays a key role in their use in agriculture, is their low toxicity. Technical specifications (TU U 24.6-35291116-001.2007) have now been applied to a large group of nanoelements based on Ag, Cu, Co, Mn, Mg, Zn, Mo, and Fe metals and their production by domestic producers has been established.[1]

Research in this direction is carried out with cultivation in agricultural crops in conditions of open ground. Thus, high efficiency as a means for pre-sowing treatment of seeds of wheat, sunflower, rapeseed, maize, amaranth, and other crops has been revealed with respect to solutions of carbon nanotubes, ultradispersed suspensions, metal nanoparticles, and nonmetals. At the same time, the increase in yields, adaptation to unfavorable climatic conditions, and improvement of the quality of plant raw materials was observed. The action of nanoelements on biosystems was of a prolonged nature. The stimulating effect of nanosuspensions when soaking the bulbs is due to the assimilation by the plants of nutrients through the bulb, leaves, and stems. Possible mechanism of influence of copper/carbon nanocomposite on the lily is interaction at the cellular level. Nanoelements are able to inform molecules of their excess energy, which increases the rate of biochemical reactions taking place in plants. In other words, nanoelements actively participate in the processes of microelement balance, that is, they are bioactive. Scientific research in the field of the use of nanotechnology in protected soil when cultivating flower crops is single.[2–7] In this connection, the aim of the work was to study the effect of the metal–carbon nanocomposite-based copper effect on lily growing under protected soil conditions.

12.2 EXPERIMENTAL RESEARCH

Scientific and industrial work was carried out in conditions of protected soil in industrial greenhouses. The experiment was laid in threefold repetition, placing the variants by the method of the randomized repetitions. The technology of forcing greenhouse lilies in sheltered soil in lily containers is generally accepted. Copper-containing nanocomposite was used in the form of nanosuspensions of various concentrations. There are two factors: factor A—hybrids of lilies and factor B—concentration of nanocomposite. The object of the research was hybrids of the greenhouse lily of the selection and seed-growing company "Van den Bose Floverbulbs" Siberia and Santander. Lily in Siberia and Santander belong to the group of Oriental hybrids. As a control, variants were used without bulb treatment and using water.

12.3 RESULTS AND DISCUSSION

In the economy, bulbs of lilies came in peat; they were packed in film bags, from a dense film that had 18 holes 1 cm in diameter for air intake. Packages with lily bulbs were stored in container boxes at a temperature of -1°C, which allow them to remain at rest. Samples of this peat were sampled and agrochemical analysis was carried out using a volumetric analysis of the preparation of aqueous extract (with recalculation of the results of the analysis for volume) and a sample with recalculation of the results of the analysis for dry weight (Table 12.1).

The moisture content of the peat corresponds to the normative indices. According to the degree of acidity, peat refers to a very acidic. The specific electrical conductivity of peat does not exceed 1.0 mS/cm. Peat has a low content of nitrogen, phosphorus, and potassium in terms of the degree of supply of food elements. Such properties of this peat contribute to the long storage and transportation of bulbs of long lilies and do not allow them to germinate.

TABLE 12.1 Characteristics of Peat During Storage of Lily Bulbs Imported from Holland.

Index	Result	Value of the horse Peat sphagnum
Moisture of peat, %	60.0	59.46
pH (suspension)	4.0	5.5–6.1
Specific conductivity of the suspension, mS/cm, at 25°C	0.093	–
Content of mobile forms of batteries		
Nitrate nitrogen N-NO$_3$	0.57	26.3
	0.1	29.0
Nitrogen ammonium N-NH$_4$	7.36	135.6
	1.5	149.0
P$_2$O$_5$	1.64	291.0
	0.3	320.0
K$_2$O	34.56	353.0
	7.0	388.0
Ca^{2+} + Mg^{2+}, mmol/g	2.98	–
	0.6	

Numerator (for dry matter)—mg/100 g; denominator—mg/L.

The lilies were cut into slices for peeling. Samples of peat mixtures were selected and agrochemical analysis was carried out (Table 12.2).

For the growth of plants on top peat, the optimum moisture content is within 78–85% of the mass. The moisture content in the peat mixture is normal.

A very important factor for the development of roots and the absorption of nutrients by them is the acidity of the soil. High acidity of soil leads to insufficient intake of such elements as phosphorus, magnesium, and iron. When growing Oriental hybrids of lilies, which include our variety, the pH should be between 5.0 and 6.5.

Lilies are sensitive to soil salinity. With a high content of salts in the soil, the roots of the lilies become hard, brittle, and acquire a yellow-brown color. Soil salinity is determined by the specific electrical conductivity. Lilies require low EU levels.

Efficiency of Metal/Carbon Nanocomposite Application 181

TABLE 12.2 Agrochemical Characteristics of the Peat Mixture at the Laying of the Experiment on the Forcing of Bulbs of Lilies.

Index	Result
Moisture of peat, %	74.3
pH (suspension)	5.10
Specific conductivity of the suspension, mS/cm, at 25°C	0.115
The content of mobile forms of batteries	
Nitrate nitrogen N-NO$_3$	0.24
	0.2
Nitrogen ammonium N-NH$_4$	6.01
	3.8
P$_2$O$_5$	3.94
	2.5
K$_2$O	32.65
	21.0
Ca^{2+} + Mg^{2+}, mmol/g	3.82
	2.4

Numerator (for dry matter)—mg/100 g; denominator—mg/L.

The content of mobile forms of nutrients is low, but the level of plant nutrition is regulated by a computer program in accordance with the needs of lily bulbs. Thanks to the inherent peat strength of buffer and high sorption capacity, mineral fertilizers are not washed out and stored in an accessible form for plants; at the same time, the danger of creating an increased concentration of salts harmful to plants is reduced.

High productivity and quality of flowers largely depend on the full provision of plants with trace elements, so when planting lily bulbs, an analysis was carried out for the content of trace elements in the peat mix (Table 12.3).

TABLE 12.3 Chemical Analysis of Peat Mixture for Micronutrients Content.

Microelements, (mg/kg dry weight)	The actual value of the test results
Zinc	15.60
Copper	3.81
Iron	126.60
Manganese	57.30
Boron	8.27

182 Nanoscience and Nanoengineering: Novel Applications

According to the content of trace elements, peat mixes contain a very low content of zinc, copper, iron, manganese, and boron.[8] By processing bulbs of lilies, metal/carbon nanocomposite on the basis of copper, we contribute to replenishing available forms of copper in lily plants. In the period of growth and development, the level of nutrition of lily plants is regulated by a computer program in accordance with the needs of macro- and microelements.

For 2 years, the reaction of the lily to the treatment of copper/carbon nanocomposite was studied. The biometric parameters of the lily were determined as the height of the flower bud in the bud budding phase (Table 12.4), the number of buds, the diameter of the open flower, and the height of the stem when cutting plants.

TABLE 12.4 Height of Flowering Shoot, cm (Average for 2015–2016, the Phase of the Beginning of Budding).

Hybrids of lilies (factor A)	Concentration of nanocomposites (factor B)					Average by factor A $HSR_{05} A = 1.7$
	Without processing (k)	Water (к)	0.01%	0.02%	0.05%	
Siberia (к)	79.6	71.9	87.7	87.9	87.2	82.8
Santander	77.9	81.6	89.1	82.2	82.3	82.6
Average by factor B	79.2	76.9	88.5	85.2	84.6	–
$HSR_{05} B = 2.7$						
$HSR_{05 \text{ private differences}}$			3.8			

When processing with nanocomposites, there was a significant increase in the height of the flowering shoot of lilies from 4.9 to 8.4 cm (at HSR = 1.7). The highest plants were Santander lily plants when treated with a 0.01% nanocomposites (Table 12.5).

The number of buds on the shoot was higher for the Siberia variety than for the Santander variety by 1.1 pcs (at HSR = 0.06 pcs). When processing with nanocomposites, there was a significant increase in the number of buds to 5.0–5.3 pieces (Table 12.6).

The length of the bud in Siberia variety averaged 10.5 cm, while the Santander variety was 1.2 cm lower. When processing with nanocomposites, a significant increase in the length of the bud occurred. The largest length was found in lilies treated with a 0.01% copper/carbon nanocomposite suspension.

Efficiency of Metal/Carbon Nanocomposite Application 183

TABLE 12.5 Number of Buds, Pcs (Average for 2015–2016, the Phase of the Beginning of Budding).

Hybrids of lilies (factor A)	Concentration of nanocomposites (factor B)					Average by factor A HSR$_{05}$ A = 0.06
	Without processing (k)	Water (к)	0.01%	0.02%	0.05%	
Siberia (к)	4.7	4.7	5.8	5.6	5.4	5.2
Santander	3.5	3.5	4.7	4.5	4.6	4.1
Average by factor B	4.1	4.1	5.3	5.1	5.0	–
HSR$_{05}$ B = 0.1						
HSR$_{05 \text{ private differences}}$			0.1			

TABLE 12.6 Length of Bud, cm (Average for 2015–2016, the Phase of the Beginning of Budding).

Hybrids of lilies (factor A)	Concentration of nanocomposites (factor B)					Average by factor A HSR$_{05}$ A = 0.06
	Without processing (k)	Water (к)	0.01%	0.02%	0.05%	
Siberia (к)	10.1	10.3	10.9	10.6	10.6	10.5
Santander	9.0	9.1	9.6	9.5	9.3	9.3
Average by factor B	9.5	9.7	10.3	10.0	9.9	–
HSR$_{05}$ B = 0.09						
HSR$_{05 \text{ private differences}}$			0.1			

The length of the bud and the number of buds depends on the photosynthetic activity, which is directly related to the width of the leaves, so the width of the lily leaf was measured during the bud budding phase (Table 12.7).

The variety Santander leaves were wider than that of Siberia, which is one of the varietal traits. The increase in the number and length of buds in both varieties when processing copper bulbs/carbon nanocomposite was associated with an increase in the width of the leaves.

The productivity of the lily (lily cut) is determined by the number of buds in the inflorescence, the diameter of the flower, and the height of the plant. In this regard, the plants of lilies count the number of buds when cutting plants (Table 12.8).

184 Nanoscience and Nanoengineering: Novel Applications

TABLE 12.7 The Width of the Leaf, cm (Average for 2015–2016, the Phase of the Beginning of Budding).

Hybrids of lilies (factor A)	Concentration of nanocomposites (factor B)					Average by factor A HSR$_{05}$ A = 0.05
	Without processing (k) (к)	Water (к)	0.01%	0.02%	0.05%	
Siberia (к)	3.0	3.1	3.4	3.3	3.1	3.2
Santander	3.4	3.5	3.7	3.6	3.6	3.6
Average by factor B	3.2	3.3	3.6	3.5	3.4	–
HSR$_{05}$ B = 0.08						
HSR$_{05\ private\ differences}$			0.1			

TABLE 12.8 Number of Buds When Cutting Plants, Pcs (Average for 2015–2016).

Hybrids of lilies (factor A)	Concentration of nanocomposites (factor B)					Average by factor A HSR$_{05}$ A = 0.04
	Without processing (k) (к)	Water (к)	0.01%	0.02%	0.05%	
Siberia (к)	4.8	4.7	6.0	5.6	5.5	5.3
Santander	3.6	3.5	4.7	4.5	4.6	4.2
Average by factor B	4.2	4.1	5.4	5.0	5.0	–
HSR$_{05}$ B = 0.06						
HSR$_{05\ private\ differences}$			0.09			

When cutting lilies, the number of buds on the plant remains the same as in the bud budding phase. There were also a greater number of buds in the Siberia lily variety than in Santander. Depending on the treatment with various concentrations, the maximum number of buds was noted at 0.01% concentration of copper/carbon nanocomposite.

The diameter of the open flower in the Santander variety was significantly larger than that of the Siberia variety (21.3 cm) (Table 12.9). Processing of bulbs with nanocomposites promoted a substantial increase in the diameter of the open flower. The largest diameter of the flower was noted when treated with 0.01% nanocomposite (22.0 cm) (Fig. 12.1).

The height of the stem when cutting lilies increased by treatment with copper/carbon nanocomposite to 131.0–141.0 cm in comparison with the control (99.0–102.3 cm). The sort of lily Siberia was slightly higher than the Santander variety.

Efficiency of Metal/Carbon Nanocomposite Application

TABLE 12.9 Diameter of the Open Flower, cm (Average for 2015–2016, Cutting Plants).

Hybrids of lilies (factor A)	Concentration of nanocomposites (factor B)					Average by factor A HSR$_{05}$ A = 0.3
	Without processing (k)	Water (к)	0.01%	0.02%	0.05%	
Siberia (к)	18.7	18.3	21.3	20.0	20.0	19.7
Santander	20.0	20.0	22.7	22.0	21.7	21.3
Average by factor B	19.3	19.2	22.0	21.0	20.9	–
HSR$_{05}$ B = 0.5						
HSR$_{05\ private\ differences}$			0.7			

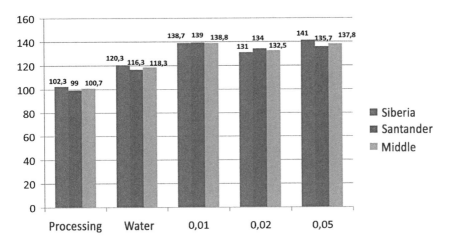

FIGURE 12.1 (See color insert.) Height of stem, cm (average for 2015–2016, cutting of plants).

Thus, the treatment of bulb lily copper/carbon nanocomposites contributes to the increase in plant height, the number, and diameter of the flowers.

Flowers—this is a product that does not bear any practical benefit in itself and its cost is rather arbitrary, but nevertheless people are ready to go to great expenses for their purchase, because any solemn event, event, congratulation, as well as simply expressing one's feelings and sympathies, cannot be imagined without flowers. Flowers—perishable products. Not all flowers manage to pass the way to their consumer.

186　　　Nanoscience and Nanoengineering: Novel Applications

For a more objective evaluation of the results obtained, it is necessary to determine the economic efficiency. For comparison with the control variant (soaking the bulbs of lilies in water), a variant was selected with treatment of lily bulbs with 0.01% copper/carbon nanocomposite under the action, which produced the highest lily productivity. The data obtained are presented in the Table 12.10.

TABLE 12.10　Cost-effectiveness of Copper/Carbon Nanocomposite in the Cultivation of a Hybrid of Lily Siberia.

Index	Option of experience	
	Water (к)	**0.01%**
Cuts in total, pcs/m²	45	45
Average number of buds on the stem, pcs.	4.8	5.9
The average price of a cut plant, rub	74.35	81.55
Cost of products, rub/m²	3345.75	3669.75
Production costs, rub/m²	2164.8	2283.4
Net profit, rub/m²	1180.95	1386.35
Level of profitability, %	54.5	60.7
Cost of production, rub/kg	48.11	50.74

The use of 0.01% copper/carbon nanocomposite during lily uplift affects the cost of production, only slightly increasing it, whereas the average selling price increased significantly (by 7 rubles) per plant due to the increase in the number of buds, which led to an increased cost of sales and, accordingly, profit and profitability of production.

12.4　CONCLUSION

According to the results of scientific and industrial experiments, processing of bulb lily copper/carbon nanocomposites promotes an increase in the height of the flower sprout, the height of the stem during the cutting of plants, the number of buds and the diameter of the open flowers.

The use of 0.01% copper/carbon nanocomposite in the cultivation of lilies led to an increase in the cost of sales and, accordingly, profit and profitability of production. Treatment of lily bulbs with a 0.01% copper/carbon nanocomposite is energetically effective since allowed to get products of higher quality.

KEYWORDS

- **copper/carbon nanocomposite**
- **suspension**
- **concentration**
- **hybrids of lilies**
- **growth stimulation**
- **protected soil**
- **buds**

REFERENCES

1. Kaplunenko, V. G.; Kosinov, N. V.; Bovsunovsky, A. N.; Chernyiy, S. A. Nanotechnology in Agriculture. *Grain.* **2008,** *4,* 47–55.
2. Lapin, A. A.; Merzlyakova, V. M.; Kodolov, V. I.; Lipotenkina, M. L.; Zelenkov, V. N. The Use of Metal. Carbon Nanocomposites in the Cultivation of Lilies in Protected Ground. Unconventional Natural Resources, Innovative Technologies and Products. Collection of Scientific Papers. RANS (Russian Academy of Natural Sciences): Moscow, 2016; Vol. 224, pp 41–46.
3. Lapin, A. A.; Merzlyakova, V. M.; Zelenkov, V. N. Calculation of the Total Antioxidant Activity of Adsorbed Structured Water in Lily Samples. Unconventional Natural Resources, Innovative Technologies and Products: Collection of Scientific Papers. RANS: Moscow, 2016; Vol. 263, pp 28–33.
4. Lapin, A. A. Merzlyakova, V. M.; Zelenkov, V. N. *Total Antioxidant Activity of Aqueous Extracts of Samples of Lilies Treated with Copper/carbon Nanocomposites. Unconventional Natural Resources, Innovative Technologies and Products: Collection of Scientific Papers*; RANS: Moscow, 2016; Vol. 263, pp 46–53.
5. Merzlyakova, V. M.; Kovyazina, O. A.; Trineeva, V. V.; Kodolov, V. I.; Lipotenkina, M. L. *Experience of Using Metal/Carbon Nanocomposites in Growing Flowers in Sheltered Soil (by the Example of a Lily). From Nanostructures, Nanomaterials and Nanotechnologies to the Nanoindustry: Abstracts*, Fifth International Conference, Izhevsk, April 2–3, 2015, Under the Total. Kodolov, I. N., Izd-IzhGTU Named after Kalashnikov, M. T. Izhevsk, 2015; Vol. 240, p 122.
6. Merzlyakova, V. M.; Kornev, V. I. *Antioxidant Activity of Aqueous Extracts of Samples of Lilies Treated with Copper/Carbon Nanocomposite,* XX Postgraduate-master's

Seminar Devoted to the Day of Power, FGBOU VO Kazan State Energy University, December 6–7, 2016.

7. Fedorenko, V. F. Ed.; Under the Society. *Nanotechnologies and Nanomaterials in the Agro-industrial Complex;* FGBNU Rosinformmagrotech: Moscow, 2011; p 312.

8. Chemical Analysis of the Peat Mixture on the Content of Trace Elements [Methodological Guidelines for Determining the Mobile Forms of Trace Elements in Greenhouse Soils, 1985].

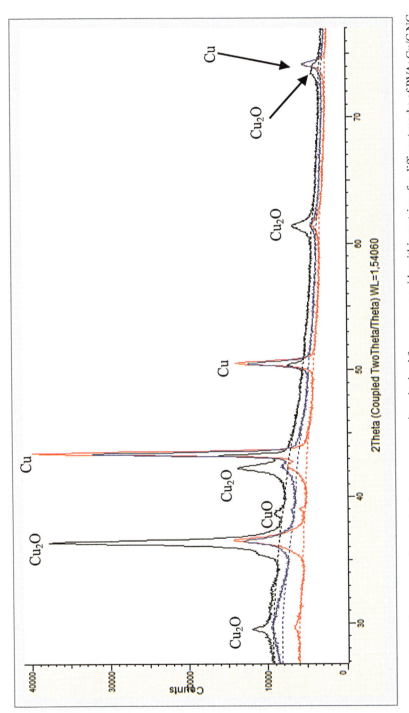

FIGURE 1.1 Diffract grams of copper/carbon nanocomposites obtained from copper oxide within matrices for different marks of PVA: Cu/C NC (red)—PVA mark BF-14, NC (blue)—BF-17, and NC (dark)—BF-24.

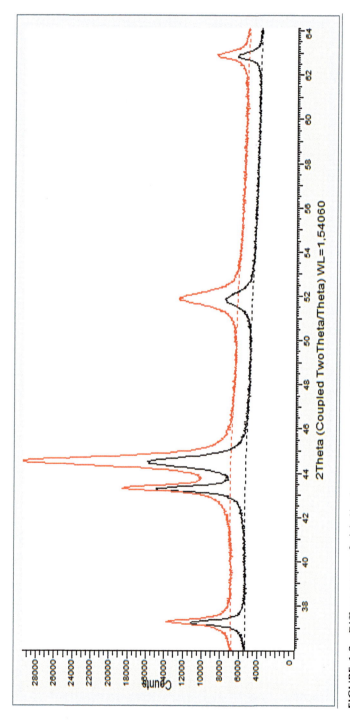

FIGURE 1.2 Diffract grams of nickel/carbon nanocomposites obtained from nickel oxide within matrices. PVA at different temperature of thermals.

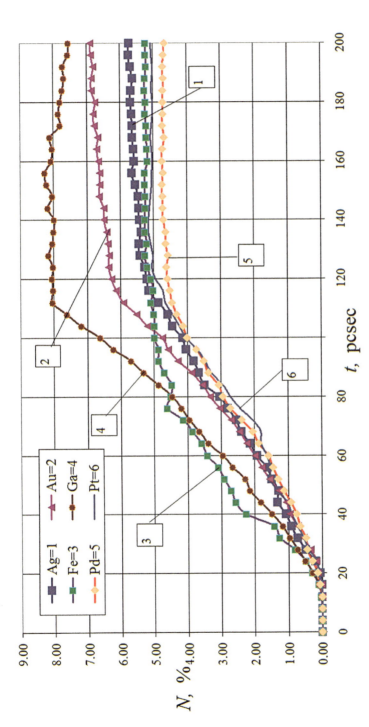

FIGURE 9.3 Percentage of precipitated atoms that got into the pore to the total number of precipitated atoms.

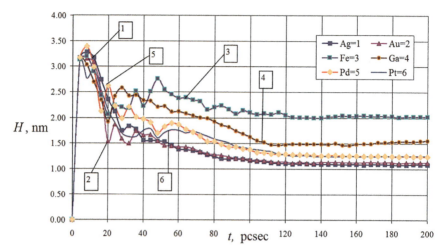

FIGURE 9.4 Depth of mass center of atoms, which penetrated the pore.

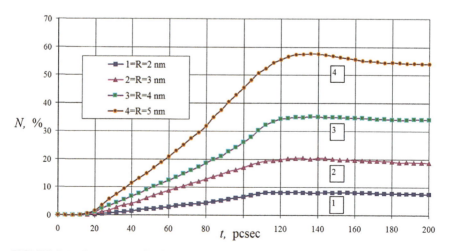

FIGURE 9.5 Percentage of gallium atoms, which got into the pore in relation to the total number of precipitated atoms.

Nanoscience and Nanoengineering: Novel Applications E

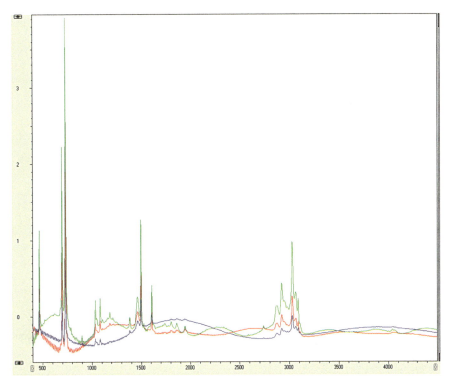

FIGURE 10.9 The peaks intensity changes in IR spectra of Cu/C nanocomposite in dependence of the duration of suspension existence. Red—in the beginning; green—the first day; and blue—the third day.

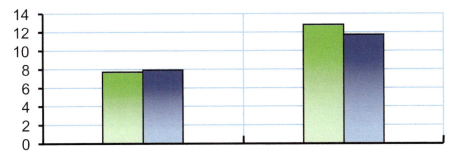

FIGURE 10.12 Comparison of adhesion strength for materials. 1 (green) and 2 (dark blue) before (A) and after (B) the modification by copper/carbon nanocomposites.

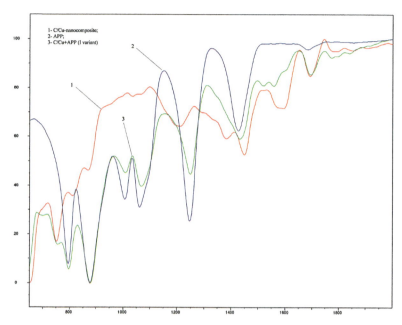

FIGURE 11.1 IR spectra of C/Cu-nanocomposites, APP and C/Cu+APP (after the second stage of modification).

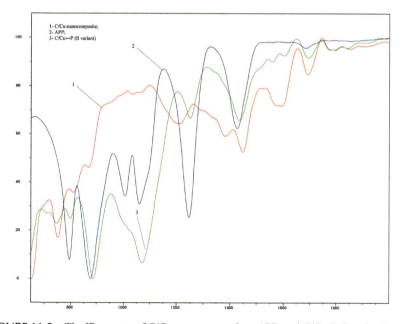

FIGURE 11.2 The IR spectra of C/Cu-nanocomposites, APP, and C/Cu-P (I variant).

Nanoscience and Nanoengineering: Novel Applications G

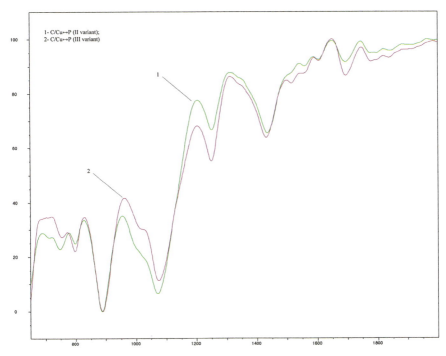

FIGURE 11.3 The IR-spectra of C/Cu-P (I variant) and C/Cu-P (II variant).

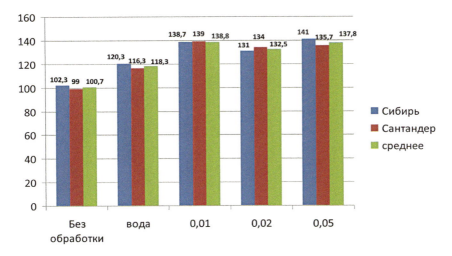

FIGURE 12.1 Height of stem, cm (average for 2015–2016, cutting of plants).

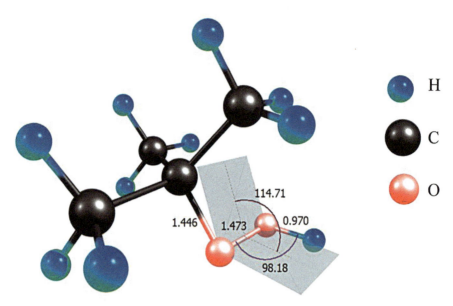

FIGURE 19.1 Structural model of the *tert*-butyl hydroperoxide molecule (equilibrium configuration obtained at MP2/6-31G(d,p) level of theory).

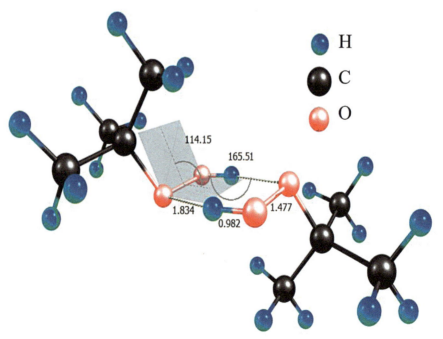

FIGURE 19.4 Structural model of the *tert*-butyl hydroperoxide dimer. Molecular geometry optimization of this homoassociate was performed at MP2/6-31G(d,p).

CHAPTER 13

BIOCIDAL ACTIVITY OF PHTHALOCYANINE–POLYMER COMPLEXES

A. V. LOBANOV[1,2*], A. B. KONONENKO[3], D. A. BANNIKOVA[3], S. V. BRITOVA[3], E. P. SAVINOVA[3], O. A. ZHUNINA[4], S. M. VASILEV[5], V. N. GORSHENEV[6], G. E. ZAIKOV[6], and S. D. VARFOLOMEEV[6]

[1]*Department of Chemical and Biological Processes Dynamics, Semenov Institute of Chemical Physics, Russian Academy of Sciences, Moscow, Russia*

[2]*Plekhanov Russian University of Economics, Moscow, Russia*

[3]*All-Russian Research Institute of Veterinary Sanitation, Hygiene and Ecology, Moscow, Russia*

[4]*Russian Scientific Center of Molecular Diagnostics and Therapy, Moscow, Russia*

[5]*Institute of Modern Standards, Moscow, Russia*

[6]*Department of Bio-Chemistry, N. M. Emanuel Institute of Biochemical Physics, Russian Academy of Sciences, Kosygina St. 4, Moscow 119991, Russia*

*Corresponding author. E-mail: avlobanov@mail.ru

ABSTRACT

The aggregation behavior of iron, manganese, and magnesium phthalocyanine (Pc) with poly-N-vinylpyrrolidone and polyethylene glycol was considered in this chapter. Biocidal activity of iron and manganese Pc complexes with polymer in relation to nine cultures of microorganisms was determined.

13.1 INTRODUCTION

Preparation of nanoscale modifications of biologically active compounds is a rapidly growing sector of nanotechnology and is directed to the synthesis of particles of different sizes, shapes, and chemical composition with enhanced or novel properties as compared with the monomolecular analogs.[1]

Iron and manganese complexes with phthalocyanines (Pcs) are homogeneous catalysts of autoxidation of the range of biogenic reductants and their analogs, such as ascorbic acid, ubiquinone, and cysteine. In this process, there is the intermediate formation of active oxygen forms—superoxide radical anion, peroxide and hydroxyl radicals, hydrogen peroxide,[2–7] whose cytostatic, and other biological activity well known.[8,9] These radical and ion-radical particles cause destructive oxidation reactions, including cleavage of nucleic acids. Less important feature of the biologically active metal complexes of Pcs is their ability to selectively accumulate in cells.[10,11] Thus, biocidal activity of Pc complexes in relation to dangerous microorganisms could be promising property for solving some problems of biological safety.

At present, the catalytic activity of Pc complexes of iron and manganese is in a monomolecular form, in the formation of reactive oxygen species has been well studied. Because of the insolubility in water of Pcs for creating of the biologically active agents to water-based widely use various solubilizers, including water-soluble, biocompatible polymers. In the composition of complexes with polymers Pcs can form aggregates of different types,[12] this can significantly affect their catalytic properties and, as a consequence, the biocidal activity. In this chapter, the aggregation behavior of iron (Fe), manganese (Mn), and magnesium (Mg) Pcs in the complexes with water-soluble polymers of poly-N-vinylpyrrolidone (PVP) and polyethylene glycol (PEG) were considered. The biocidal activity of the Pc–polymer complexes was determined.

13.2 EXPERIMENTAL

FePc, MnPc, and MgPc (Acros Organics, USA), PVP and PEG (Sigma, USA) with a molecular weight of 26,500 and 1000 g/mol, respectively, were used in this work. The purity and individuality of metal complexes of Pcs were confirmed by method of MALDI-mass spectrometry on the appliance Thermo DSQ II (USA). Base solution of the metal complex of Pc was prepared by dissolving the dry sample of substance in the

Biocidal Activity of Phthalocyanine–Polymer Complexes 191

concentrated sulfuric acid. Further, 1% (vol) of the solution was added in dimethylformamide to a concentration of Pc $6\cdot10^{-3}$ mol/L. Concentration was monitored by electronic absorption spectra. The solutions were kept in the dark at 4°C. Then, 2% (wt) solution of PVP and PEG in bidistilled water were separately prepared. Solutions of polymers and Pc are mixed in a ratio of 100:1 to a concentration of Pc $6\cdot10^{-5}$ mol/L. The appliance HACH DR-4000V (USA) was used for registration of the electronic spectra. An aliquot of 2 mL of a solution of polymer and metal complex Pc was placed in a quartz cuvette of 1 cm in width. For comparison experiments, FePc and MePc were adsorbed onto silica nanoparticles with a diameter of 60 nm according to the procedure described by Udartseva.[13] To prepare solutions of sodium dodecyl sulfate (SDS) to a concentration of 0.0082 mol/L (M = 288.4 g/mol) a sample of dry SDS (0.0021 g) was dissolved in 10 mL of distilled water. To obtain solutions of Triton (X-100) with a concentration of 0.00025 mol/L (M = 625 g/mol), sample of TX-100 (0.0016 g) was dissolved in 10 mL of distilled water. These solutions were used in comparative experiments in some cases.

Biocidal activity was determined by the method of diffusion in agar containing the test-culture by determining the diameter of the bacteria growth inhibition zones. Test-cultures of *Staphylococcus aureus* P209 and *Escherichia coli* 1257 were used in experiments with various dilutions of the preparation. Test-cultures of microorganisms were cultivated on meat-peptone agar at 37°C for 24 h, then wash out with physiological solution was made, and the initial concentration of suspension 10^9 microbial cells/mL was detected by bacterial turbidity standard. Serial dilutions of cultures were prepared in sterile physiological solution, in this case, final concentration of bacteria was 10^7 microbial cells/mL. Melted and cooled to 45°C meat-peptone agar of 10 mL in volume was mixed with 1 mL of suspension of each culture and placed in Petri plates. Then, the wells of 3 mm in diameter were made by special sterile stencil and solutions of the complexes of FePc and MnPc with polymers in different dilution were added. Culturing was incubated for 24 h at 37°C and accounting of zones of growth inhibition of test-cultures were produced.

13.3 RESULTS AND DISCUSSION

The complexes of FePc and MnPc are not dissolved in water. The use of PVP and PEG polymers as 1% additive allows solubilizing metal complexes of Pcs to form clear aqueous solutions. It is known that the absorption band of monomolecular Pcs lies in the range 670–690 nm.[14] According to the data of electron spectroscopy in the composition of the complexes of polymers, FePc has absorption bands near 655 and 850 nm (Fig. 13.1). The form of the spectrum is almost independent of the concentration of FePc and type of polymer. Structure of the spectrum allows considering that FePc is represented mainly by molecular aggregates whose size can be up to a few nanometers. Changing of the initial degree of oxidation of the iron ion (II) in such structures cannot be excluded.[15]

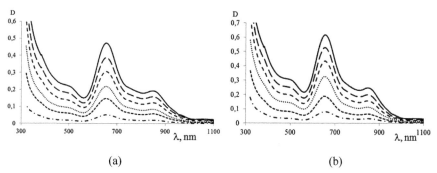

FIGURE 13.1 The absorption spectra of FePc in complex with the PVP (*a*) and PEG (*b*) in water, corresponding concentrations of FePc 0.5×10^{-6}, 1.0×10^{-6}, 1.5×10^{-6}, 2.0×10^{-6}, 2.5×10^{-6} mol/L.

Type of the electronic spectra of MnPc in the polymer complexes varies little depending upon the nature of the polymer (Fig. 13.2). The absorbance with a maximum of 620 nm is observed in the spectra. The presence of the absorption indicates the presence of MnPc in the form of H-aggregate stacking structure in which, it is likely, dimers of type $(Mn^{III}Pc)O_2$ can be contained.[16] The absorption of low intensity, which manifests itself in the form of a shoulder at 690 nm, corresponds to a small fraction of monomolecular form MnPc. Thus, both the metal complex by reacting with the polymers PVP and PEG self-assemble into nanoscale aggregates.

Pc complex of magnesium in the multicomponent systems can be in three states:[17–22] in the form of a monomer with narrow absorption band in the 675 nm, H-aggregates with their absorption band shifted relative to the monomer hypsochromically to ~630 nm, and J-aggregates with an absorption band bathochromically shifted relative to the monomer to wavelengths of 830 nm (Fig. 13.3). H- and J-aggregates have a different structure. H-aggregates are stacking structure type and J-aggregates are brickwork type construction.[23–26] The absorption spectrum of MgPc in TX-100 solutions has a narrow absorption band in the range 675 nm characteristic for a monomer state of metal complex and no aggregate bands.

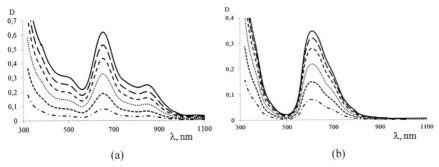

FIGURE 13.2 The absorption spectra of MnPc in complex with the PVP (*a*) and PEG (*b*) in water, corresponding concentrations of MnPc 0.5×10^{-6}, 1.0×10^{-6}, 1.5×10^{-6}, 2.0×10^{-6}, 2.5×10^{-6} mol/L.

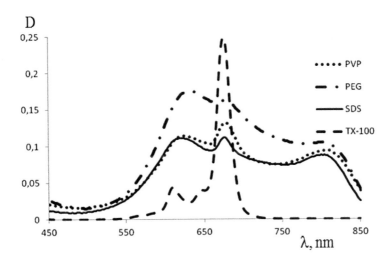

FIGURE 13.3 The absorption spectra of MgPc (2.5×10^{-6} mol/L) in complex with the PVP, PEG, SDS, and TX-100 in water.

To assess the biocidal activity of the polymer complexes, their action against a number of pathogenic microorganisms was tested (Tables 13.1 and 13.2). Solutions of the polymers and Pc with a concentration $6 \cdot 10^{-5}$ mol/L have a bactericidal action, persisting at a dilution by 16 times. The magnitude of biocidal action, estimated from diameter of zone of growth inhibition of microorganisms, was more than activity of monomolecular form of FePc and MnPc by two to three times which persists in the case of adsorption of Pcs on silica nanoparticles.

TABLE 13.1 The Biocidal Activity of Metal Phthalocyanine Complexes with PVP and PEG Against *Escherichia coli* 1257 and *Staphylococcus aureus* P209, Estimated from the Diameter of Zone of Growth Inhibition of Test-culture (mm).

Complex	Dilution				
	–	1:2	1:4	1:8	1:16
	Escherichia coli 1257				
FePc-PVP	24	22	18	14	10
FePc-PEG	38	32	20	15	10
MnPc-PVP	28	25	20	15	10
MnPc-PEG	29	20	14	9	9
MgPc-PVP	0	0	0	0	0
MgPc-PEG	0	0	0	0	0
	Staphylococcus aureus P209				
FePc-PVP	22	18	16	14	9
FePc-PEG	42	36	24	20	13
MnPc-PVP	32	24	18	15	10
MnPc-PEG	26	20	17	14	8
MgPc-PVP	0	0	0	0	0
MgPc-PEG	0	0	0	0	0

The specificity of the action of the type of polymer in the experiment was not marked as on the aggregation behavior of FePc and MnPc as on evaluation of their action against microorganisms. The biocidal activity of polymeric complexes of FePc is similar or slightly higher than the corresponding value for MnPc. Thus, the likely structural differences of nanoaggregates of FePc and MnPc in the composition of polymer complexes have almost no effect on their biocidal properties that allows speaking about universal mechanism for the generation active oxygen forms and their action on microorganisms for the considered systems.

An analysis of biocidal activity of solutions of FePc and MnPc complexes with polymers showed that they have bacteriostatic and bactericidal effect against members of genus *Enterobacteriaceae*, including *Salmonella*,

Biocidal Activity of Phthalocyanine–Polymer Complexes 195

playing a significant role in the development of toxic-food human infections. The polymer complexes of FePc and MnPc are active against *S. aureus* culture. This demonstrates the viability of development of controlling means of contamination of environmental objects against pathogenic and conditionally pathogenic microorganisms on the basis of polymers and complexes of iron and manganese.

TABLE 13.2 The Spectrum of Activity of Solutions of FePc and MnPc Complexes with PVP and PEG at a Dilution of 1:4, Estimated from the Diameter of Zone of Growth Inhibition of Test-culture (mm).

Microorganism	Zone of growth inhibition	
	PVP-FePc	**PEG-FePc**
Proteus mirabilis	15	17
Proteus vulgaris	20	20
Citrobacter freundii	19	20
Salmonella infantis	16	20
Salmonella typhimurium	14	16
Salmonella london	15	16
Salmonella enteritidis	14	13
	PVP-MnPc	**PEG-MnPc**
Proteus mirabilis	12	10
Proteus vulgaris	15	14
Citrobacter freundii	15	12
Salmonella infantis	14	11
Salmonella typhimurium	12	10
Salmonella london	11	13
Salmonella enteritidis	12	11

13.4 CONCLUSION

The results of this work show that the aggregates of Pc complexes of iron and manganese, due to more expressed biocidal properties as compared with the monomolecular form, are perspective compounds for the development of means of biological protection, sanitary processing, and preparations for medical purposes.

ACKNOWLEDGMENTS

The work was supported by RFBR (project no. 15-03-03591).

KEYWORDS

- phthalocyanines
- aggregation
- biocidal activity
- microorganisms
- monomolecular
- metal/Pc complexes
- polyethylene glycol

REFERENCES

1. Mohanpuria, P.; Rana, N. K.; Yadav, S. K. Biosynthesis of Nanoparticles: Technological Concepts and Future Applications. *J. Nanopart. Res.* **2008,** *10,* 507–517.
2. Batinic-Haberle, I.; Rajic, Z.; Tovmasyan, A.; Reboucas, J. S.; Ye, X.; Leong, K. W.; Dewhirst, M. W.; Vujaskovic, Z.; Benov, L.; Spasojevic, I. Diverse Functions of Cationic Mn (III) N-substituted Pyridylporphyrins, Recognized as SOD Mimics. *Free Radical Biol. Med.* **2011,** *51* (5), 1035–1053.
3. Pasternack, R. F.; Halliwell, B. Superoxide Dismutase Activities of an Iron Porphyrin and Other Iron Complexes. *J. Am. Chem. Soc.* **1979,** *101,* 1026–1031.
4. Ilan, Y.; Rabani, J.; Fridovich, I.; Pasternack, R. F. Superoxide Dismuting Activity of an Iron Porphyrin. *Inorg. Nucl. Chem. Lett.* **1981,** *17,* 93–96.
5. Weinraub, D.; Peretz, P.; Faraggi, M. Chemical Properties of Water-soluble Porphyrins. 1. Equilibriums between Some Ligands and Iron (III) Tetrakis(4-N-methylpyridyl) Porphyrin. *J. Phys. Chem.* **1982,** *86,* 1839–1842.
6. Solomon, D.; Peretz, P.; Faraggi, M. Chemical Properties of Water-soluble Porphyrins. 2. The Reaction of Iron (III) Tetrakis(4-N-methylpyridyl) Porphyrin with the Superoxide Radical Dioxygen Couple. *J. Phys. Chem.* **1982,** *86,* 1842–1849.
7. Weinraub, D.; Levy, P.; Faraggi, M. Chemical Properties of Water-soluble Porphyrins. 5. Reactions of Some Manganese (III) Porphyrins with the Superoxide and Other Reducing Radicals. *Int. J. Radiat. Biol.* **1986,** *50,* 649–658.
8. Aust, S. D.; Morehouse, L. A.; Thomas, C. E. Role of Metals in Oxygen Radical Reactions. *J. Free Radicals Biol. Med.* **1985,** *1,* 3–25.

9. Li, B.; Gutierrez, P. L.; Amstad, P.; Blough, N. V. Hydroxyl Radical Production by Mouse Epidermal Cell Lines in the Presence of Quinone Anti-Cancer Compounds. *Chem. Res. Toxicol.* **1999**, *12*, 1042–1049.

10. van Hillegersberg, R.; Kort, W. J.; Wilson, J. H. P. Current Status of Photodynamic Therapy in Oncology. *Drugs* **1994**, *48*, 510–527.

11. Amato, I. Cancer Therapy. Hope for a Magic Bullet that Moves at the Speed of Light. *Science* **1993**, *262*, 32–33.

12. Lobanov, A. V.; Dmitrieva, G. S.; Sultimova, N. B.; Levin, P. P. Aggregation and Photophysical Properties of Phthalocyanines in Supramolecular Complexes. *Russ. J. Phys. Chem. B* **2014**, *8*, 272–276.

13. Udartseva, O. O.; Lobanov, A. V.; Andreeva, E. R.; Dmitrieva, G. S.; Mel'nikov, M. Ya.; Buravkova, L. B. Photophysical Properties and Photodynamic Activity of Nanostructured Aluminum Phthalocyanines. *Biophysics* **2014**, *59*, 856–862.

14. Volpati, D.; Alessio, P.; Zanfolim, A. A.; Storti, F. C.; Job, A. E.; Ferreira, M.; Riul, A. Jr.; Oliveira, O. N. Jr.; Constantino, C. J. L. Exploiting Distinct Molecular Architectures of Ultrathin Films Made with Iron Phthalocyanine for Sensing. *J. Phys. Chem. B* **2008**, *112*, 15275–15282.

15. Grodkowski, J.; Dhanasekaran, T.; Neta, P.; Hambright, P.; Brunschwig, B. S.; Shinozaki, K.; Fujita, E. Reduction of Cobalt and Iron Phthalocyanines and the Role of the Reduced Species in Catalyzed Photoreduction of CO_2. *J. Phys. Chem. A* **2000**, *104*, 11332–11339.

16. Dolotova, O.; Yuzhakova, O.; Solovyova, L.; Shevchenko, E.; Negrimovsky, V.; Lukyanets, E.; Kaliya, O. Water-soluble Manganese Phthalocyanines. *J. Porphyrins Phthalocyanines*. **2013**, *17*, 881–888.

17. Yi, J.; Chen, Z.; Xiang, J.; Zhang, F. Photocontrollable J-Aggregation of a Diarylethene_ Phthalocyanine Hybrid and Its Aggregation-Stabilized Photochromic Behavior. *Langmuir* **2011**, *27*, 8061–8066.

18. Bayrak, R.; Dumludag, F.; Akcay, H. T. Synthesis, Characterization and Electrical Properties of Peripherally Tetra-aldazine Substituted Novel Metal Free Phthalocyanine and Its Zinc (II) and Nickel (II) Complexes. *Mol. Biomol. Spectrosc*. **2013**, *105*, 550–556.

19. Nyokong, T. Effects of Substituents on the Photochemical and Photophysical Properties of Main Group Metal Phthalocyanines. *Coord. Chem. Rev.* **2007**, *251*, 1707–1722.

20. Chauke, V.; Durmus, M.; Nyokong, T. Photochemistry, Photophysics and Nonlinear Optical Parameters of Phenoxy and Tert-butylphenoxy Substituted Indium (III) Phthalocyanines. *Chemistry* **2007**, *192*, 179–187.

21. Youssef, T. E.; Mohamed, H. H. Synthesis and Photophysicochemical Properties of Novel Mononuclear Rhodium (III) Phthalocyanines. *Polyhedron* **2011**, *30*, 2045–2050.

22. Gouloumis, A.; Gonzalez-Rodriguez, D.; Vazquez, P. Control Over Charge Separation in Phthalocyanine-anthraquinone Conjugates as a Function of the Aggregation Status. *J. Am. Chem. Soc.* **2006**, *128*, 12674–12684.

23. Yanika, H.; Aydina, D.; Durmus, M. Peripheral and Non-peripheral Tetrasubstituted Aluminium, Gallium and Indium Phthalocyanines: Synthesis, Photophysics and Photochemistry. *J. Photochem. Photobiol. A Chem.* **2009**, *206*, 18–26.

24. Akpe, V.; Brismar, H.; Nyokong, T.; Osadebe, P. O. Photophysical and Photochemical Parameters of Octakis (Benzylthio)Phthalocyaninato Zinc, Aluminium and Tin:

Red Shift Index Concept in Solvent Effect on the Ground State Absorption of Zinc Phthalocyanine Derivatives. *J. Mol. Struct.* **2010,** *984,* 1–14.

25. Staicu, A.; Pascu, A.; Boni, M. Photophysical Study of Zn Phthalocyanine in Binary Solvent Mixtures. *J. Mol. Struct.* **2013,** *1044,* 188–193.

26. Agirtas, M. S.; Çelebi, M.; Gümüs, S. New Water Soluble Phenoxy Phenyl Diazenyl Benzoic Acid Substituted Phthalocyanine Derivatives: Synthesis, Antioxidant Activities, Atypical Aggregation Behavior and Electronic Properties. *Dyes Pigments* **2013,** *99,* 423–431.

CHAPTER 14

SOLUTIONS AND FILMS OF SILVER NANOPARTICLES PRODUCED BY PHOTOCHEMICAL METHOD AND THEIR BACTERICIDAL ACTIVITY

A. B. KONONENKO[1], D. A. BANNIKOVA[1], S. V. BRITOVA[1],
E. P. SAVINOVA[1], O. A. ZHUNINA[2], A. V. LOBANOV[3,4*],
S. M. VASILEV[5], V. N. GORSHENEV[6], G. E. ZAIKOV[6], and
S. D. VARFOLOMEEV[6]

[1]*All-Russian Research Institute of Veterinary Sanitation, Hygiene and Ecology, Moscow, Russia*

[2]*Russian Scientific Center of Molecular Diagnostics and Therapy, Moscow, Russia*

[3]*Department of Chemical and Biological Processes Dynamics, Semenov Institute of Chemical Physics, Russian Academy of Sciences, Moscow, Russia*

[4]*Plekhanov Russian University of Economics, Moscow, Russia*

[5]*Institute of Modern Standards, Moscow, Russia*

[6]*Department of Bio-Chemistry, N. M. Emanuel Institute of Biochemical Physics, Russian Academy of Sciences, Kosygina St. 4, Moscow 119991, Russia*

Corresponding author. E-mail: avlobanov@mail.ru

ABSTRACT

Photochemical synthesis of silver nanoparticles was proposed. Nanoparticle water formulations and films can be used for disinfection treatment, preventing the multiplication of dangerous microorganisms in the premises, equipment of the food industry, and in medicine.

14.1 INTRODUCTION

At present, environment friendly, non-toxic domestic disinfectant absent in the domestic market. In the industrial and residential applications, chlorinated products (bleach, chloramine, and sodium hypochlorite) are used, but they are not perfect in terms of toxicity, environmental safety, and durability over time. In addition, traditional chlorinated disinfectants pose a certain risk for human health, the environment, cause corrosion of equipment contacted with them, different limited duration.[1]

The preparations of silver nanoparticles with stabilizers capable of forming a film on the surface of a biocide preserving protective effect for a long time, and also preparations of silver nanoparticles in a liquid form in aqueous base have prolonged antimicrobial activity and decreased toxicity and volatility.[2] Thus, the development of simple and easy way to create of bactericide silver nanoparticles is important.

14.2 EXPERIMENTAL

Silver nanoparticles were obtained by photochemical synthesis in according to modified method[3] including interaction of the silver ions with the stabilizing agent in an aqueous solution at room temperature under the action of visible light. As stabilizers, polyvinyl alcohol (MW 10,000 g/mol), starch, polyvinylpyrrolidone (MW 26,400 g/mol), sodium dodecyl sulfate, Triton X-100, oleic acid, or binary mixture of these substances was used. In the standard experiment in 50 mL of distilled water was dissolved in 40 mg of the stabilizer-reducing agent, and then added 30 mg of silver nitrate and the obtained clear solution was irradiated with light of a halogen lamp 150 W with intensive stirring of solution on a magnetic stirrer. The distance between the solution and the light source was 30 cm that corresponds to the power density of light flux is 10 mW/cm^2. The irradiation time under the described conditions was 30–120 min. The size and the statistical size distribution of stabilized silver nanoparticles were determined by transmission electron microscopy with the use of the device "Hitachi-11" (Japan). The size of silver nanoparticles for all solutions is 20–30 nm. Seventy percent of the amount of nanoparticles corresponds to this size. The exception is sodium dodecyl sulfate solution, wherein particles sizes of 60 nm predominate. The remaining part is particles from 2 to 200 nm.

The bactericidal activity of the preparations of the silver nanoparticles with different stabilizers was determined by agar diffusion. In thickness of

Solutions and Films of Silver Nanoparticles 201

meat-peptone agar, containing a daily culture of microorganisms in dose of 107 microbial cells/mL, made sterile wells with diameter of 4 mm. In the wells, add working dilution of the preparation and placed it in an incubator at 37°C for 18–20 h. The results were evaluated according to the zone of growth inhibition of the test-cultures around the well.

14.3 RESULTS AND DISCUSSION

As test-cultures *Escherichia coli, Staphylococcus aureus, Salmonella enteritidis, Salmonella dublin, Salmonella choleraesuis,* and *Salmonella typhimurium* were selected. In order to study the spectrum of silver nanoparticles, their activity has also been studied in relation to bacteria of the genera *Citrobacter, Providencia, Hafnia, Proteus, Morganella,* and *Listeria.* All preparations have a bactericidal effect in relation to the studying microorganisms when silver content of 4–35 mg/L. The obtained results for the example of action of solutions of silver nanoparticles stabilized with polyvinylpyrrolidone are shown in Tables 14.1 and 14.2.

TABLE 14.1 The Bactericidal Activity of Preparations of Silver Nanoparticles Stabilized with Polyvinylpyrrolidone.

Name of microorganism	Zone of growth inhibition (mm)			
	Silver content in the preparation (mg/L)			
	35	17	8	4
E. coli O 1257	18	18	12	10
S. aureus 209 P	18	16	10	10
S. enteritidis	16	14	11	8
S. dublin	17	14	11	9
S. choleraesuis	14	12	10	10
S. typhimurium	10	10	10	10
Citrobacter freundii	12	11	10	10
Providencia rettgeri	11	9	8	8
Proteus mirabilis	20	19	18	14
Proteus vulgaris	12	12	11	10
Y. enterocolitica	17	14	9	8
Y. pseudotuberculosis	15	11	8	8
Listeria monocytogenes	18	12	10	10

As can be seen from Table 14.1, any differences in the sizes of zones of growth retardation depending on the nature of the microorganisms were observed. Preparations of silver nanoparticles during storage at room temperature away from light retained level of bactericidal activity against the test-cultures of *E. coli* and *Salmonella* spp. during 6 months. However, very little turbidity of agar was noted in the areas of growth retardation of Staphylococcus, this indicated about the bacteriostatic effect. During storage of the solutions, the amount of nanoparticle is not decreased by more than 10%. Particle size distribution was maintained at the initial level. Bactericidal activity of the silver nanoparticles preparations containing as stabilizers and film-forming components, such as polyvinyl alcohol (the synthesis of silver nanoparticles was 30 min and 2 h), sodium dodecyl sulfate, Triton X-100, and oleic acid was determined. Disinfecting activity of preparations of the silver nanoparticles was studied on smooth surfaces (ceramic tile and flooring tile) contaminated suspension of daily culture of test-microorganism using a protein protection. Exposition of action, in this case, was 30 min and 3 h. The preparations were applied at the rate of 500 mL/m^2. The neutralization of the drug was carried out by 10-fold dilution of the washouts from the processed object. The objects processed by water were controlled. The results of determination of the bactericidal activity of these drugs are shown in Table 14.2. The presented data suggest that all preparations have bactericidal activity against the test-cultures at the silver content at least 4 mg/L. The disinfectant activity against *E. coli* and *Salmonella* spp. was detected in preparations containing silver is not less than 17 mg/L and 3 h of exposition. This concentration of silver nanoparticles in the preparations does not allow to achieve disinfection effect on test-objects contaminated with *S. aureus*. In culturing in the beef-extract broth (BEB) washouts with such objects the growth of the cultures appears at 2–3 days. Complete decontamination of the objects contaminated with *Staphylococcus aureus*, was achieved under the action of the preparations with a silver content of 35 mg/L and exposition for 3 h.

The morphology of Salmonella populations when exposed to silver nanoparticles at doses causing bacteriostatic and bactericidal effects was studied by electron microscopy. The situation in which were monitored the dynamics of the Salmonella population on processed surfaces of the silver solution was modulated in the experiments.

In the control samples, colonies of Salmonella looked like a population of rod-shaped cells, combined extracellular matrix and coated with a surface dense cover (biofilm). Separating rod-shaped cells were observed in some sites. On the surfaces processed with the solutions of the silver nanoparticles

Solutions and Films of Silver Nanoparticles   203

with a silver content of 2 mg/L, partial destruction of the outer covering and the extracellular matrix between separate bacterial cells, and as a result, the violation of the spatial orientation of the cells occurred in the cell population of Salmonella.

TABLE 14.2 Bactericidal Activity of the Preparations of the Silver Nanoparticle Containing Film-forming Component.

Preparation	Silver content (mg/L)	Zone of growth inhibition (mm)		
		Name of microorganism		
		E. coli	*Salmonella* spp.	*S. aureus*
1	35	17	17	17
	17	16	15	14
	8	13	12	10
	4	10	10	8
2	35	18	13	12
	17	15	10	9
	8	11	7	8
	4	9	7	8
3	35	16	14	18
	17	15	13	16
	8	14	12	12
	4	11	9	9
4	35	17	15	17
	17	16	14	15
	8	15	13	12
	4	11	10	9
5	35	20	17	17
	17	19	14	14
	8	16	11	10
	4	11	9	8

Components and preparations conditions: 1—polyvinyl alcohol (synthesis of 30 min), 2—polyvinyl alcohol (synthesis of 2 h), 3—polyvinyl alcohol and sodium dodecyl sulfate, 4—polyvinyl alcohol and Triton X -100, and 5—polyvinyl alcohol and oleic acid.

On the surfaces processed with the solutions of the silver nanoparticles with a silver content of 4 mg/L, cells of heteromorphic type were developed in Salmonella populations. Formation of rounded cells (spheroplasts, often forming the characteristic bands) was observed.[4] The reversibility of these processes in the population was proved by the appearance of the typical culture growth of Salmonella after additional rearing in the fresh nutrient media. Scanning electron microscopy of such crops identified the formation powerful accumulations of reversed cells with typical crimped structure. Primary education of outer covering (biofilm produced by only cells having a complete cell wall) was observed in some sites.

The interaction of the solutions of silver nanoparticles with a silver content of 40 mg/L caused irreversible destruction of Salmonella population on processed filters. The absence of the growth of the culture at secondary rearing in fresh nutrient medium as agar and broth, indicate the impossibility of reversion of the population. It is noteworthy that while the typical bacterial microcolonies presented rod-shaped cells under the covers are clearly visible on the scans. Is obvious that these microcolonies hitting on the surfaces processed the bactericidal silver solutions cannot develop, "die away," while retaining the native morphology. Probably, the high silver concentrations cause instantaneous blockage of enzyme systems in the bacterial cells, making impossible their future development.

The preparations of silver nanoparticles were tested in model experiments with the aim of reducing of contamination level of objects of veterinary supervision in poultry farms (surfaces of equipment and tools for cutting, poultry carcasses). For this purpose, the listed objects were artificially contaminated with the suspension of the daily cultures of *E. coli* and *Salmonella* spp. The silver content in the preparations was 35 mg/L. The surfaces were processed at the rate of 500 mL/m^2 and exposition for 3 h, tools were dipped into a solution of the preparations of the silver nanoparticles for 30 min, and poultry carcasses were placed into a container with cold water containing the preparation of the silver nanoparticles for cooling (30 min). After the expiration of the exposition, the washouts with the sterile cotton swabs in the test tubes with the physiological solution were made and the bacteriological analysis for the presence of the test-cultures used in the experiment was carried out. The test-objects processed with water were controlled.

The continuous growth of the test-cultures was detected in the control culturing of the washouts with the surfaces while the growth of test-cultures was absent on surfaces processed with the preparations of the silver nanoparticles. In the study of the tools (scissors and tweezers) processed with the preparations of the silver nanoparticles the growth of microorganisms in the culturing of the washouts was presented to a single colony, and in the control (processing of water) reached up to hundreds of colonies in some samples. After cooling of the poultry carcasses without processing of water (control), abundant growth of test-cultures in the culturing in all samples washouts with the surface was observed. In the washout from carcasses processed with the solution of the silver nanoparticles, enterobacteria growth was observed in the number of 20–40 colonies in some samples.

14.4 CONCLUSION

Thus, the method of disinfection, including dilution of the preparations of the silver nanoparticles to a content of 35 mg/L, the application of preparations at the rate of 500 mL/m^2 to previously prepared (mechanically cleaned) objects to be decontaminated (ceramic tiles and flooring tile, the surfaces of the equipment and tools for cutting, and poultry carcasses), exposition for 3 h and disinfecting of tools for cutting and poultry carcasses—30 min, control of the disinfection using the washouts of the sterile cotton swabs into test-tubes with the physiological solution and the subsequent bacteriological analysis, can be offered. The means can be recommended for use in medicine, antimicrobial protection for food products, as antimicrobial additives in paint and varnish compositions.

ACKNOWLEDGMENTS

This work was supported by RFBR, research project No. 15-03-03591.

KEYWORDS

- **silver nanoparticles**
- **bactericidal activity**
- **residential applications**
- **polyvinyl alcohol**
- **antimicrobial activity**
- **films**
- **stabilizers**

REFERENCES

1. Monisov, A. A.; Shandala, M. G., Eds.; *Disinfectants. Handbook, Part 1. Disinfectants;* TOO Rarog: Moscow, 1996; p 176.
2. Baranova, E. K.; Mulyukin, A. L.; Kozlov, A. N.; Revina, A. A.; El-Registan, G. I. Science Intensive Technologies. **2005,** *6* (5), 33–37.
3. Sergeev, B. M.; Kirukhin, M. V.; Prusov, A. N.; Sergeev, V. G. Bulletin of Moscow University. 2. Chemistry Series. **1999,** *40* (2), 129–133.
4. Pitzurra, M.; Szybalski, W. *J. Bacteriol.* **1959,** *77,* 614–620.

CHAPTER 15

POLYMER COMPOSITES WITH GRADIENT OF ELECTRIC AND MAGNETIC PROPERTIES

JIMSHER N. ANELI[1*], L. NADARESHVILI[2], A. AKHALKATSI[3], M. BOLOTASHVILI[1], and G. BASILAIA[1]

[1]*Institute of Machine Mechanics, Mindeli St. 10, Tbilisi 0186, Georgia*

[2]*Institute of Cybernetics of Georgian Technical University, 5 S. Euli St., Tbilisi 0186, Georgia*

[3]*I. Javakhishvili Tbilisi State University, 3 I. Chavchavadze Ave., Tbilisi 0176, Georgia*

Corresponding author. E-mail: JimAneli@yahoo.com

ABSTRACT

The character of variations of the local electric resistance of film polymer composites on the basis of polyvinyl alcohol with graphite powder from one side and the magnetic susceptibility of the same polymer with nickel nanoparticles from another one has been studied. It is established that the changes of these parameters essentially depend on both initial shape of the films and on direction of their orientation. It is concluded that the films of gradient anisotropic polymer composites may be used in electronics.

15.1 INTRODUCTION

It is well known that there are several methods for obtaining materials with anisotropic properties by chemical methods (copolymerization, polymer-analogous transformation, radiation–chemical modifications, etc.).[1,2] At present, for obtaining of such structures, one of the best methods is the orientation of polymer films in the definite direction and environment conditions. It is known also that at stretching of thermoplastic polymers above glass temperature, the material in orientation state is formed. Such polymers are characterized by mono-axis crystal symmetry. In this state, the principal direction of macromolecules coincides with the direction of stretching. If the polymer filled with different dispersive fillers, particularly with electric conductive and magnetic materials (graphite, carbon black, and metal powders), the particles distribution of lasts interacting with macromolecules transform from chaotic state to orientation one. The change of polymer microstructure significantly defines the material electric and magnetic properties.[3,4]

In this work, the character of change of electric conductivity and magnetic susceptibility of polymer composite films based on polyvinyl alcohol (PVA), graphite, and nickel at their mechanical stretching has been investigated.

15.2 BASIC PART

The films were prepared by using following technology:

The water solution of fine-grained graphite (average diameter of grains less than 10 μm) or nickel (average diameter of grains less than 20 nm) suspensions in PVA were prepared. The mixture was filtrated and the film was formed on the dryer table.

The specific volumetric electric resistance of the polymer films was changed in the interval 10–50 kΩ.cm. The selection of such interval of the composite resistance was dictated by preliminary selection of conducting composites effectively reacted on the mechanical deformations.[4] The composites contained the magnetic filler are characterized simultaneously both electrical and magnetic properties.

The experiments were carried out on the basis of polymer composite films with rectangle and trapezoidal shape. The thickness of films was not more than 0.2 mm. The deviation of the values of resistance for any local region of the film was not more than 10%. These films were fixed in special clamps, placed to the heater and were stretched on 50–150% at rate 50 cm/min and temperatures 100–120°C. Stretching was conducted for rectangle form sample along big side and for trapezoidal sample in parallel to bases direction (Figs. 15.1 and 15.2).

After stretching of the deformed films, local ohmic resistances were measured. First of all, it was necessary to mark the film with square grid. In our case, the length of square side was equal to 5 mm. The local resistances were measured by using twin needles after touching them to the film. The measuring of resistance of elementary cages was performed several times and then the average values of resistances were calculated.

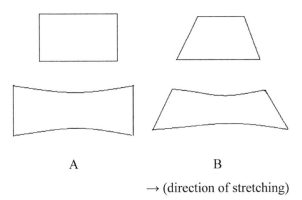

→ (direction of stretching)

FIGURE 15.1 Rectangle (A) and trapezoidal (B) shape films before (top) and after (bottom) stretching along big sides.

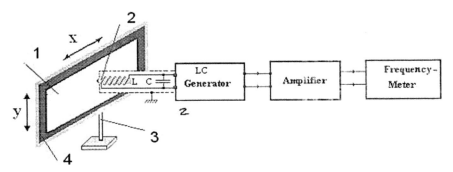

FIGURE 15.2 Scheme of the measuring of the magnetic characteristics of the polymer films. 1—Sample; 2—ferrite tip; 3—support; 4—frame.

Another series of investigations proceeded on establishment of local magnetic characteristics of gradient anisotropic magnetic polymeric composites with magnetic powder fillers like nickel powder.

The films from magnetic polymer composites were subjected to the similar stretching procedure as it was previously described.

The resulting distribution of magnetic particles density in polymer composites was recorded by the method of LC-generator similar to the one used by us in work[5] for NMR detection in europium garnet at low temperatures.

It is also interesting to note that a similar method was used for the first precision determination of the magnetic field penetration length in superconductors.[6]

For assessment of magnetic susceptibility distribution over the surface of gradient magnetic films, it was used an LC generator of sinusoidal oscillations.[5]

The procedure of measurements of the magnetization changes is realized through the sensitive detection of the change of inductance of the LC generator oscillatory contour coil supplied with a tipped ferrite rod core at scanning by it over the investigated surface (Fig. 15.2). In whole, the susceptibility measurement system consists of the following stages: LC generator, emitter follower for cascade matching, sinusoidal oscillation amplifier, and frequency meter.

The frequency of LC generator oscillatory contour is defined by its fixed total capacitance and inductance. The oscillatory contour is an isolated system and tuned on a reference frequency providing its operation on the maximal sensitivity as result of which the slightest contour inductance changes result in the significant changes in its frequency.

Let us note that the induction coil of oscillatory contour was brought away on 30–40 cm from the LC generator by a screened transmission line and the contour inductance with its spurious capacitance enters into contour effective parameters.

The frequency detuning is possible using a 2-mm-thick and 15-mm-long tipped ferrite core put into the coil during its scanning over the film. Namely, this principle is used for the definition of the contour quality factor change expressed in the frequency units. During the displacement of coil having a tipped ferrite core near the surface of the gradient magnetized film surface the change of coil inductance takes place which is followed by the corresponding change of LC generator frequency in a quite significant range giving one possibility to evaluate the change of value of the real part of susceptibility related to frequency.

The experimental set-up is presented in Figure 15.2. In the inductive coil of the resonance contour of LC-generator, it is placed cylindrical tipped ferrite rod used as probe. The investigated rectangular shape magnetic polymer composite film is displaced relatively the immovable ferrite tip. The scanning of the film surface is realized along the previously marked net contour (Fig. 15.3).

Measurements of the dependences F-l were provided with using of the ferromagnetic film, the scheme of which is presented in the Figure 15.3.

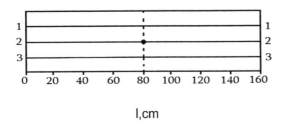

FIGURE 15.3 Scheme of ferromagnetic film for investigation of the local magnetic susceptibility along directions 1–1, 2–2, and 3–3; dotted line—the middle of the film.

The change of magnetic particle concentration causes the change of inductance δL of the resonance contour of LC-generator resulting in the frequency displacement of LC-generator f related with δL by relation $\delta f/f = 1/2 \delta L/L$. This frequency displacement could be precision measured what stipulates the high sensitivity of the method.

At the natural frequency of used LC-generator near ~2 MHz the observed range of the frequency change δf was about ~1000 Hz at the precision of the frequency measurement ~1 Hz.

As a result of measuring of local resistances oriented along the parallel to long side of the rectangle shape samples, it was established that maximum change of this parameter was noticed along symmetry axes of the rectangle along orientation direction. This change has an extreme character (the maximum is at the central part of the film) and its full shape has Gaussian form. Figure 15.4 shows that the maximum heights of the local resistances depend on the value of stretching. This result is in good agreement with the known conception on the mechanism of conductivity of conductive polymer composites.[3] Investigation of obtained films using of the metallographic microscope shows that the average optical density of penetrated light through the film nearly exponentially depends on the ordinates of the elementary squares. Therefore, it may be proposed that the dependence of concentration of the conducting particles in the local regions of the film has the same character. However, the basis of such distribution of the filler particles in the uniaxially stretched polymer composites is not yet clear.

By analogical shape of the same dependence is characterized, one of local resistances in rectangle to stretching directions, although these dependencies are somewhat weaker.

It was interesting to establish the character of considered above functional dependences of the local resistances to the concentration of the electric conducting particles. Figure 15.4 shows that the increase of filler concentration leads to reduction of the intensity of resistance change at stretching. This phenomenon may be described by the following processes. It is known that the filler particles at stretching of polymer composites commit the mutual transition in the polymer matrix initiated by interacting with them macromolecules segments, in result of which the average distances between these particles and consequently the charge transition change accordingly.[7] Here, the inverse processes—approach and removing of the particles and consequently the probability of creation of the conducting chains changes, respectively. However, in case of composites with high concentrations of the conducting particles, the probability of arising of new contacts between ones is higher than in opposite case, since the frequency of these elementary processes is higher than in case of composites with relatively low concentrations of these particles. Here, the described process is analogical to reserving of the switching conducting chains in the complex electrical engineering schemes. The obtained results show that the process of displacement of the conducting particles and consequently the change of conductivity take place more intensively in the middle part of the stretched polymer film than in other ones, namely, this change has gradient character with increasing from grips till middle of the film. The amount of this gradient is the higher and the lower is filler concentration (Fig. 15.5).

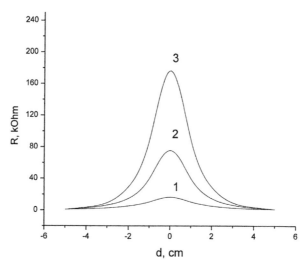

FIGURE 15.4 Dependence of local resistances of polymer film on the value of stretching parallel to long side of the rectangle on 50 (1), 100 (2), and 150% (3). The curves maximums are located on the central line with abscissa coordinate "0."

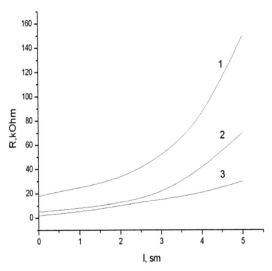

FIGURE 15.5 Dependence of local resistances of films on the stretching degree for composites based on PVA containing 30 (1), 40 (2), and 50 (3) mas% of graphite powder (curves correspond to left half side of central strip of the film).

The following series was fulfilled on the trapezium shape conducting films. Here, it was created the mechanical stretch gradient is perpendicular to base direction, along which the stretching was realized. This gradient was increased from big base and was ended at more stressed small base with maximum (Fig. 15.6).

The experimental data on the definition of character of the dependence of distribution of the local resistances on stretched both rectangle and trapezoidal form films on their coordinates show that the shape of these dependences essentially is defined mainly by two factor: (1) the content of electrical conducting fillers in the selected part of the stretched film and (2) distribution of conducting particles in the polymer matrix after stretching along selected direction. Nonlinearity of the dependence of local electrical conductivity both along and perpendicular direction of stretching is explained mainly by the nature of the conductivity of the polymers filled with conducting particles. Namely, it is well known that the charge transfer in such systems obeys to tunnel mechanism—nonlinear dependence of the conductivity on the distance between neighbor particles.

In Figure 15.7, the results of measurements of generator frequency change along contour lines 1–1, 2–2, and 3–3 are presented. Similar results are obtained by measuring of magnetic susceptibility of film along rectangular to 1–1, 2–2, and 3–3 directions.

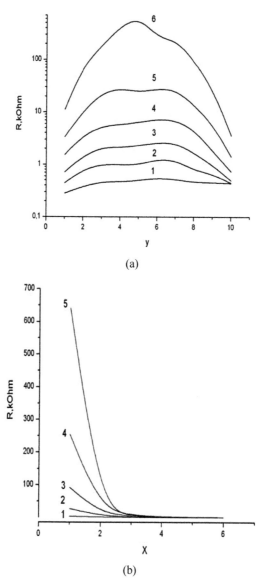

FIGURE 15.6 Dependences of local resistances on the film (PVA with 25 mas.% graphite) coordinate in the strips stretched on 150% parallel to bases of trapezoidal shape films in perpendicular to stretching direction from big base to small one (a) and parallel to stretching direction from small base to big one (b). The numbers on the curves indicate the numbers of stripes in perpendicular (a) and parallel (b) to stretching direction. The curve number 5 of b corresponds to central strip of trapezoidal film and others—to side from it ones (4-3-2-1). The asymmetry of the curves on the 6(a) is due to certain inhomogeneous distribution of the filler particles in polymer matrix.

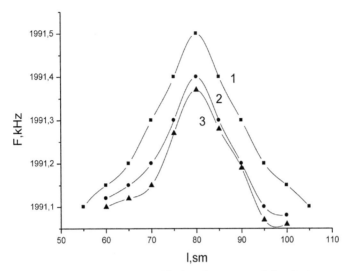

FIGURE 15.7 Dependences of the oscillation frequency of the ferrite sensor on the coordinates of film made from polymer composite along directions 2–2 (1), 1–1 (2), and 3–3 (3).

The presented results show that for magnetic polymer composite films with rectangle shape the dependences F-l have similar form, as was obtained for conducting polymer composites (R-l) with same shape. Such resemblance is based on similar character of distribution of the filler particles—conducting ones in rectangle electrical conducting polymer composites and magnetic particles in the magnetic analogues after their uniaxial stretching, because the maximums of curves of dependences of the local resistance on film coordinates correspond to rarefied regions of these films and the maximums of the curves F-l correspond to those regions—frequency increases, when magnetic particle concentration decreases and, consequently, local susceptibility, and inductance decrease.

Results of detailed measurements will be published elsewhere.

15.3 CONCLUSION

1. Gradient anisotropic structures are formed after orientation (stretching) in spatial conditions of thin polymer composites based on PVA, graphite, and nickel powders with electrical conducting and magnetic properties. Structure anisotropy leads to anisotropy both electrical conductivity and magnetic properties of these films.

2. After orientation of the rectangle films parallel to it, any side forms the film, electrical resistance of which parallel and rectangle directions to orientation axis changes by Gauss low.
3. Gradient distribution of the local resistance in the stretched films along and rectangle to stretching direction is due to gradient of local deformations in the same directions.
4. The gradient distribution of magnetic particles in the stretched direction leads to equivalent change of magnetic properties of these films.
5. The experiments described above open the perspectives in the field of creation of the films with desirable anisotropy of electric conductance and magnetic properties. The perspective is in application of these materials for creation of the so-called printed schemes. In electronics, these films will be useful for preparing of multifunctional micro-schemes.

KEYWORDS

- polymer films
- composites
- stretching
- anisotropy
- local electrical resistance magnetic susceptibility

REFERENCES

1. Lekishvili, N. G.; Nadareishvili, L. I. *Polymers and Polymeric Materials for the Fibre and Gradient Optics;* VSP: Utrecht, Boston, Köln, Tokyo, 2002; p 230.
2. Nadareishvili, L. GB-optics—A New Direction of Gradient Optics. *J. Appl. Pol. Sci.* **2004,** *91,* 489–493.
3. Aneli, J. N.; Khananashvil, L. M.; Zaikov, G. E. Structuring and Conductivity of Polymer Composites. Nova Science Publishers: Hauppauge, NY, 1998; p 326.
4. Aneli, J. N.; Khananashvil, L. M.; Zaikov, G. E. Effects of Mechanical Deformations on Structurization and Electric Conductivity of Polymer Composites. *J. Appl. Pol. Sci.* **1999,** *74,* 601–621.
5. Pavlov, G. D.; Chekmarev, V. P.; Mamniashvili, G. I.; Gavrilko, S. I. Method of NMR Recording in Magnetoordering Materials. USSR Patent № 279893, 1988.
6. Schawlow, A. L.; Devlin, G. E. *Phys. Rev.* **1959,** *113* (1), 120–126.
7. Aneli, J. N.; Zaikov, G. E.; Mukbaniani, O. V. *Chem. Chem. Technol.* **2011,** *5,* 75–82.

CHAPTER 16

MODERN ASPECTS OF TECHNOLOGIES OF ATOMIC FORCE MICROSCOPY AND SCANNING SPECTROSCOPY FOR NANOMATERIALS AND NANOSTRUCTURES INVESTIGATIONS AND CHARACTERIZATIONS

VICTOR A. BYKOV[1,2*], ARSENY KALININ[1,2], VYATCHESLAV POLYAKOV[1], and ARTEM SHELAEV[1,2]

[1]*NT-MDT-Spectral Instruments Companies Group, Moscow 124460, Russia*

[2]*Moscow Institute of Physics and Technology, Zhukovsky, Russia*

*Corresponding author. E-mail: spm@ntmdt-si.ru

ABSTRACT

New trends in development of scanning microscopy and atomic force microscopy are considered. Micro- and nanoelectronics with extra high-level metrology requirements and up to material science, biology, and ecology with requirements to the side of simplification in operation procedures, possibility of the materials and molecules recognitions are discussed. Perspective ways of nanoscale methods development are proposed.

16.1 INTRODUCTION

During last years, the development of scanning probe microscopy (SPM) technology was transformed to the side of specialization. The application field was increased very wide—from one side micro- and nano-electronics with extra high-level metrology requirements and up to material science, biology, and ecology with requirements to the side of simplification in operation procedures, possibility of the materials and molecules recognitions.

SPM gives an opportunity to carry out studies of spatial, physical, and chemical properties of objects with the typical dimensions of less than a few nanometers. Owing to its multifunctionality, availability, and simplicity, atomic force microscope (AFM) has become one of the most prevailing "tools for nanotechnology" nowadays. NTEGRA platform has been designed as the special base for the constantly developing options of SPM that combines them with various other modern research methods. Integration of SPM and confocal microscopy/luminescence/Raman scattering spectroscopy/infrared apertureless near-field spectroscopy and microscopy (ASNOM) take place. Owing to the effect of giant amplification of Raman scattering (tip-enhances Raman scattering, TERS), it allows carrying out spectroscopy studies and obtaining images with 10 nm resolution.

16.2 MICROMECHANICS MANUFACTURING

New generation of AFM control electronics now allows a real-time cantilever deflection tracking and analyzing. Based on a fast force-distance measurement, we developed a new group of non-resonant AFM methods of SPM—called hybrid mode. Hybrid mode is the most proposal AFM mode since it summarizes all advantages of amplitude modulation and contact modes allowing simultaneously: free of share force topography measurement with direct tip-sample interaction control, real-time quantitative nanomechanical measurements,[1] conductivity,[2] piezoresponse, and electrostatic imaging, all with conventional scanning speed. Hybrid mode is also very helpful for liquid measurements because it utilizes the issue with cantilever Eigen frequency detection.

Progress in micromechanics manufacturing resulted in significant increase of the cantilever yield rate (to practically 100%) with repeatability of resonant characteristics at 10% level, thus preconditioning implementation of the concept of multi-probe cartridges for AFM.

A cartridge of this type is a multi-probe contour-type sensor with 38 cantilevers. The cantilevers can be either of the same type or "colored" with predefined coverings and rigidities. Depending on AFM system type, the cartridge rotation to select working cantilever can be manual or software-controlled and takes only a few seconds.

A whole cartridge can be exchanged manually through a simple procedure without the risk of damage cantilevers. The cartridges operate in dedicated measuring heads, which are designed for integration in the latest instruments by the Company (Titanium, NEXT, SOLVER-NANO, and VEGA-SPM).

For fully software-controlled AFMs Titanium and NEXT the cantilever setup procedure was motorized and automated including precise cartridge rotation to the user-selected cantilever, optical beam deflection (OBD) system adjustment, lock-in amplifier tuning, and sample positioning (Fig. 16.1). This approaches us to the concept of ease-of-use AFM where routine system adjustment before scanning is proceeded automatically in a few tens of seconds.

The new Revolution Cartridge with multi-probe technology for automated replacement of cantilevers makes a breakthrough in AFM usability:
- 38 tips on Cartridge
- Fast tip exchange
- Fully automated operation

FIGURE 16.1 Part of multi-probe cartridge and SPM head with multi-probe cartridge for scanning probe microscope NEXT and Titanium.

16.3 FULLY AUTOMATED TOPOGRAPHY IMAGING

Ease-of-use is not the only feature of automated multi-probe cartridge. One of the most demanding applications of modern AFM is routine and repeatable atomic and molecular resolution. This requires extra-low tip-sample thermal drift assumed us[3] lower than 1 Å/min. Development of thermally stabilized cabinet with 0.01°C temperature control accuracy and drift-minimized mechanical design of Titanium AFM helped us to achieve mentioned drift level and repeatable atomic/molecular resolution imaging.[3] But conventional cantilever exchange procedure requires opening the cabinet and manipulating with AFM therefore, destabilizing perfect temperature conditions. So the concept of automated multi-probe cartridge together with active thermal stabilization and drift-minimized mechanical design can be a perfect tool for routine high-resolution AFM imaging (Fig. 16.2).

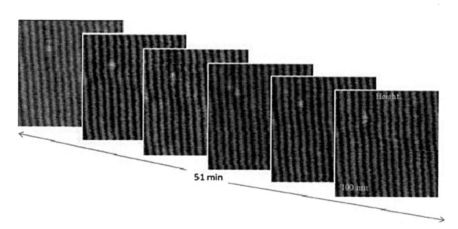

FIGURE 16.2 Height images of semi-fluorinated alkanes on graphite collected in sequential scans with 1 Hz scanning rate (measured by Titanium AFM in thermally stabilized cabinet).

AFM is a candidate to solve some of "Metrology Difficult Challenges" proposed by The International Technology Roadmap for semiconductors like "Structural and elemental analysis at device dimensions and measurements for beyond CMOS," "Nondestructive, production worthy wafer and mask-level microscopy for critical dimension measurement for 3D structures, overlay, defect detection, and analysis." A rapid development of polymer[4] and single-molecular electronics[5] also requires AFM to measure and control the topography, nanomechanical, conductivity, temperature, and other properties at the nanoscale.

To summarize, future electronics development and manufacturing can be a wide field for AFM application, especially for large-sample AFMs. But the biggest drawback of AFM technology to overcome is low throughput.

Throughput of AFM is limited by system adjustment time before scanning (OBD system and lock-in adjustment, area of interest searching time, etc.), scanning parameters adjustment time, scanning speed, and amount of data gathered after one scanning session. So to develop the next-generation AFM all these limits should be overcome.

To minimize system adjustment and scanning parameters tuning time, we develop and improve new software algorithms allowing fully automated topography imaging. New high-speed control electronics together with Hybrid mode allow more data points and different properties to be recorded per one scanning session. We also develop new AFM-scanner control algorithms to increase a topography imaging speed noticeably.

These developments implemented to the fully motorized large-sample AFM is a promising tool for nanotechnology industry.

16.4 APERTURELESS SCANNING NEAR-FIELD OPTICAL MICROSCOPE

To use AFM-cluster technology in the portable SPM, such as Solver-NANO (http://www.ntmdt.com/practical-afm/solver-nano) can open the road for the using of this unit right on the space stations for material quality control in the space and space station conditions.

Development of modes for scanning spectroscopy combined with SPM in the instruments NTEGRA-SPECTRA-II provides new options of confocal laser luminescence spectroscopy and Raman spectroscopy as well as higher reliability of detection for TERS and high-resolution scanning probe-optical microscopy and spectroscopy. Probes with diamond nanocrystals containing N-V defects are capable to detect magnetic states as microscopic as single spins and so they are promising for studies of surface catalytic activity and for detection of free radicals, including applications in biology and medicine.

ASNOM probe induces light scattering to give the possibility to investigate infrared as chemical nature of surface functional groups and to measure the doped impurity implantations in microelectronic structures[6] that it is impossible to observe in electron microscopy (Fig. 16.3).

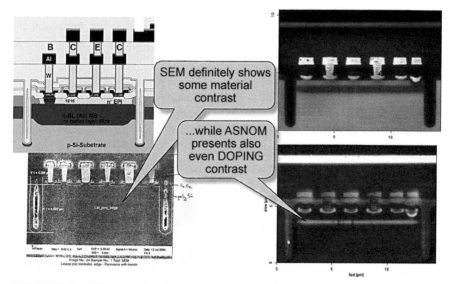

FIGURE 16.3 Silicon semiconductor structure chip in SEM (left) and ASNOM images (D. Kazantsev, ITEF RAS, and NT-MDT-SI).

16.5 CONCLUSION

New nanoscale equipment and techniques give the possibility to investigate and control modern processes of nanostructures formation as well as to prognosis the different nanosystems properties.

KEYWORDS

- **scanning electron microscopy**
- **atomic force microscopy**
- **microelectronic structure**
- **single spins**
- **NTEGRA-SPECTRA-II**
- **ASNOM**

REFERENCES

1. Magonov, S.; Belikov, S.; Surtchev, M.; Leesment, S.; Malovichko, I. High-resolution Mapping of Quantitative Elastic Modulus of Polymers. *Microsc. Microanal.* **2015,** *21* (3), 2183–2184.
2. Montenegro, J.; Vazquez-Vazquez, C.; Kalinin, A.; Geckeler, K. E.; Granja, J. R. Coupling of Carbon and Peptide Nanotubes. *J. Am. Chem. Soc.* **2014.** DOI:10.1021/ja410901r
3. Alexander, J.; Magonov, S. High-resolution Imaging in Different Atomic Force Microscopy Modes. NT-MDT Application note, Vol. 88, 2015. http://www.ntmdt.com/data/media/files/products/general/high-resolution_imaging_in_afm_an088_a4_full.pdf.
4. Karpov, Y.; Erdmann, T.; Raguzin, I.; Al-Hussein, M.; Binner, M.; Lappan, U.; Stamm, M.; Gerasimov, K. L.; Beryozkina, T.; Bakulev, V.; Anokhin, D. V.; Ivanov, D. A.; Günther, F.; Gemming, S.; Seifert, G.; Voit, B.; Di Pietro, R.; Kiriy, A. High Conductivity in Molecularly p-Doped Diketopyrrolopyrrole-based Polymer: The Impact of a High Dopant Strength and Good Structural Order. *Adv. Mater.* **2016,** *28,* 6003–6010.
5. Xiang, D.; Wang, X.; Jia, C.; Lee, T.; Guo, X. Molecular-scale Electronics: From Concept to Function. *Chem. Rev.* 2016. http://doi.org/10.1021/acs.chemrev.5b00680.
6. Huber, A.; Kazantsev, D.; Keilmann, F.; Wittborn, J.; Hillenbrand, R. Simultaneous Infrared Material Recognition and Conductivity Mapping by Nanoscale Near-field Microscopy. *Adv. Mat.* **2007,** *19* (17), 2209–2212.

CHAPTER 17

X-RAY PHOTOELECTRON INVESTIGATION OF THE CHEMICAL STRUCTURE OF POLYMER MATERIALS MODIFIED WITH CARBON METAL-CONTAINING NANOSTRUCTURES

I. N. SHABANOVA[1*], V. I. KODOLOV[2,3], and N. S. TEREBOVA[1]

[1]*Laboratory of Nanostructures, Physico-Technical Institute, Ural Branch of the Russian Academy of Sciences, Kirov St. 132, Izhevsk 426000, Russia*

[2]*Basic Research-High Educational Centre of Chemical Physics and Mesoscopy, Ural Division of Russian Academy of Sciences, Izhevsk, Udmurt Republic, Russia*

[3]*Deparment of Chemistry and Chemical Technology, Kalashnikov Izhevsk State Technical University, Studencheskaya St. 7, Izhevsk, Russia*

Corresponding author. E-mail: xps@ftiudm.ru

ABSTRACT

In this work, polymer materials modified with carbon metal-containing (C/M) nanostructures have been studied by X-ray photoelectron spectroscopy on a device equipped with a magnetic energy-analyzer. The optimal composition and concentration of C/M nanostructures have been determined for the maximal change of the structure and the improvement of the service properties of the studied materials.

17.1 INTRODUCTION

Polymer materials find different industrial application. The methods for improving the properties of polymer materials by their modification with nanostructures have been intensely developed recently. Nanosized carbon metal-containing (C/M) nanostructures have a large specific surface which is responsible for their properties. The use of C/M nanostructures as modifiers is a promising way for obtaining materials with good physical-mechanical and improved thermophysical characteristics. The mechanism of the modifying action of nanosized C/M nanostructures on polymers is still unclarified.

The goal of this work is the XPS investigation of the influence of the content of C/M nanostructures on the degree of the polymer modification.

17.2 EXPERIMENT

The X-ray photoelectron spectra were obtained by using an X-ray electron magnetic spectrometer. The spectrometer resolution was 10^{-4}, luminosity 0.085%, the excitation with AlKα line (1486.5 eV), and vacuum was 10^{-8} to 10^{-10} Pa.[1,2]

The samples were polymethylmethacrylate (PMMA), polyvinyl alcohol (PVA), and polycarbonate (PC); the modification was conducted by adding C/M nanostructures (10^{-1} to 10^{-5}, 1, 3, 5, 15, 30, 60, and 90%).

The polymer modification with minute additions of C/M nanostructures (10^{-1} to 10^{-5} and 1.3%) was conducted as follows: the C/M nanostructure suspension was prepared in the methylene chloride medium with the help of immersible ultrasonic devices. The required amount of the fine-dispersed suspension was added into 10% polymer solution in methylene chloride; the composition was mechanically mixed. After that, films were prepared by the solvent evaporation at raising the temperature to 90°C. The PVA modification with C/M nanostructures (5, 15, 30, 60, and 90%) was conducted by the mechanochemical method in a triturating machine with the addition of an active medium (distilled water). After that, the powder was annealed and investigated.[3]

The reference samples were polymers without nano additions, C/M nanostructures in the form of carbon fiber nanofilms associated with metal-containing clusters formed in the nanoreactors of polymer matrices.[4] The samples were prepared in the Science and Education Centre of Chemical Physics and Mesoscopy, Udmurt Science Centre of the Ural Branch of the Russian Academy of Sciences.

17.3 RESULTS AND DISCUSSION

We have studied the variations of the structure of organic glass, PC, and PVA modified with 10^{-5}, 10^{-4}, 10^{-3}, 10^{-2}, and 10^{-1}% of C/M nanostructures.

Figure 17.1 shows the C1s-spectrum of a C/M nanostructure consisting of three components: C—C (sp^2)—284 eV, C—H—285 eV, and C—C (sp^3)—286.2 eV. The presence of a small content of the C—H component indicates that the synthesis of nanostructures from the polymer matrix is incomplete. The intensity ratio of the maxima of C—C (sp^2) and C—C (sp^3) depends on the size of a nanostructure: the larger the surface area is in comparison with the volume, the larger the intensity of the C1s-spectrum component with the sp^3-hybridization of the valence electrons according to Makarov et al.[5]

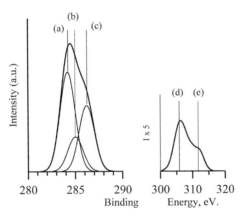

FIGURE 17.1 The XPS C1s-spectrum of C/M nanostructures consisting of three components: (a) C—C (sp^2)—284 eV; (b) C—H—285 eV; (c) C—C (sp^3)—286.2 eV and the satellite structure; (d) satellite (sp^2); and (e) satellite (sp^3).

Figure 17.2 presents the C1s-spectra of polymethylmethacrylate (PMMA).

In the reference sample organic glass film, the bonds C—H (285eV) and C—O (287 eV) (Fig 17.2a) are characteristic of the organic glass structure. When the content of the C/M nanostructure in the organic glass is being increased, starting with the content of 10^{-4}%, the C1s-spectrum structure changes. The structure characteristic of the C/M nanostructures appears consisting of C—C (sp^2) and C—C (sp^3) components. The C—H component indicates the remains of the organic glass structure. It should be noted that the C—O component is absent in the C1s-spectrum. During the organic glass modification, with the increase of the nanostructures concentration to 10^{-3}%, in the C1s-spectrum, the C—H component decreases, and the C—C (sp^2), and C—C (sp^3) components characteristic of the nanostructures grow, that is, the degree of the polymer modification grows. When the concentration of the nanostructures is further growing (10^{-2}%), the degree of the modification decreases and, when the content of the nanostructures is increased to 10^{-1}%, the modification does not take place. The C1s-spectrum shape becomes similar to that of the C1s-spectrum of the non-modified organic glass. Thus, the ratio of the C—C and C—H bonds in the C1s-spectrum can indicate the degree of the polymer nanomodification. At the content of the nanostructures 10^{-3}% the modification is maximal.

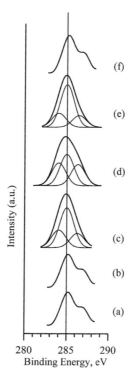

FIGURE 17.2 The XPS C1s-spectra of polymethylmethacrylate: (a) reference sample; (b) sample nanomodified with 10^{-1}% of C/M nanostructures; (c) sample nanomodified with 10^{-2}% of C/M nanostructures; (d) sample nanomodified with 10^{-3}% of C/M nanostructures; and (e) sample nanomodified with 10^{-4}% of C/M nanostructures; and (f) sample nanomodified with 10^{-5}% of C/M nanostructures.

Similar results are obtained for the PC nanomodification (Fig. 17.3). For the reference sample of PC containing a larger amount of oxygen, the

C1s-spectrum also consists of two components C—H (285 eV) and C—O (287 eV); however, the relative intensity of the C—O component is significantly larger than that in the organic glass. When the nanostructure content is in the range of 10^{-5} to 10^{-2}%, the peculiarities characteristic of the C/Cu-nanoform C1s-spectrum appear in the PC C1s-spectrum. The maximal change in the C1s-spectrum structure is observed when the nanostructures content in the polymer is 10^{-3}%. In contrast to the organic glass, the PC structure variation starts at the nanoforms content of 10^{-5}%. In this case, the C—O component decreases in the C1s-spectrum which is indicated by the shift of the high-energy maximum from 287 to 286.2 eV in the C1s-spectrum, which corresponds to the binding energy of the C—C (sp^3) component. When the content of the nanostructures is in the range of 10^{-4} to 10^{-2}%, the C—O bonds are absent in the C1s-spectrum. When there are 10^{-1}% of nanostructures in the PC, no changes are observed in the polymer structure.

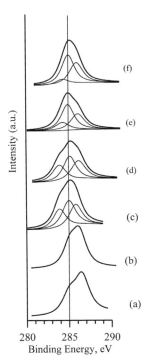

FIGURE 17.3 The XPS C1s-spectra of polycarbonate: (a) reference sample;(b) sample nanomodified with 10^{-1}% of C/M nanostructures; (c) sample nanomodified with 10^{-2}% of C/M nanostructures; (d) sample nanomodified with 10^{-3}% of C/M nanostructures; and (e) sample nanomodified with 10^{-4}% of C/M nanostructures; and (f) sample nanomodified with 10^{-5}% of C/M nanostructures.

Figure 17.4 shows the C1s-spectra of PVA having the smallest content of oxygen in its structure in comparison with the other studied polymers. Similar to the C1s-spectrum of the PC and organic glass films, the C1s-spectrum of the PVA reference sample contains two components C—H and C—O, however, the C—O component is less intensive than C—H. The variation of the PVA structure is observed only at the modification with C/Cu nanostructures in the amount from 10^{-3} to 10^{-2}%. When the nanostructures are added in the PVA polymer solution, the components characteristic of the C1s-spectrum of C/M nanostructures appears in the polymer C1s-spectrum (Fig. 17.1). When the content of the nanostructures is 10^{-1}% in PVA, no changes are observed in the polymer structure similar to the above polymers; this can be explained by the processes of the coagulation of the nanostructures due to their high content in the polymer.

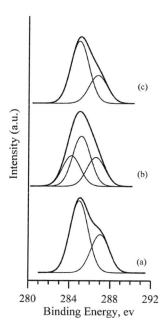

FIGURE 17.4 The XPS C1s-spectra of polyvinyl alcohol: (a) reference sample; (b) sample nanomodified with 10^{-3}% of C/M nanostructures; and (c) sample nanomodified with 10^{-4}% of C/M nanostructures.

When the content of C/M nanostructures is more than 1% (5, 15, 30, 60, and 90%) (Fig. 17.5), which have been incorporated into PVA by mechanochemical method preventing coagulation, the mechanism of nanomodification changes. The absence of both oxygen atoms (there is no the C—O component in the C1s-spectrum)⁶ and of the component characteristic of the C—H bond is observed. Thus, the complete PVA carbonization is observed in the presence of nanostructures as catalysts of the reaction in which the bonds of carbon and oxygen are broken, they are removed, and the formation of the C—C bonds with sp-, sp²-, and sp³-hybridization of the valence electrons takes place.

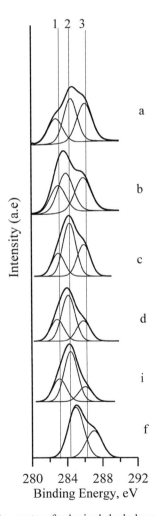

FIGURE 17.5 The XPS C1s-spectra of polyvinyl alcohol modified with C/M nanostructures: (a) PVA + 90% C/Ni; (b) PVA + 60% C/Ni; (c) PVA + 15% C/Ni; (d) PVA + 15% C/Ni; (e) PVA + 5% C/Ni; and (f) reference sample PVA: C—H—285 ± 0.2 eV.; C—O—287 ± 0.2 eV where (1) C—C (sp)—283.0 ± 0.2 eV.; (2) C—C (sp²)—284.3 ± 0.2 eV.; and (3) C—C (sp³)—286.2 ± 0.2 eV.

In the C1s-spectrum, the ratio of the components bonds C—C (sp)/C—C (sp^2) of the carbon atoms changes when the content of the nanostructures in the polymer is in the range of 15–90%: the content of the C—C (sp) component increases in relation to the C—C (sp^2) component. When the C/Ni nanostructures content is 30 and 60% in PVC, the ratio of the carbon components C—C (sp) is maximal (Table 17.1). Heating up to 300–400°C does not lead to the change in the C1s-spectrum shape.

TABLE 17.1 The dependence of the carbyne formation in the polymer on the content of the nanostructures in PVA ($\Delta \sim\pm 0.2$).

The content of the nanostructures in PVA, %	The ratio of the components of the C1s-spectrum C—C (sp)/C—C (sp^2)
90	0.60
60	0.70
30	0.70
15	0.50
5	0.40

17.4 CONCLUSION

In this work, it is shown that the degree of the interaction of the nanostructures with the polymer depends on the content of the nanostructures and their activity in the given medium. The temperature growth prevents the development of self-organization in the medium. Thus, for describing the process of the medium structurization under the action of the nanostructures it is necessary to introduce some critical parameters, such as the content of nanoparticles, activity of the nanoparticles, and critical temperature.

The modification of the studied polymers, that is, the variation of their structure, takes place at the content of the nanoparticles determined for each particular polymer. The XPS studies show that the smaller number of oxygen atoms are bound to carbon atoms in the polymer, the larger amount of nanoparticles is required for the variation of the polymer structure; the minimal content of nanoparticles, 10^{-5}%, is required for PC, 10^{-4}% for organic glass, and 10^{-3}% for PVA. When the content of the nanoparticles in the polymer is from the minimal content to 10^{-4}%, the C1s-spectrum changes: the component C—O (277 eV) disappears, and the components C—C (sp^2), and C—C (sp^3) characteristic of C/M nanostructures appear. The comparison of the structures of the studied polymers indicates a high reactivity of the C—O bond in the PC CO_3 group; in the organic glass CO_2 group the reactivity is smaller, and it is the smallest in the PVA CO group. The disappearance of the component characteristic of the C—O bond in the C1s-spectrum indicates the possibility of the replacement of this group of atoms by the nanoparticles, that is, the formation of strong bonds between the polymer atoms and the atoms of the nanostructure surface. In this case, the polymer obtains the structure-forming activity of the nanostructures, which are the centers of the appearing new structure. The breakage of the chemical bond between the C—O group and the nearest surrounding of the polymer atoms because of the different reactivity of the C—O bond occurs at different minimal content of the nanoparticles in the polymers.

Based on the investigation results, it can be suggested that the larger the content of the oxygen atoms bound to carbon is, the larger the variation of the polymer structure and the formation of the regions similar to the structure of the C/Cu nanoform in the polymer structure are. The smaller the oxygen atoms are in the initial polymer, the larger content of the nanostructures is required for the structurization of the polymers, and for the structure of the polymers to become more similar to that of the nanoform.

The comparison of the PVA modified with 5, 15, 30, 60, and 90% of C/M nanostructures shows that at the content of the nanostructures in the range of 30–60%, the largest amount of carbyne is formed from the carbon atoms of the polymer and the nanostructures. It should be noted that when the PVA is replaced by the PC, in the C1s-spectrum, the C—C component with sp-hybridization of the valence electrons of the carbon atoms disappears, and the M-C component appears, which is characteristic of most nanomodified polymers. It is connected with the absence of the dehydration process in the PC. Thus, for the carbyne formation, the simultaneous presence of a certain type and amount of polymer and nanostructures is required, which provide the polymer carbonization.

It is known[7] that carbyne consisting of the set of carbon chains is unstable. For obtaining more stable crystalline carbyne, it is necessary to bind separate carbon chains with the impurity atoms of the transition metals. In the studied samples, the d-metal atoms can be such atoms, the main part of which are deep in the substance and are not detected by the XPS. The d-metal atoms are most likely the catalysts of the carbyne crystallization.

The variation of the polymer structure is accompanied by the change of their technological properties; the tensile strength of the studied films increases by 13%, the electrical resistance decreases by a factor of 3.3, the optical density of the films increases causing an increase in the heat capacity of the polymers.

Carbyne has high conducting properties; therefore, the carbyne formation in the nanomodified polymer PVA leads to an increase in its electrical conduction.

ACKNOWLEDGMENTS

This work was supported by the Program of the FANO Russia (State Registration No. AAAA-A17-117022250040-0.

KEYWORDS

- polycarbonate
- polymethylmethacrylate
- polyvinyl alcohol
- carbon metal-containing nanostructures
- X-ray photoelectron spectroscopy
- modification

REFERENCES

1. Siegbahn, K.; Nordling, K.; Falman, A.; Nordberg, R. *Electron Spectroscopy;* Borovsky, I. B., Ed.; Mir: Moscow, 1971; p 493.
2. Shabanova, I. N.; Dobysheva, L. V.; Varganov, D. V.; Karpov, V. G.; Kovner, L. G.; Kl'ushnikov, O. I.; Manakov, Yu. G.; Makhonin, Ye. A.; Khaidarov, A. V.; Trapeznikov, V. A. New Automated X-ray Electron Magnetic Spectrometers: A Spectrometer with Technological Adaptors and Manipulators; a Spectrometer for Studying Melts. *Izv. Akad. Nauk SSSR, Ser. Fiz.* **1986,** *50* (9), 1677–1682.
3. Shabanova, I. N.; Kodolov, V. I.; Terebova, N. S.; Trineeva, V. V. X-ray Photoelectron Spectroscopy in the Investigation of Carbon Metal-containing Nanosystems and Nanostructured Materials. ISBN 978-5-4312-0128-8. Izd. Izhevsk State University: Izhevsk, Russia, 2012; p 250.
4. Kodolov, V. I.; Khokhryakov. N. V. Chemical Physics of the Processes of the Formations and Transformations of Nanostructures and Nanosystems. ISBN 978-5-9620-0151-7, RIO FGOU VPO Izhevskaya GSHA: Izhevsk, Russia, 2009.

5. Makarov, L. G.; Shabanova, I. N.; Terebova, N. S. The Use of the X-ray Photoelectron Spectroscopy Method for Studying the Structure of Carbon Nanostructures. *Zavodsk. Lab. (Diagnos. Mater).* **2005,** *71* (5), 26–28
6. Shabanova, I. N.; Terebova, N. S.; Sapozhnikov, G. V. *J. Electron. Spectr. Rel. Phen.* **2014,** *195,* 43.
7. Sladkov, A. M.; Kudryavtsev, Yu. P. Diamond, Graphite, Carbyne – Allotropic Forms of Carbon. *Priroda* **1969,** *5,* 37–44.

CHAPTER 18

SMALL-SIZED TECHNOLOGICAL ELECTRON SPECTROMETER WITH MAGNETIC ENERGY-ANALYZER

YU. G. MANAKOV[1], I. N. SHABANOVA[2*], V. A. TRAPEZNIKOV[2], and YE. A. MOROZOV[3]

[1]*Udmurt State University, Universitetskaya St., Izhevsk 426034, Russia*

[2]*Physico-Technical Institute, Ural Branch of the Russian Academy of Sciences, 132 Kirov St., Izhevsk 426000, Russia*

[3]*Izhevsk State Technical University, 7, Studentcheskaya St., Izhevsk, Russia*

[*]*Corresponding author. E-mail: xps@ftiudm.ru*

ABSTRACT

In this chapter, the description is given about the development of a unique small-sized portable X-ray electron magnetic spectrometer for the serial production and application in industry. Due to the growth of the role of surface analysis for the development of the technologies for obtaining materials of a new generation, it is necessary to develop small-sized and low-cost X-ray electron magnetic spectrometers for building them in an industrial production line and for researches in the sphere of high nanotechnologies accompanied by the factors leading to fouling an energy-analyzer in an electrostatic spectrometer.

The development of a small-sized X-ray electron magnetic spectrometer is attained by the decrease of the spectrometer orbit radius to 10–15 cm, the creation of a new system of the compensation of external magnetic fields, a new energy analyzer, and an improved protection against powerful process effects; by changing the shape of the toroidal chamber for the attachment to the production line, by using technological adapters for investigations at the temperatures in the range of 80–2000 K, etc. Thus, the small-sized X-ray electron magnetic spectrometer allows conducting investigations in the fields where the use of electrostatic analyzers leads to worsening their parameters and the spectrometer destruction. The use of the small-sized X-ray electron magnetic spectrometer will allow the development of new technologies.

18.1 INTRODUCTION

The progress in surface physics, chemistry, and mechanics is first of all due to the appearance and quick improvement of the electron spectroscopy methods in which the depth of the analyzed layer is determined by the mean free path of electrons and the sensitivity is fractions of a monoatomic layer. Among the electron spectroscopy methods, X-ray photoelectron spectroscopy (XPS) is the most promising method for investigating the electronic structure and the chemical bond of the elements of outer and inner surfaces.

In XPS, different types of monochromators[1,2] such as magnetic and electrostatic are used for focusing electrons. Most modern electron spectrometers are devices with electrostatic focusing. It is serial products.

The double focusing of electrons by magnetic field offered by Siegbahn and Edvarson[3] and Siegbahn et al.[4] is more advantageous in comparison with the electrostatic focusing on other equal conditions.

At present, Siegbahn's idea has been further developed only in Russia by Siegbahn's student Professor Trapeznikov.

The analysis of the work on the creation of magnetic spectrometers shows a significant advantage of a magnetic energy-analyzer in comparison with an electrostatic one:

1. The possibility of obtaining higher contrast and resolution.
2. Due to the focal plane, it is possible to use a multianode and multichannel mode of recording electrons.
3. The magnetic analyzer is constructively separated from the spectrometer vacuum chamber which allows using different methods of action on a sample (from the liquid helium temperature to the temperatures of metal melting) accompanied by the intensive release of aggressive gaseous fluxes without the deterioration of the device focusing properties. At the use of electrostatic devices, the defocusing of the electrodes, which are inside the spectrometer

Small-Sized Technological Electron Spectrometer

chamber, is likely to take place due to technological waste appearing at heating or spattering of samples, that is, magnetic spectrometers are more advantageous than electrostatic spectrometers.

4. Magnetic spectrometers have constant luminosity and resolution over the entire energy region including spectra of valence bands.

In the Physicotechnical Institute of the Ural Branch of the Russian Academy of Sciences, first Russian electron magnetic spectrometers with an automated system of control have been developed and built. They are not inferior to the best foreign spectrometers in their main parameters; however, they are much better equipped with technological adapters such as an evaporation chamber equipped with resistance evaporators, an impact testing machine for fracture and impact tests of standard samples in vacuum, a chemical chamber for studying the processes of oxidation, reduction, adsorption, etc., adapters for the laser radiation, ion, and electron action on a sample, adapters for the thermal action in the temperature range of 80–2000 K, for sputtering, spalling, scraping, fracturing, layer-by-layer mechanical removal (with a certain step) of the sample surface in vacuum, manipulators, and mass-spectrometers.

A unique automated electron magnetic spectrometer is created for investigating melts up to 2000 K. The creation of the spectrometer significantly widens the sphere of the application of the XPS method and allows studying the electronic structure of amorphous metal melts in the solid and liquid state.

A unique 100 cm electron magnetic spectrometer with a large radius of the cyclotron orbit has very high sensitivity (by two orders of magnitude higher than that of the 30 cm spectrometers); this allows investigating minute doses of radiation at fast processes.

18.2 RESULTS

At present, a unique small-sized portable electron magnetic spectrometer is being developed for the serial production and application in industry.

The spectrometer is being developed on the basis of the created and used 30 cm electron magnetic spectrometers.

The main disadvantages which have prevented the creation of serial devices of the magnetic type are large sizes, sensitivity to the action of external magnetic fields and interferences, and high cost of the device. As the result of the above disadvantages, electron magnetic spectrometers were built in a single exemplar, and at present, their number in the world is less than 10. However, the above disadvantages can be obviated.

Due to the growth of the role of surface analysis for the development of the technologies of a new generation, it is necessary to develop small-sized and low-cost magnetic spectrometers for applications in industry and researches in the field of high nanotechnologies. It is connected with that new technologies are accompanied by the factors leading to fouling an energy-analyzer in an electrostatic spectrometer by reaction by-products or causing the deformation of the energy-analyzer at high temperatures.

The development of the small-sized electron magnetic spectrometer becomes possible due to the decrease of the spectrometer orbit radius to 10–15 cm; the creation of a new system for the compensation of external magnetic fields replacing the Helmholtz rings;[5] the creation of a new magnetic energy-analyzer;[6] the improvement of the protection against various inter-ferences allowing the device to operate in the presence of powerful process effects; the change of the toroidal chamber shape, which will facilitate the attachment of the spectrometer to the production line, and the use of the adapters (one or several) required for researches such as

1. an evaporation chamber with resistance evaporators;
2. a chemical chamber for studying the processes of oxidation, reduc-tion, adsorption, etc.;
3. an adapter for thermal action on a sample:
   a) from 80 K (for studying the reasons for cold brittleness, inter-atomic interactions in high-temperature superconductors);
   b) up to 2000 K (metal melts and amorphous systems);
4. adaptors for sputtering, spalling, scraping, shearing, fracturing, layer-by-layer mechanical removal (with a certain step) of the sample surface in vacuum.

Computer programs have been developed for calculating uniform magnetic fields created by the systems based on rectangular current-carrying circuits.

Based on the computer experiment, the optimal systems of the rectangular circuits have been found for using them as the systems for the compensation of the external magnetic field for the electron magnetic spectrometers.

The above programs allow to investigate fields created by circular and rectangular circuits of the same sizes and to perform the comparative analysis of them.

For the electron magnetic spectrometers, high-performance systems of compensation have been calculated which increase the degree of the protection against alternating and static magnetic fields. The extension of the compensating magnetic fields with the specified degree of homogeneity, which is created by the above systems, is several times longer than the existing extension (e.g., the extension of the field with homogeneity of ~99.9% increases by a factor 1.5–5).

The calculated dynamic systems of the compensation of the magnetic field have been used as a basis for the compensation systems of the 12 cm electron magnetic spectrometer (Physicotechnical Institute, Izhevsk).

The suggested resolution of the spectrometer is 0.1 eV, luminosity is 0.1%, sensitivity is a fraction of a monoatomic layer. The time of the determination of the element concentration is a fraction of a second. In addition, the cost of a small-sized electron magnetic spectrometer is less and it may have extensive applications.

At present, there is no serial production of spectrometers with a magnetic energy-analyzer in the world. The developments and investigations in the field of the creation of spectrometers with a magnetic analyzer and the comparison of the researches performed with the use of electrostatic and magnetic spectrometers show that the main scope of works with the use of an electrostatic analyzer has been fulfilled, and the conduction of investigations in the extreme conditions such as high temperatures, intensive gas release, processes of material vaporization, fast processes, chemical reactions, etc., is economic and possible only with the use of a magnetic analyzer.

Spectrometers with a magnetic energy-analyzer will allow active conduction of investigations in the fields where the use of an electrostatic spectrometer will lead to the deterioration of the device metrological performance: at gas release from a sample at high temperatures, at melting of samples, at evaporation or sputtering of a substance, etc. The use of the new device will allow eliminating spoilage in production, to decrease the production price, and to develop new technologies.

18.3 CONCLUSION

The portable technological X-ray electron magnetic spectrometer for the analysis of the composition of outer and inner surfaces is designed for the quantitative and qualitative analysis of ultrafine surface layers (10–20 A) and a concentration in depth of solid and liquid materials; for the control over the formation of the coating-substrate interface in the development of the technology of the fabrication of elements in microelectronics; for the study of the grain boundaries and ruptured surfaces of the fractures of embrittled materials; for the work with adapters for sputtering, layer-by-layer removal of the surface from 10 A to several microns, melting, cooling down to liquid nitrogen temperature, fracture in vacuum for studying the grain boundaries, laser radiation, mechanical and ion beam action on samples, the scientific purposes including the investigation of the electronic structure, chemical bond and nearest surroundings of atoms in solid and liquid materials.

The main specific features of the device are portability and small weight and the possibility of using different adapters for the action on a sample during spectra obtaining.

The spectrometer can be used for the development of new technologies. It can be built on the production line. It can find its application in the institutes of higher education for conducting workshops, laboratory works, and scientific investigations with the use of adapters for sputtering, layer-by-layer removal of surface from 10 A to several microns, melting, cooling down to liquid nitrogen temperature, fracture in vacuum for studying grain boundaries, for mechanical, laser, and ion beam action on samples.

The spectrometer technical characteristics: resolution is 1–0.1 eV; luminosity is 0.5–0.1%; sensitivity—to a fraction of a monoatomic layer; time of the concentration determination—in the range of 10 s a fraction of a second. The sizes of the energy-analyzer are 500 × 500 × 250 mm^3; total weight is 30 kg (Fig 18.1).

FIGURE 18.1 The X-ray electron magnetic spectrometer with the electron orbit radius of 10 cm.

KEYWORDS

- **X-ray photoelectron spectroscopy (XPS)**
- **unique small-sized portable X-ray electron magnetic spectrometer**
- **magnetic spectrometers**
- **superconductors**
- **magnetic energy-analyzer**

REFERENCES

1. Afanas'yev, V. P.; Ya Yavor, S. *Electrostatic Analyzers for Charged Particle Beams;* Atomizdat: Moscow, 1978; p 224.
2. Fridrikhov. S. A. *Energy Analyzers and Monochromators for Electron Spectroscopy;* LGU: Leningrad, Russia, 1978; p 158.
3. Siegbahn, K.; Edvarson, K. X-Ray Spectroscopy in the Precision Range of 1:105. *Nucl. Phus.* **1956,** *1,* 137–147.
4. Siegbahn, K.; Nordling, K.; Falman, A., et al. *Electron Spectroscopy;* Mir: Moscow, 1971; p 490.
5. Isupov, N. Yu.; Manakov, Yu. G.; Manakova, E. Yu.; Nurullina, R. A.; Trapeznikov, V. A.; Shabanova, I. N. The System for Regulating the Magnetic Field in the Annular Chamber. Patent 2,314,549, June 20, 2006.
6. Shabanova, I. N.; Nurullina, R. A.; Trapeznikov, V. A.; Manakov, Yu. G. Electron Magnetic Spectrometer. Patent 2,338,295, January 30, 2007.

PART IV
Structure and Properties of Nanostructures and Nanosystems

CHAPTER 19

NMR ^{1}H, ^{13}C, AND ^{17}O SPECTROSCOPY OF THE *TERT*-BUTYL HYDROPEROXIDE

N. A. TUROVSKIJ[1*], YU. V. BERESTNEVA[2], E. V. RAKSHA[2], and G. E. ZAIKOV[3]

[1]*Physical Chemistry Department, Donetsk National University, Universitetskaya St. 24, Donetsk 83001, Ukraine*

[2]*L. M. Litvinenko Institute of Physical Organic and Coal Chemistry, R. Luxemburg St. 24, Donetsk 83114, Ukraine*

[3]*Department of Bio-Chemistry, N. M. Institute of Biochemical Physics, Russian Academy of Sciences, Kosygin St. 4, Moscow 119991, Russia*

*Corresponding author. E-mail: Turovskij@gmail.com

ABSTRACT

NMR ^{1}H and ^{13}C spectra of *tert*-butyl hydroperoxide ((CH_3)$_3$COOH) in acetonitrile-d$_3$ (CD_3CN), chloroform-d ($CDCl_3$), and dimethyl sulfoxide-d$_6$ (DMSO-d$_6$) have been investigated by the NMR method. The calculation of magnetic shielding tensors and chemical shifts for ^{1}H, ^{17}O, and ^{13}C nuclei of the (CH_3)$_3$COOH molecule in the approximation of an isolated particle and considering the influence of the solvent in the framework of the continuum polarization model was carried out. Comparative analysis of experimental and computer NMR spectroscopy results revealed that the gauge-including atomic orbital (GIAO) method with MP2/6-31G (d,p) level of theory and the PCM approach can be used to estimate the parameters of NMR ^{1}H and ^{13}C spectra of (CH_3)$_3$COOH.

19.1 INTRODUCTION

Reactivity of hydroperoxide compounds is of vital importance for a wide range of chemical processes. They are one of the effective origins of the highly reactive oxygen species in the oxidative modification of hydrocarbons,[1] PAH,[2] as well as carbon nanoparticles (graphene, nanotubes, nanoscrolls, etc.).[3–5] Hydroperoxide moiety is one of reactive group on the graphene oxide surface.[6,7]

Quantum chemical methods are widely used for investigating the structural features of the organic peroxides and their associates with various classes of compounds and to study the reactivity of hydroperoxides.[8,9] Computational chemistry is effective tool to get more structural information about peroxide bond activation by enzymes,[10] transition metals compounds,[11] sulfide,[12] and quaternary ammonium salts.[13] The existing semiempirical, ab initio, and density functional theory (DFT)—methods can reproduce the peroxides molecular geometry with sufficient accuracy.[14] Molecular modeling of the peroxide bond homolytic decomposition and hydroperoxides association processes is an additional information source of the structural effects that accompany these reactions. One of the criteria for the quantum-chemical method selection to study the hydroperoxides reactivity can be NMR ^{1}H and ^{13}C spectra parameters reproduction with sufficient accuracy. It should be noted that the parameters of the NMR spectra are very sensitive to slight changes in spatial and electronic structure of the molecule. The joint use of experimental and computational methods can be an information source of the structural features of the molecule caused by intra- and inter-molecular interactions.[15] Calculations within the DFT and perturbation theory (MP2) in the gauge-including atomic orbital (GIAO) approximation for NMR spectra modeling of organic compounds are most often used, since they provided a good relationship between the computational cost and accuracy.[15,16]

Tert-butyl hydroperoxide is commercially available one and has a convenient structure for the model investigations. This work deals with the NMR ^{1}H spectroscopy of the *tert*-butyl hydroperoxide ((CH_3)$_3$COOH) as a model compound.

19.2 EXPERIMENTAL

(CH_3)$_3$COOH was purified according to Hock and Lang.[17] Experimental NMR ^{1}H and ^{13}C spectra of the hydroperoxide solutions were obtained by using the Bruker Avance II 400 spectrometer (NMR ^{1}H—400 MHz and NMR ^{13}C—100 MHz) at 298K. Solvents—acetonitrile-d$_3$ (CD$_3$CN), chloroform-d

NMR ^{1}H, ^{13}C, and ^{17}O Spectroscopy of the *Tert*-Butyl Hydroperoxide

(CDCl$_3$), and dimethyl sulfoxide-d$_6$ (DMSO-d$_6$) were Sigma-Aldrich reagents and used without additional purification but were stored above molecular sieves before use. Tetramethylsilane (TMS) was used as internal standard. The hydroperoxide concentration in solution was 0.03 mol·dm^{-3}. Molecular geometry and electronic structure parameters, thermodynamic characteristics of the $(CH_3)_3COOH$ molecule were calculated using the Gaussian 03[18] software package. The hydroperoxide molecular geometry optimization and frequency harmonic vibrations calculation were carried out on the first step of investigations. The nature of the stationary points obtained was verified by calculating the vibrational frequencies at the same theory level. To choose the optimal method for the $(CH_3)_3COOH$ geometry calculation, the hydroperoxide structure parameters were estimated on the MP2 theory level with 6–31 G(d,p) basis set. The solvent effect was considered in the PCM approximation.[19,20] The magnetic shielding tensors (χ, ppm) for ^{1}H, ^{13}C, and ^{17}O nuclei of the $(CH_3)_3COOH$ molecule were calculated with the MP2/6-31G(d,p) and MP2/6-31G(d,p)/PCM optimized geometries by standard GIAO approach.[21] Optimization of the molecular geometry in the same approximation and calculation of magnetic shielding tensors in the framework of the GIAO approach were also performed for the TMS and H$_2$O molecules. Obtained χ values for TMS and H$_2$O (Table 19.1) were used for the hydroperoxide ^{1}H, ^{13}C, and ^{17}O nuclei chemical shifts calculations.

TABLE 19.1 The Magnetic Shielding Tensors for ^{1}H and ^{13}C Nuclei of the Tetramethylsilane and ^{17}O Nuclei of the H$_2$O Calculated Within GIAO/MP2/6-31G(d,p) Approach.

Nuclei	χ, ppm				
	–	CDCl$_3$	CD$_3$CN	(CD$_3$)$_2$SO	H$_2$O
^{1}H	31.958	31.951	31.953	31.940	–
^{13}C	207.541	207.814	207.925	207.858	–
^{17}O	351.740	–	–	–	358.771

Note: χ values are averaged over corresponding nuclei.

19.3   RESULTS AND DISCUSSION

Experimental NMR ^{1}H and ^{13}C studies of $(CH_3)_3COOH$ were carried out in the following solvents: CD$_3$CN, CDCl$_3$, and DMSO-d$_6$ at 298K. The concentration of the hydroperoxide in all samples was 0.03 mol·dm^3. TMS was used as the internal standard. Parameters of the experimental NMR ^{1}H and ^{13}C spectra of the $(CH_3)_3COOH$ are listed in Table 19.2.

TABLE 19.2 Experimental Parameters of the *Tert*-butyl Hydroperoxide NMR ^{1}H and ^{13}C Spectra.

	δ (^{1}H), ppm		δ (^{13}C), ppm	
	$-CH_3$	$-CO-OH$	$-CH_3$	$-CO-OH$
$CDCl_3$	1.27	7.24	25.71	80.87
CD_3CN	1.18	8.80	26.06	80.44
$(CD_3)_2SO$	1.12	10.73	26.01	76.56

Table 19.2 reveals the following features. The shift of the $-CH_3$ and $-COOH$ groups signals in the NMR ^{1}H spectrum of the hydroperoxide with the solvent polarity increasing is observed. The signal of the hydroperoxide group proton appears at 7.24 ppm in $CDCl_3$, and in the more polar DMSO-d_6 it was found at 10.73 ppm. When passing from $CDCl_3$ to DMSO-d_6, the signal of the $-CH_3$ group's protons is shifted to the strong field. Another effect is observed in the case of the NMR ^{13}C spectrum, the signal of a tertiary carbon atom is shifted to a strong field of 4.34 ppm in DMSO-d_6 as compared to $CDCl_3$.

Experimental data for the NMR ^{17}O spectrum of the hydroperoxide are taken from Barieux and Schirmann.[22] Chemical shifts relative to H_2O of the $(CH_3)_3CO^xO^yH$ are 246(O^x) and 206(O^y) as well as 249(O^x) and 206(O^y) in hexane.

The first step of the $(CH_3)_3COOH$ NMR spectra calculation was the estimation of the hydroperoxide molecular geometry parameters and electronic structure to choose method and basis set for the further investigations. The $(CH_3)_3COOH$ molecular geometry optimization was carried out by MP2 method, the 6-31G(d,p) basis set was used in calculations. Peroxide bond O—O is a reaction center in this type of chemical initiators and thus the main attention was focused on the geometry of $-COOH$ fragment. The calculation results were compared with experimental values[23] and a good agreement between calculated and experimental parameters can be seen (Table 19.3). Calculations with solvent effect accounting were also carried out for the $(CH_3)_3COOH$ molecule by MP2/6-31G(d,p) method within PCM approximation. Obtained geometry parameters are listed in Table 19.3. The solvent does not affect the O—O bond length. Solvent effect is noticeable for O—H and C—O bonds, O—O—H and C—O—O bond angles, and C—O—O—H torsion angle. Significant changes of O—H and C—O bonds, O—O—H and C—O—O bond angles are observed in the case of CH3CN and DMSO, while CHCl3 has less influence.

NMR ^{1}H, ^{13}C, and ^{17}O Spectroscopy of the *Tert*-Butyl Hydroperoxide 251

TABLE 19.3 Molecular Geometry Parameters of the *Tert*-butyl Hydroperoxide Molecule Calculated Within MP2/6-31G(d,p)/PCM Approximation.

Parameter	Solvent					Experiment [23]
	–	CHCl$_3$	CH$_3$CN	(CH$_3$)$_2$SO	H$_2$O	
r_{O-O}, Å	1.473	1.472	1.472	1.472	1.471	1.473
r_{O-H}, Å	0.970	0.980	0.986	0.986	0.989	0.990
r_{C-O}, Å	1.446	1.449	1.450	1.450	1.450	1.443
$\angle$O–O–H,°	98.2	98.9	99.4	99.3	99.3	100.0
$\angle$C–O–O,°	107.7	107.9	108.2	108.1	108.3	109.6
$\angle$C–O–O–H,°	114.7	116.3	113.8	114.5	111.9	114.0

The GIAO calculations for the (CH$_3$)$_3$COOH molecule were performed at MP2/6-31G(d,p) level of theory in the approximation of the isolated molecule as well as with solvent effect accounting within PCM approach. Figure 19.1 illustrates the structural model of the (CH$_3$)$_3$COOH molecule (equilibrium configuration obtained at MP2/6-31G(d,p) level of theory). To estimate the hydroperoxide NMR ^{1}H, ^{13}C, and ^{17}O spectra parameters the magnetic shielding constants (χ, ppm) for corresponding nuclei were calculated by GIAO method. Obtained χ values for magnetically equivalent nuclei were averaged. On the base of the obtained χ values, the chemical shift values (δ, ppm) of the ^{1}H, ^{13}C, and ^{17}O nuclei in the hydroperoxide molecule were evaluated. TMS and H$_2$O were used as standard, for which the molecular geometry optimization and χ calculation were performed using the same level of theory and basis set. Values of the ^{1}H and ^{13}C chemical shifts were found as the difference of the magnetic shielding tensors of the corresponding TMS and hydroperoxide nuclei. Values of the ^{17}O chemical shifts were found as the difference of the χ of the H$_2$O and hydroperoxide nuclei.

For the MP2/6-31G(d,p) calculated ^{1}H and ^{13}C chemical shifts, there was no full conformity with the experimental data. Experimental δ values for –CO–OH moiety proton in all solvents are higher than the δ = 6.805 ppm calculated in the isolated molecule approximation. The labile protons signals are usually shifted to lower field in DMSO-d$_6$ solution. Difference between δ values of –CO–OH moiety proton for CH$_3$CN and CHCl$_3$ solutions may also be due to the influence of the solvent. Formation of hydroperoxides self-associates is possible in the low-polarity solvents[24] that may affect the magnitude of the proton chemical shift of the –CO–OH fragment.

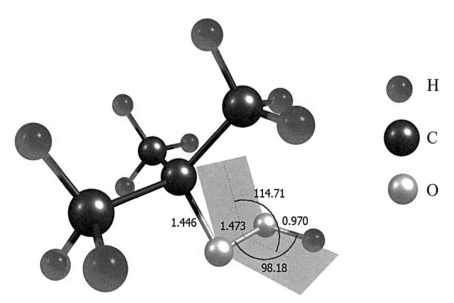

FIGURE 19.1 **(See color insert.)** Structural model of the *tert*-butyl hydroperoxide molecule (equilibrium configuration obtained at MP2/6-31G(d,p) level of theory).

Calculations of the χ values for hydroperoxide moiety atoms were carried out on MP2/6-31G(d,p)/GIAO level at different values of valent angle of carboxyl group ($\angle$COOH). Dependences of the χ and δ from the $\angle$COOH value were obtained (Figs. 19.2 and 19.3). $\angle$COOH value was varied within 0°–360° with step of 15°.

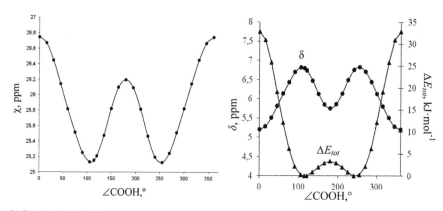

FIGURE 19.2 Change of the *tert*-butyl hydroperoxide total energy, χ and δ of the —CO—OH moiety with $\angle$COOH value changing.

NMR ^{1}H, ^{13}C, and ^{17}O Spectroscopy of the *Tert*-Butyl Hydroperoxide 253

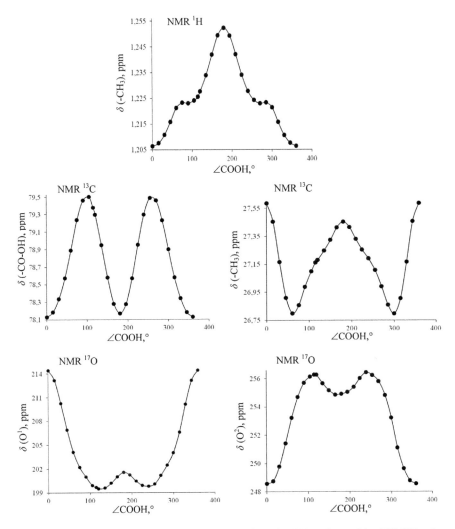

FIGURE 19.3 Change of the chemical shifts of —CO—OH moiety with ∠COOH value changing.

Chemical shift of the hydroperoxide moiety proton depends on this fragment configuration and is between 5.21 and 6.83 ppm (Fig. 19.2). The χ dependence character is similar to the change in the total energy (ΔE_{tot}) of (CH$_3$)$_3$COOH with the variation of ∠COOH value (Fig. 19.2). For the chemical shift of the —CO—OH group proton, the dependence is reversed— hydroperoxide configurations at the minimum points on the curve correspond to the maximum values of δ, thus the change in the ∠COOH torsion angle to 180° or 0° results in increase the total energy of the system and a decrease in the value of δ. Three minima are observed for chemical shift of the $-CH_3$ group's protons at values of $\angle COOH$ 0° (δ = 1.21 ppm), 90°, and 270° (δ = 1.22 ppm). At $\angle COOH$ = 180°, the chemical shift of the $-CH_3$ group's protons has a maximum value of 1.25 ppm (Fig. 19.3).

The character of δ versus $\angle COOH$ dependence for the carbon of the hydroperoxide group is similar to the proton $-CO-OH$ group. The nature of the O^x and O^y chemical shifts dependences is opposite (Fig. 19.3): minimum points for O^x correspond to the maximum points for O^y.

Figure 19.4 presents one of many possible configurations of the $(CH_3)_3COOH$ molecule dimer. Molecular geometry optimization of this homoassociate was performed at MP2/6-31G(d,p) theory level without solvent effect accounting. This associate is stabilized by the formation of two intermolecular hydrogen bonds: O...HO distance is 1.834 Å, bond angle O...H–O is 161.51°, and O–H bond length increases up to 0.982 Å. However, the configuration of the hydroperoxide fragment is not largely changed. For this associate, χ values were calculated at (MP2/6-31G(d,p) theory level, isolated molecule approximation) and the chemical shift values of the ^1H and ^{13}C nuclei were evaluated. The δ value of 10.541 ppm has been obtained for the $-CO-OH$ moiety proton. This is significantly higher than the experimental values observed in $CDCl_3$ and CD_3CN solutions.

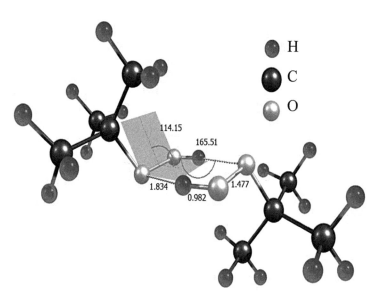

FIGURE 19.4 (See color insert.) Structural model of the *tert*-butyl hydroperoxide dimer. Molecular geometry optimization of this homoassociate was performed at MP2/6-31G(d,p).

NMR ^{1}H, ^{13}C, and ^{17}O Spectroscopy of the *Tert*-Butyl Hydroperoxide

Study the hydroperoxide concentration effect on the position of the signals in the NMR ^{1}H spectrum in CD$_3$CN and CDCl$_3$ solutions have been carried out. The hydroperoxide concentration was ranged within $(2.1–500.0) \times 10^{-3}$ and $(9.0–20.0) \times 10^{-3}$ mol·dm^{-3} in CD$_3$CN and CDCl$_3$, respectively. Changing in the hydroperoxide concentration in these ranges does not lead to a change in the signal position in the spectrum. Hence, the chemical shift of the hydroperoxide group proton is independent of the hydroperoxide concentration in the system in experimental conditions. Thus, calculation and NMR ^{1}H spectroscopy results showed that the hydroperoxide dimers do not form in the system. And observed chemical shift values for the —CO—OH moiety proton are due to the solvent effect. In order to account for the solvent effect in the calculation of magnetic shielding tensors, the PCM approach was used. Magnetic shielding tensors for ^{1}H and ^{13}C nuclei of the(CH$_3$)$_3$COOH calculated by GIAO method at MP2/6-31G(d,p) theory level with PCM solvent effect have been used for the chemical shift values of the ^{1}H and ^{13}C nuclei estimation.

Table 19.4 illustrates the magnetic shielding tensors (χ, ppm) and chemical shift values (δ, ppm) for ^{1}H and ^{13}C nuclei of the (CH$_3$)$_3$COOH calculated at MP2/6-31G(d,p) level of theory with solvent effect accounting.

TABLE 19.4 NMR ^{1}H and ^{13}C Parameters of the *Tert*-butyl Hydroperoxide (MP2/6-31G(d,p)/PCM).

Solvent	NMR ^{1}H				NMR ^{13}C			
	—CO—OH		—CH$_3$		—CO—OH		—CH$_3$	
	χ, ppm	δ, ppm	χ, ppm	δ, ppm	χ, ppm	δ, ppm	χ, ppm	δ, ppm
–	25.15	6.81	30.73	1.23	128.16	79.38	180.38	27.16
CHCl$_3$	23.82	8.13	30.71	1.24	127.84	79.98	180.42	27.39
CH$_3$CN	23.18	8.77	30.70	1.25	127.61	80.31	180.47	27.46
(CH$_3$)$_2$SO	23.16	8.78	30.70	1.24	127.64	80.22	180.47	27.39

Concerning the spectral patterns of a —CH$_3$ group protons, inspection of Table 19.4 reveals the following features: the pattern of NMR ^{1}H spectra of (CH$_3$)$_3$COOH is rather correctly reproduced at selected computational level for all solvents; the best agreement between the experimental and calculated ^{1}H and ^{13}C chemical shifts of the —CO—OH moiety is observed for acetonitrile-d$_3$ solution; for all cases, solvent effect accounting leads to a better result compared to the isolated molecule approximation; calculated and experimental values of δ for —CO—OH group proton decrease symbatically with the solvent polarity increasing. But in DMSO-d$_6$ solution, a significant shift to lower field region is observed for —CO—OH group proton as compared with other solvents. Calculated δ value for this proton in DMSO is very close to the experimentally observed one in CD_3CN solution. This shift can be explained by the formation of hydroperoxide-DMSO-d_6 hetero-associates in experimental conditions. One should note that similar values of $\delta = 10.77 \div 10.33$ ppm have a —CO—OH group proton of $(CH_3)_3COOH$ complex-bonded with tetraalkylammonium bromides.[25]

Calculation of the hydroperoxide NMR ^{17}O spectrum was also carried out. Chemical shifts relative to H_2O of the $(CH_3)_3CO^xO^yH$ are $256.2(O^x)$ and $199.6(O^y)$ in isolated molecule approximation and $250.7(O^x)$ and $207.1(O^y)$ in H_2O. Solvent accounting in the framework of PCM model leads to the good agreement between experimental and calculated NMR ^{17}O chemical shifts.

19.4 CONCLUSION

Experimental and computational NMR spectroscopy investigations of the $(CH_3)_3COOH$ were carried out. The influence of the solvent on the NMR spectra parameters of the $(CH_3)_3COOH$ was studied. It was shown that with increasing polarity of the solvent signal of the hydroperoxide group proton shifts toward weak fields. On the basis of the complex data analysis of the spectroscopic studies and molecular modeling shows that the GIAO method with the MP2/6-31G (d,p) level of theory and the PCM approximation can be used to estimate NMR ^1H, ^{13}C, and ^{17}O spectra parameters of $(CH_3)_3COOH$. Chemical shifts for the hydroperoxide group atoms have revealed to be sensitive to the hydroperoxide configurations.

KEYWORDS

- NMR spectroscopy
- *tert*-butyl hydroperoxide
- chemical shift
- magnetic shielding constant
- GIAO
- molecular modeling

REFERENCES

1. Xiao, Y.; Liu, J.; Kaihong, X.; Wang, W.; Fang, Y. Aerobic Oxidation of Cyclohexane Catalyzed by Graphene Oxide: Effects of Surface Structure and Functionalization. *Mol. Catalysis* **2017**, *431,* 1–8.
2. Chaparala, S. V.; Raj, A. Reaction Mechanism for the Oxidation of Zigzag Site on Polycyclic Aromatic Hydrocarbons in Soot by O_2. *Combust. Flame* **2016**, *165,* 21–33.
3. Czech, B.; Oleszczuk, P.; Wiącek, A. Advanced Oxidation (H_2O_2 and/or UV) of Functionalized Carbon Nanotubes (CNT-OH and CNT-COOH) and its Influence on the Stabilization of CNTs in Water and Tannic Acid Solution. *Environ Pollut.* **2015,** *200,* 161–167.
4. Weydemeyer, E. J.; Sawdon, A. J.; Peng, C. A. Controlled Cutting and Hydroxyl Functionalization of Carbon Nanotubes through Autoclaving and Sonication in Hydrogen Peroxide. *Chem. Commun.* **2015,** *51,* 5939–5942.
5. Datsyuk, V.; Kalyva, M.; Papagelis, K.; Parthenios, J.; Tasis, D.; Siokou, A.; Kallitsis, I.; Galiotis, C. Chemical Oxidation of Multiwalled Carbon Nanotubes. *Carbon* **2008,** *46,* 833–840.
6. Papaianina, O. S.; Savoskin, M. V.; Vdovichenko, A. N.; Rodygin, M. Yu.; Abakumov, A. A.; Nosyrev, I. E.; Popov, A. F. Graphite Oxide – Stages of Formation and a New View on its Structure. *Theor. Exp. Chem.* **2013,** *49,* 88–95.
7. Papaianina, O. S.; Savoskin, M. V.; Vdovichenko, A. N.; Lebedeva, Yu. P.; Nosyrev, I. E.; Kompanets, M. A.; Opeida, I. O. New Approach to the Reduction of Graphite Oxide. *Theor. Exp. Chem.* **2014,** *50,* 35–38.
8. Turovskij, N. A.; Raksha, E. V.; Gevus, O. I.; Opeida, I. A.; Zaikov, G. E. Activation of 1-Hydroxycyclohexylhydroperoxide Decomposition in the Presence of Alk4NBr. *Oxid. Commun.* **2009,** *32* (1), 69 – 77.
9. Turovsky, M. A.; Raksha, O. V.; Opeida, I. O.; Turovska, O. M.; Zaikov, G. E. Molecular Modelling of Aralkylhydroperoxides Homolysis. *Oxid. Commun.* **2007,** *30* (3), 504–512.
10. Siegbahn, P. E. M. Modeling Aspects of Mechanisms for Reactions Catalyzed by Metalloenzymes. *J. Comput. Chem.* **2001,** *22,* 1634–1645.
11. Ryan, P.; Konstantinov, I.; Snurr, R. Q.; Broadbelt, L. J. DFT Investigation of Hydroperoxide Decomposition Over Copper and Cobalt Sites Within Metal-organic Frameworks. *J. Catalysis* **2012,** *286,* 95–102.
12. Litvinenko, S. L.; Lobachev, V. L.; Dyatlenko, L. M.; Turovskii, N. A. Quantum-chemical Investigation of the Mechanisms of Oxidation of Dimethyl Sulfide by Hydrogen Peroxide and Eroxoborates. *Theor. Exp. Chem.* **2011,** *1,* 2–8.
13. Turovskij, N. A.; Raksha, E. V.; Berestneva, Yu. V.; Pasternak, E. N.; Zubritskij, M. Yu.; Opeida, I. A.; Zaikov, G. E. *Supramolecular Decomposition of the Aralkyl Hydroperoxides in the Presence of Et₄NBr;* Pethrick, R. A., Pearce, E. M., Zaikov, G. E., Eds.; Apple Academic Press, Inc.: Toronto, NJ, 2013; p 322.
14. Antonovskij, V. L.; Khursan, S. L. *Physical Chemistry of Organic Peroxides;* PTC AKADEMKNIGA: Moscow, 2003; p 391.
15. Belaykov, P. A.; Ananikov, V. P. Modeling of NMR Spectra and Signal Assignment Using Real-time DFT/GIAO Calculations. *Russ. Chem. Bull.* **2011,** *60* (5), 783–789.
16. Vaara, J. Theory and Computation of Nuclear Magnetic Resonance Parameters. *Phys. Chem. Chem. Phys.* **2007,** *9,* 5399–5418.

17. Hock, H.; Lang, S. Autoxydation Von Kohlenwasserstoffen, IX. Mitteil.: Über Peroxyde von Benzol-Derivaten (Autoxidation of Hydrocarbons IX. Msgs. about Peroxides of Benzene Derivatives). *Chem. Ber.* **1944,** *77,* 257–264.

18. Gaussian 03, Revision B.01, Frisch, M. J.; Trucks, G. W.; Schlegel, H. B.; Scuseria, G. E.; Robb, M. A.; Cheeseman, J. R.; Montgomery, J. A. Jr., Vreven, T.; Kudin, K. N.; Burant, J. C.; Millam, J. M.; Iyengar, S. S.; Tomasi, J.; Barone, V.; Mennucci, B.; Cossi, M.; Scalmani, G.; Rega, N.; Petersson, G. A.; Nakatsuji, H.; Hada, M.; Ehara, M.; Toyota, K.; Fukuda, R.; Hasegawa, J.; Ishida, M.; Nakajima, T.; Honda, Y.; Kitao, O.; Nakai, H.; Klene, M.; Li, X.; Knox, J. E.; Hratchian, H. P.; Cross, J. B.; Adamo, C.; Jaramillo, J.; Gomperts, R.; Stratmann, R. E.; Yazyev, O.; Austin, A. J.; Cammi, R.; Pomelli, C.; Ochterski, J. W.; Ayala, P. Y.; Morokuma, K.; Voth, G. A.; Salvador, P.; Dannenberg, J. J.; Zakrzewski, V. G.; Dapprich, S.; Daniels, A. D.; Strain, M. C.; Farkas, O.; Malick, D. K.; Rabuck, A. D.; Raghavachari, K.; Foresman, J. B. Ortiz, J. V.; Cui, Q.; Baboul, A. G.; Clifford, S.; Cioslowski, J.; Stefanov, B. B.; Liu, G.; Liashenko, A.; Piskorz, P.; Komaromi, I.; Martin, R. L.; Fox, D. J.; Keith, T.; Al-Laham, M. A.; Peng, C. Y.; Nanayakkara, A.; Challacombe, M.; Gill, P. M. W.; Johnson, B.; Chen, W.; Wong, M. W.; Gonzalez, C.; Pople, J. A. Gaussian, Inc.: Pittsburgh, PA, 2003.

19. Mennucci, B.; Tomasi, J. Continuum Solvation Models: A new Approach to the Problem of Solute's Charge Distribution and Cavity Boundaries. *J. Chem. Phys.* **1997,** *106,* 5151–5158.

20. Cossi, M.; Scalmani, G.; Rega, N.; Barone, V. New Developments in the Polarizable Continuum Model for Quantum Mechanical and Classical Calculations on Molecules in Solution. *J. Chem. Phys.* **2002,** *117,* 43–54.

21. Wolinski, K.; Hinton, J. F.; Pulay, P. Efficient Implementation of the Gauge-independent Atomic Orbital Method for NMR Chemical Shift Calculations. *J. Am. Chem. Soc.* **1990,** *112* (23), 8251–8260.

22. Barieux, J. J.; Schirmann, J. P. ^{17}O-enriched Hydrogen Peroxide and T. butyl Hydroperoxide: Synthesis, Characterization and Some Applications. *Tetrahedron Lett.* **1987,** *28* (51), 6443–6446.

23. Kosnikov, A. Yu.; Antonovskii, V. L.; Lindeman, S. V.; Antipin, M. Yu.; Struchkov, Yu. T.; Turovskii, N. A.; Zyat'kov, I. P. X-ray Crystallographic and Quantum-chemical Investigation of *Tert*-butyl Hydroperoxide. *Theor. Exp. Chem.* **1989,** *25* (1), 73–77.

24. Remizov, A. B.; Kamalova, D. I.; Skochilov, R. A.; Batyrshin, N. N.; Kharlampidi, Kh. E. FT-IR Study of Self Association of Some Hydroperoxides. *J. Mol. Struct.* **2004,** *700,* 73–79.

25. Turovskij, N. A.; Berestneva, Yu. V.; Raksha, E. V.; Pasternak, E. N.; Zubritskij, M. Yu.; Opeida, I. A.; Zaikov, G. E. ^{1}H NMR Study of the *tert*-butyl Hydroperoxide Interaction with Tetraalkyl Ammonium Bromides. *Polym. Res. J.* **2014,** *8* (2), 85–92.

CHAPTER 20

ON PROCESSES OF CHARGE TRANSFER IN ELECTRIC CONDUCTING POLYMERIC COMPOSITES

TAMAZ A. MARSAGISHVILI[1] and JIMSHER N. ANELI[2*]

[1]*Institute of the Inorganic Chemistry and Electrochemistry, Mindeli St. 11, Tbilisi 0186, Georgia*

[2]*Institute of Machine Mechanics, Mindeli St. 10, Tbilisi 0186, Georgia*

[*]*Corresponding author. E-mail: jimaneli@yahoo.com*

ABSTRACT

Analysis of the processes of charge transfer in electric conducting solid polymer systems has been conducted. Processes occurred in conducting polymer materials are divided into two categories: first—charge transfers between electrodes and second—particles and charge transfers between conductive particles. There are used Green temperature functions (GFs) in terms of polarization operators. It allows to take into account effects of frequency and spatial distribution. Analytical expressions for kinetic parameters characterizing charge transfer processes have been obtained. General formula for the dependence of the current on the average distance between conducting particles has been obtained too. The formula is applied to some real materials.

20.1 INTRODUCTION

In recent time so-called non-traditional electrical conducting materials, particularly composites, containing a dielectric basis (ceramics, polymers), and conducting dispersive fillers (carbon-graphite and metallic powders), display a growing competition to traditional conductors (e.g., metals and classical semiconductors). This situation is due to many positive characteristics of these materials, among which are the high corrosion stability, accessible technology of production, and low cost. Now there are many scientific and technical works in the sphere of the noted materials with both theoretical and experimental character (see for example, Ref. 1–6).

Growth of conductivity of the electrical conducting polymer composites (ECPC) with increase of conducting filler content is the rule without exclusions. The specific feature of this dependence is an increase of specific volumetric electrical conductivity γ or, which is the equivalence to decrease of specific volumetric electrical resistance ρ_v at definite (for a particular composite) threshold filler concentration, induced by an insulator–conductor transition. This transition conforms to the so-called threshold of proceeding or percolation. In this case γ value jump, which may reach several decimal degrees, is stipulated by formation of a continuous chain of filler particles in the polymer matrix—the infinite cluster.[7,8]

At present, the problem of the conductivity mechanism of ECPC is still to be discussed. As to the opinion of some investigators, the charge transfer is conducted by chains, consisted of filler particles having direct electric contact.[9,10] In the opinion of other authors conductivity of ECPC is caused by thermal emission of electrons through spaces between particles.[11,12] It is another opinion that currently exists in ECPC with air gaps of polymer films between filler particles. In this case electrons, which energy is lower than the potential barrier value may be tunneled through it, if their own wave-length is comparable with space width of insulating film.[13,14] Wessling[15] has used non-equilibrium thermodynamics to develop a model based on consideration of a process of formation of conductive chains. At present, the percolation theory[16] is widely used for calculations of γ for conducting composites (with both organic and inorganic binders). Accordingly to this theory, the composites consist non-interacting phases. It was shown that a type and concentration of fillers mostly define the intensity of the conducting channels in polymer matrix.[17,18] The analysis of works on the investigations of electrically conducting properties of ECPC induces one general conclusion: despite a variety of the above-considered models of electrically conducting ECPC, no one could unfortunately pretend for the versatility. Each model includes one or several approximations and suppositions, which aggravate the correctness of estimations of ECPC conducting properties. That is, why the comparison of theoretically calculated data with the experimental results usually gives deviations, which reach several orders in some cases. The

On Processes of Charge Transfer 261

coincidence is rarely reached definite concentrations of conducting filler and specific conditions of the composite production.

In this work, a new approach to charge transfer processes in the polymer materials contained the conducting fillers has been proposed.

20.2 THE SYSTEM HAMILTONIAN

The processes in the systems with electrical conducting polymer media may be divided into several types. The first type for the process of the electron transfer from the electrode to the electrical conducting admixture, second one—the electron transfer from one conducting particle to another in the polymer matrix, and third type of the process—electron transfer from conducting particle to the electrode in the polymer.

The processes of charge transport from the electrode to particle may be described analytically by choosing the concrete system for which it is possible to write the system Hamiltonian. A principal difficulty of investigation of charge transfer processes is connected with the necessity of using of quantum approach. In the frame of this approach, there is no model generally accepted for the processes of charge transfer in the irregular systems. The using of such apparatus of mathematical physics as the apparatus of Green temperature function (GF) allows a describing of the complex condensed system and processes of charge transfer in them. This technique allows the unification of theoretical approaches, using widely the different models for describing the effects of the system of frequency and space dispersion.[19-22]

20.3 CURRENT DENSITY FOR THE PROCESS OF CHARGE TRANSFER FROM METAL OR SEMICONDUCTOR ELECTRODE TO THE CONDUCTING PARTICLES IN THE POLYMER MATRIX

The calculation of the current density connected with process of charge transfer from electrode to particle may be carried out on the basis of quantum-mechanical calculations for process of transition between two electron states.

The density of cathode current may be presented as:

$$i_c = e \frac{N_0}{S} \int d\varepsilon \rho(\varepsilon) \rho(\varepsilon_1) n_F(\varepsilon) n_F(\varepsilon_1) W_c(\varepsilon, \varepsilon_1),$$

where e is the electron charge, S is the electrode surface, N_0 is the number of particles in the system volume of particles V_0, $\rho(\varepsilon)$ is the density of single-electron states in the electrode, $\rho(\varepsilon_1)$ is the density of single-electron states in the particle, $n_F(\varepsilon)$ is Fermi function of the electron distribution, $W_c(\varepsilon,\varepsilon_1)$ is the probability of the electron transfer from given energetic level of the electrode ε on the given energetic level of particle ε_1.

The electron transfer may be both electron-non-adiabatic and electron adiabatic. Usually in the wide range of the parameters, the process of charge transfer of the electron is electron-non-adiabatic transfer and below first of all, we will consider namely electron-non-adiabatic transfer of the electron.

At integration on the energy, it must be noted that usually the contribution to the integral on energies possess a small range of energies and formal integration on energies may be carried out in the infinite limits.

The probability of the electron-non-adiabatic transfer of the electron is expressed as:

$$W_c(\varepsilon,\varepsilon_1) = \frac{\beta}{i}\exp(\beta F_i)\int d\theta Sp\left[\exp((-\beta(1-\theta)H^i)L\exp(-\beta\theta H^J)L\right].$$

Omitting the cumbersome calculations, we present the quantum expression for density of cathode current of heterogeneous process with participation of metallic or semiconductor electrode with Fermi distribution:

$$i_c = e\pi\left|L_{fi}(\vec{R}^*,\psi^*)\right|^2 \int d\varepsilon d\varepsilon' kT\rho(\varepsilon)\rho(\varepsilon')\exp(-2\ln(\sin\pi\theta^*))\Phi(\vec{R}^*,\psi^*)U(\vec{R}^*,\Psi^*)$$
$$\exp\left\{-\beta\theta^*e\eta - \beta\theta^*\Delta F - \Psi^m(\vec{R}^*,\psi^*;\theta)\right\},\beta = 1/kT.$$

Here, ΔF is free energy of the process. The star in the designation marks means the value of this coordinate in the point of maximum at calculation of corresponding integration by given coordinate, as a rule, by the saddle-point method.

The saddle-point θ^* may be found from equation:

$$e\eta + \beta\Delta F + \frac{\partial\psi^m(R^*,\psi^*;\theta)}{\partial\theta} + 2\pi ctg(\pi\theta) = 0.$$

In these formulae, L_{fi} is the resonance integral from interaction of particle with surface of semiconductor or metal electrode. Matrix element is calculated by using wave functions in the frames of concrete model for particle.

The resonance integral L_{fi} may be considered as some phenomenological parameter. Arguments of this resonance integral characterize the geometrical characteristic of process, distance to surface (R) and spatial orientation of particle (Ψ) at transfer of charge. Function $U(R^*, \psi^*)$ is calculated for concrete processes allowing for the geometry of electrode and particles. Function $\Phi(\vec{R}^*, \psi^*)$ presents the function of distribution of admixture particles. This function may be model one, connected with particles concentration. Function $\Psi^m(\vec{R}^*, \psi^*; \theta)$ is one of reorganization of the polymer medium. Its formal expression takes the form:

$$\Psi^m(\vec{R}^*, \psi^*, \theta) = \frac{1}{\pi} \int d\vec{r} d\vec{r}' \Delta E_i(\vec{r}, \vec{R}^*, \psi^*) \Delta E_k(\vec{r}', \vec{R}^*, \psi^*)$$

$$\int_{-\infty}^{\infty} d\omega \operatorname{Im} g_{ik}^R(\vec{r}, \vec{r}'; \omega) \frac{sh\dfrac{\beta\omega(1-\theta)}{2} sh\dfrac{\beta\omega\theta}{2}}{\omega^2 sh\dfrac{\beta\omega}{2}},$$

where ΔE is the change of the electrical field voltage of admixture particle and electrode at process of charge transfer.

This function in general describes both the processes of tunneling of an electron and classic reorganization of medium at charge transfer.

Let us introduce the energy of reorganization of the polymer medium by means of relation:

$$E_r^m(\vec{R}, \psi) = -\frac{1}{2} \int d\vec{r} d\vec{r}' \Delta E_i(\vec{r}; \vec{R}, \psi) g_{ik}^R(\vec{r}, \vec{r}'; \omega = 0) \Delta E_k(\vec{r}'; \vec{R}, \psi).$$

In the factorization approximation for function g^R over spatial and time coordinates the medium reorganization function may be presented in form:

$$\Psi^m(\vec{R}^*, \psi^*, \theta) = E_r^m \frac{2}{\hbar} \int_{-\infty}^{\infty} d\omega f(\omega) \frac{sh\dfrac{\beta\omega(1-\theta)}{2} sh\dfrac{\beta\omega\theta}{2}}{\omega^2 sh\dfrac{\beta\omega}{2}}.$$

At integration over r and r', it is necessary to take into account the structure of medium and the situation, when both effects of spatial dispersion of medium (function $g(r, r')$) and the effects of its frequency dispersion (function $f(\omega)$) will be described by different model functions allowing for existing of definite modes of polymer polarization.

The calculations show that complete fulfilling of the analytical calculations is impossible and it is necessary to carry out of the numerical integration.

The activation energy may be defined by formula:

$$E_a = -2\ln\ (\sin \pi\theta^*) + \theta^* \left(1-\theta^*\right) E_r^m - \theta^* e\eta - \Delta F\theta^*.$$

Presented above in Section II, relations are valid also for processes of the charge transfer of the electron from particle to electrode at corresponding change of marks of energetic parameters.

The calculations of the kinetic parameters for processes of electron-adiabatic processes of electron may be considered in detail analogically with non-adiabatic processes. However, it may be used as another method of estimation of the parameters for adiabatic processes.

The probability of transfer may be presented in view:

$$W_{ad} = A_{ad} \exp\ (-\beta E_a)$$

Pre-exponent in formula multiplayer may be obtained from analogical expression for non-adiabatic process by substituting of electron resonance integral on the critical value by formula:

$$L \rightarrow \frac{L_c}{\sqrt{2}},$$

where critical value for single-frequency model of medium (with frequency ω_m) has a form:

$$L_c = (\frac{kTE_r^m \omega_m^2}{\pi^3})^{1/4}.$$

The activation energy of the process is defined by the way in which the adiabatic term is designed from potential energies of the channel terms of initial and final states U_i and U_f:

$$U = 0.5(U_i + U_f) - (0.25(U_i - U_f)^2 + L^2)^{0.5}.$$

Further on the points of minimum and the maximum of the system energy and the activation energy are defined as difference between values corresponding to maximum and minimum at initial state.

20.4 PROCESSES OF CHARGE TRANSFER BETWEEN ADMIXTURE PARTICLES IN THE POLYMER

The processes of charge transfer in the conducting polymer systems are defined by a number of factors. First of all, there are the geometrical parameters of conducting particles, their shape, mutual orientation of interacted particles, possibility of adsorption of polymer molecules on the surface of conducting particles, effects of "solvatation" of conducting particle with polymer molecules. Similar details of such systems are mainly defined by method of preparation of conducting polymer material and properties of components. In dependence on peculiarities of structure of composite material, it is necessary to use one or another model for calculation of kinetic parameters. So, if the polymer molecule is chemically adsorbed on the surface of conducting particle actually instead of the charge transfer process between metal particles in the polymer matrix, it is necessary to consider the process with participation of adsorbed particles. So, the adsorbed particle becomes the essential element of the process.

In all kinetic parameters, it will be presented intramolecular reorganization of adsorbed particles. All calculations will be essentially complicated, but execution of them in sufficiently correct form is possible.

One of the serious problems at investigation of such systems is the distribution of electromagnetic field near particle surfaces. This distribution often plays the defining role at calculation of kinetic parameters. The picture of distribution of the field essentially depends on "solvating" capability of the polymer molecules. For solvated particles, also it is necessary for the use of the model of particles with oscillation subsystem in which not at all cases, it is possible to describe what namely chemical bonds participate in the oscillation, however, it can be foreseen some generalized oscillated modes. There are possibilities of analytical calculations, although in this case the definite number of parameters arises, which are the characteristics of the system and numerical estimation of which are very approximate.

The process of charge transfer between admixture particles in the polymer is considered as a process of transfer between two metal electrodes in the condensed medium. The model for such processes may be presented on the basis of above-conducted calculations for processes of transfer between electrodes and particles. For conducting calculations, it is necessary to detail of the system Hamiltonian. It may be used the same model approximations for describing of particles 1 and 2 and medium, which were used in the last section. In result for initial state the Hamiltonian one gets the form:

$$H^i = H^i_{p1} + H^i_m + H^i_{p2} + H^{int}_{p1,m} + H^{int}_{p2,m} + H^{int}_{p1,p2}.$$

Analogically expressed the Hamiltonian of the final state.

20.5 THE RATE FACTOR OF CHARGE TRANSFER PROCESS BETWEEN CONDUCTING PARTICLES

The current density for the process of electron transfer from particle 1 to 2 in the polymer matrix has a few parameters:

$$i_c = e\pi \left|V_{fi}(\vec{R}^*,\psi^*)\right|^2 \int d\varepsilon d\varepsilon' kT \rho_1(\varepsilon)\rho_2(\varepsilon') \exp(-2\ln(\sin \pi\theta^*))\Phi(\vec{R}^*,\psi^*)A(\vec{R}^*,\Psi^*)$$
$$\exp\left\{-\beta\theta^* e\eta_{12} - \beta\theta^* \Delta F_{12} - \Psi_{12}{}^m(\vec{R}^*,\psi^*;\theta)\right\}.$$

Here, the parameters have sense analogical to formula (47) and critical point for θ^* is defined from equation:

$$e\eta_{12} + \beta\Delta F_{12} + \frac{\partial \psi^m(R^*,\psi^*;\theta)}{\partial \theta} + 2\pi ctg(\pi\theta) = 0.$$

The pre-exponent multiplayer is calculated for concrete particles allowing for the particles geometry, space orientation, and the distance of electron transfer.

For activation energy of transfer process we have:

$$E_a = -2\ln(\sin \pi\theta^*) + \theta^*\left(1-\theta^*\right)E_r^m - \theta^* e\eta_{12} - \Delta F_{12}\theta^*.$$

The presented expressions allow one to conduct simple estimations of process kinetic parameters in frames of strongly simplified models. In this way, one could calculate the electron kinetic parameters using of average magnitudes of admixture concentrations and, consequently, to get average magnitudes of parameters.

At estimation of kinetic parameters, it must be foreseen that direct charge transfer takes place really in the limits of 30 Å, and that because of electron tunneling. The charge transfer through an intermediate state, naturally, is possible, but the probability of the final process will be equal to the product of probabilities of all elementary acts of electron transfer.

On Processes of Charge Transfer 267

20.6 SCHEME FOR CARRYING OUT OF QUANTITATIVE ESTIMATION FOR THE PROCESSES

For estimation of kinetic parameter values of different processes, it is presented below the schemes of calculations:

1. For calculation of the kinetic parameters, first of all, it is necessary to make more exact namely, what particles react and what is the reaction medium, and what electrodes are used at real measurements. It is necessary also to have maximal information about electrode, its zone structure, a bending of zones near the surface, etc.
2. To introduce the degrees of freedom of the reaction and divide them into classic and quantum for medium. Classical degrees of freedom may be ones, which satisfy the condition:

$$th\left(\frac{\omega_i(1-\theta^*)}{2kT}\right) \approx \frac{\omega_i(1-\theta^*)}{2kT} \ll 1, th\frac{\omega_f\theta^*}{2kT} \approx \frac{\omega_f\theta^*}{2kT} \ll 1.$$

At fulfilling of preliminary calculations, the number 0.5 may be taken as value of θ^*.

3. To fulfill the estimation of the medium reorganization energy by simple formulas, for example.[15] However, it is necessary to take into account here, that in the dependence of medium only definite modes of polarization are reorganized and, consequently, instead of static-dielectric penetration, it must be used the values of the dielectric permittivity from left and right sides of corresponding pick of absorption. The allowing for the effects of frequency and spatial dispersion may be conducted lately at exact calculations. After preliminary estimation for final calculations, it must be selected a model for describing of spatial dispersion of medium.
4. To define or estimate the electron resonance integral or dipole moment of transition using quantum-chemical methods, sums law, and fulfilled calculations for similar systems. It must be foreseen that quantum-chemical calculations allow one to conduct calculations either in vacuum or in medium with static-dielectric background, but not in medium with complex structure. Therefore, at final calculations of the kinetic parameters of processes, it is often necessary to introduce the corresponding corrections in the results, for example, in the characteristic value of decay of the resonance integral.

5. To define the character of charge transfer process if it is electron-non-adiabatic or adiabatic. For electron-non-adiabatic process, it must be used above given methods and for adiabatic ones—the transfer rule, in which is the method of calculation of pre-exponent, rate factor, and definition of activation energy of the process after calculation of transition configuration and the finding of the coordinates of minimum of initial system state and maximum in the transition state. The calculation of the Landau–Zener parameter may be carried out by using of formula:

$$\gamma_e = 2\pi L^2 / (|v||\Delta U|),$$

where v is the rate of movement over the reaction coordinate of the system near transition configuration, ΔU is the difference of inclinations in terms of potential energy near point of their crossing. If the condition $\gamma_e \gg 1$ is fulfilled then the process has an electron-adiabatic character. At calculation of the kinetic parameters, it must be used the adiabatic terms.[22] The transmission factor in this case is $\kappa \approx 1$. At fulfilling of reciprocal condition, the process has electron-non-adiabatic character.

6. To define of transition configuration. The definition of transition configuration over spatial and rotational coordinates of reaction is more complex task, there are no general methods. It is necessary to have information about distribution of the reacting particles in the medium for using of right function for their distribution. The information is necessary about possible mutual orientation of reacted particle. At writing of the potentials of reagent interactions all, this information is necessary for introduction of reaction coordinates and calculation of transition configuration by them.

7. To calculate the transition coefficient for heterogeneous reaction θ^*.

8. To estimate of the activation energy for classic degree of freedoms and tunneling factor—for quantum ones.

9. After conducting of preliminary calculations to make more exact the models of processes for each stage of electron transfer process, to coordinate the models of separate stages, and fulfill of the exact calculations on the full scheme for all systems.

As it was mentioned above, experimental data treatment requires the exacting of parameters entered to the results of analytical calculations obtained in the frames of theoretical models. However, some conclusions may be made on the basis of semi-classical estimations of the parameters. We will provide estimations for metallic electrode and electrical conducting particles.

One of the defining factors at calculation of the process rate constant is the distance of transition. In the same time, this parameter defines often the character of process if it is adiabatic or non-adiabatic. It may be shown qualitatively that if the character of the charge transfer process is adiabatic the rate constant weakly depends on transfer distance because the main dependence in the process rate constant on the distance is presented through electron resonance integral. In the same time, for adiabatic processes, it is necessary to change the electron resonance integral on its critical value, which is not presented by the function of the transfer distance. The calculations in different systems show that in the condensed medium, the electron transfer process has adiabatic character only on rather small distances. Consequently, on small distances, the rate constant of the process will not be dependent on distance. Moreover, the value of rate constant will mainly be dependent on the medium properties and for very rough model of homogeneous, isotropic local medium in the frames of single-mode model, it will be presented by frequency of this mode ω_m and by energy of the reorganization of medium E_r^m, and the square of the electron resonance integral must be changed on the following value:

$$\frac{(kTE_r^m)^{1/2}}{2\pi^{3/2}}\omega_m.$$

In the last expression, the energy of medium reorganization depends on the electron transfer distance, but this dependence has relatively week character. At variation of distance of transfer on 60–70%, the reorganization energy is changed on about 25–40%, in dependence on the model.

The electron resonance integral for adiabatic processes of the electron transfer is exponentially decreasing function of the distance.

$$L(R) = L_0 \exp(-R/\Delta).$$

For many processes of non-adiabatic electron transfer the characteristic distance of the decay of electron resonance electron Δ (it is proposed that the distance R is measured in angstroms).

Qualitatively, the current density in the polymer system with electrical conducting filler is the function of distance between conducting particles and graphically corresponding dependence has the graphical view presented in Figure 20.1 ($\Delta = 0.4$ Å).

For comparison of the theoretically calculated dependences of electrical current density on the average distance between conductive particles in the polymer matrix with experimentally obtained results for three polymer composites, containing the nanopowders of nickel and silver, all curves are presented on the one diagram (Fig. 20.1), showing that characters of these curves are qualitatively the same. The deviation between numerical values of the theoretical and experimental data is defined by inhomogeneous distribution of conducting particles in the polymer matrix.

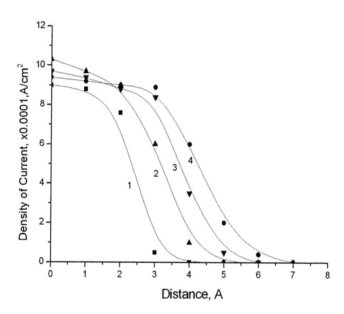

FIGURE 20.1 Dependence of density of current on the average distance between conducting particles in ECPC. 1—theoretically calculated; 2—ECPC based on polydimethylvinylsiloxane with silver nanoparticles (50–100 nm); 3—ECPC based on polydimethylvinylsiloxane with silver nanoparticles (200–250 nm); 4—ECPC based on polyvinyl alcohol with nickel nanoparticles.

20.7 CONCLUSION

In result of the analysis of charge transfer processes in the electrical conducting solid polymer systems using quantum mechanical approach in modeling of conductivity of the conducting of polymer composites, it is established that generally the processes of the charge transfer in these systems may be divided into two types: (1) the processes of charge transfer between electrodes and particles and (2) the processes of the charge transfer between conductive particles.

Application of GFs of polarization operators for the medium molecules permits to obtain the expressions for kinetic parameters of the charge transfer processes from electrodes to particles and between particles in condensed matter. The comparison of the general theoretical dependence of the current in the ECPC on the average distance between conducting particles with analogical dependence for some real systems shows that the theoretical approach carried out in this work is very useful for describing of the charge transfer processes in the conducting polymer systems.

KEYWORDS

- **charge transfer**
- **vector-potential**
- **Green temperature functions**
- **Hamiltonian**
- **polymer composite**
- **electrical conductivity**

REFERENCES

1. Norman, R. N. *Conductive Rubber and Plastics;* Elsevier: Amsterdam, Netherlands, 1970.
2. Donnet, A.; Voet, A. *Carbon Black;* Marcel Dekker: New York, NY, 1976.
3. Zhu, D.; Bin, Y.; Matsuo M. *J. Polym. Sci. Part B Polym. Phys.* **2007,** *45,* 1037–1045.
4. Aneli, J. N.; Zaikov, G. E.; Mukbaniani, O. V. *Chem. Chem. Technol.* **2011,** *5,* 75–88.
5. Aneli, J. N.; Khananashvili, L. M.; Zaikov, G. E. *Structuring and Conductivity of Polymer Composites;* Novo-Science. Publication: New York, NY, 1998; p 326.
6. Sheng, P. *Phys. Rev. B* **1980,** *21,* 2180–2195.
7. Beaucage, G.; Rane, S.; Shaffer, D. W.; Fisher, G. D. *J. Polym. Sci. Part B Polym. Phys.* **1999,** *37,* 1105–1112.
8. Bin, J.; Kitanake, M.; Zhu, D.; Matsuo. M. *Macromolecules* **2003,** *36,* 6213–6219.
9. Bin, J.; Xu, C.; Matsuo, M. *Carbon* **2002,** *40,* 195–199.
10. Benguigu, L.; Jakubovich, J.; Narkis, M. *J. Polym. Sci. Part B Polym. Phys.* **1987,** *25,* 27.
11. Balberg, I. *Carbon* **2002,** *40,* 139–143.
12. Heaney, M. B. *Physica A* **1997,** *241,* 296–300.
13. Lee B. *Polymer Eng. Sci.* **1992,** *32* (1), 36–37.

14. Bridge. W. B.; Folkes, M. J.; Wood, B. R. *J. Phys. D* **1990**, *23* (7), 890–898.
15. Wessling, B. *Macromol. Chem.* **1984**, *185,* 1265–1275.
16. Zainutdinov, A. K.; Kasimov, A. A.; Magrupov. M. A. Pisma, V. J. T. F. *Lett. J. Tech. Phys.* **1992**, *18* (2), 29–34.
17. Strumplen, R.; Glatz-Reichenbach, J. J. *Electroceramics* **1999**, *3* (4), 329–346.
18. Thongruang, W.; Balik, C. H.; Spontak, R. J. *Polymer* **2002**, *43,* 2279–2286.
19. Abrikosov, A. A.; Gorkov, L. P.; Dzialoshinskii, I. E. *Methods of Quantum Field Theory in Statistical Mechanics;* Dover: New York, NY, 1975.
20. March, N. H.; Young, W. H.; Sampanthar, S. *The Many Body Problem in Guantum Mechanics;* Dover: New York, NY, 1995.
21. Dogonadze, R. R.; Marsagishvili, T. A. *The Chemical Physics of Solvation. Part A;* Elsevier: Amsterdam, Netherlands, 1985; p 39.
22. Dogonadze, R. R.; Kuznetsov, A. M.; Marsagishvili, T. A. Present State of the Theory of Charge Transfer Processes in Condensed Phase. *Electrochim. Acta* **1980**, *25* (1), 1–28.

CHAPTER 21

ROUTES TO CHEMICAL MODIFICATION OF THE SURFACE OF NANOSIZED FLUORIDES

A. V. SAFRONIKHIN[*], H. V. EHRLICH, and G. V. LISICHKIN

Lomonosov Moscow State University, Moscow, Russia

[*]Corresponding author. E-mail: safronikhin@yandex.ru

ABSTRACT

In the investigation EuF_3 nanoparticles modification method is proposed. Ligand-assisted synthesis of RE fluorides makes it possible to obtain stable colloids of surface-modified nanoparticles. This method gives products with higher values of coordinated ligand surface density and can be also used to control the nanoparticle size and shape. Formation of a chemical bond between the ligands and metal ions is shown by IR and luminescence spectroscopies. It is shown that surface modification is a way to enhance the luminescence. EuF_3 nanoparticles modified with Dbm and Phen have intense luminescence, with red emission at ~612 nm prevailed. It is shown that energy transfer from the coordinated Phen and Dbm to Eu^{3+} ions in the surface complexes makes a main contribution to the luminescence intensity increase.

21.1 INTRODUCTION

A key objective of state-of-the-art chemistry is the creation of new functional materials. This can be performed, in particular, through the change of surface layer composition and structure, that is, by chemical modification of the surface. Chemical modification involves bonding of a modifier (usually an organic compound) to a surface (Fig. 21.1).

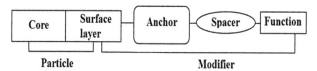

FIGURE 21.1 Generalized structure of surface-modified particle.

Surface modification plays a key role in production of such functional materials, such as sorbents, sensors, immobilized metal complex catalysts, etc. It has been successfully used for years to protect the surfaces from the effect of the external environment, to regulate wetting and surface lyophility, to control particle size and form (at modification during nanoparticle synthesis, in situ), and to prevent nanoparticles aggregation. On the other hand, modifier molecules can induce new properties of surface modified materials.

Today, the methods of chemical modification of the surfaces of metals, nonmetals, metal oxides, nonmetal oxides (primarily SiO_2), zeolites, and organic polymers have been worked out in detail. An exception is one large class of chemical compounds—metal salts, which until recently have not been considered as objects for surface modification. Interest in them has emerged in recent years—in particular, in the context of the progress of nanotechnology. Methods of synthesis of nanoscale particles of ionic compounds and their composites have been developed, and the properties of such systems and possibilities of their practical application have been investigated.

Among different metal salts, nanosized rare earth (RE) fluorides and composites on their basis have attracted increasing attention in recent years. This is due to their luminescent (Eu^{3+}, Tb^{3+}, Sm^{3+}, and Dy^{3+}), magnetic (Gd^{3+}) and other properties. Materials containing RE ions are widely used as components of lighting devices (including light emitting diodes), optical fibers, and lasers. Nanosized RE containing materials find also applications in medicine and bioassay as luminescent labels for biological imaging, as contrast agents for magnetic resonance imaging (MRI), and as agents for radiotherapy, photodynamic therapy, and neutron capture therapy of cancer. A new challenge is the creation of multifunctional materials on the basis of the RE compound nanoparticles which can be applied, for example, as multiplexed imaging agents.

In this work, we present approaches to surface modification of RE fluoride particles and compare them. Two principally different approaches can be used to produce surface modified ionic crystals. First one is modification

Routes to Chemical Modification   275

during the synthesis of ionic crystals (ligand assisted synthesis, in situ surface modification) when solutions of the reagents are simultaneously added drop by drop into a modifier solution (so-called double-jet precipitation technique).[1] Second one is modification of preliminarily synthesized ionic crystals (post-synthesis surface modification).[2]

21.2 EXPERIMENTAL RESEARCH

RE fluoride nanoparticles were synthesized by the technique described in our previous work.[3] The post-synthesis modification of EuF_3 nanoparticles was made by different methods including impregnation, sorption from aqueous solutions, and mechanochemical treatment. Acetylacetone (Acac), dibenzoylmethane (Dbm), o-phenanthroline (Phen), and citric acid (Citr) were used as modifiers.

In situ surface modification was carried out in the following way. RE salt and NaF solutions were simultaneously added dropwise into a modifier solution. For in situ modification, citric and amino acids (glycine (Gly), aspartic acid (Asp), and tryptophan (Trp)) were used as modifiers.

X-ray diffraction (XRD) measurements were carried out using a DRON-3M powder X-ray diffractometer. Morphology and size of the synthesized particles were investigated by transmission electron microscopy (TEM) on a LEO912 AB OMEGA instrument. The specific surface of the products was measured on an ASAP 2010 analyzer (Micromeritics) by the Brunauer–Emmett–Teller (BET) method using low-temperature nitrogen adsorption. Chemical composition of the samples was studied by IR and luminescence spectroscopes and elemental analysis. IR spectra were collected on an IR200 Thermo Nicolet instrument using KBr pellet technique. The room-temperature luminescence spectra of the powders were obtained on LS 55 luminescence spectrometer (PerkinElmer). Elemental analysis was performed on a CHN-2400 analyzer (PerkinElmer).

21.3 RESULTS AND DISCUSSION

The XRD data obtained indicate formation of respective RE fluorides as well in the case of synthesis without a modifier as in the case of ligand-assisted synthesis. The positions of the peaks are in good agreement with literature values. No peaks of any other phases or impurities were detected that indicated every sample is single phase and of high purity.

IR spectroscopy proved that the ligands form complexes with RE ions on the particle surface. IR spectra of EuF_3 modified with β-diketones (Acac@EuF_3 and Dbm@EuF_3) have characteristic bands of the enol form of the β-diketones. In IR spectra of EuF_3 modified with Phen, the spectral shift from 1600 to 1591 cm^{-1} is observed that can be caused by coordination of Eu^{3+} ions with Phen through N atoms. Reasoning from these facts it can be assumed that the surface complexes have the mononuclear chelate structures which are probably close to the structures of the respective molecular complexes (Fig. 21.2).

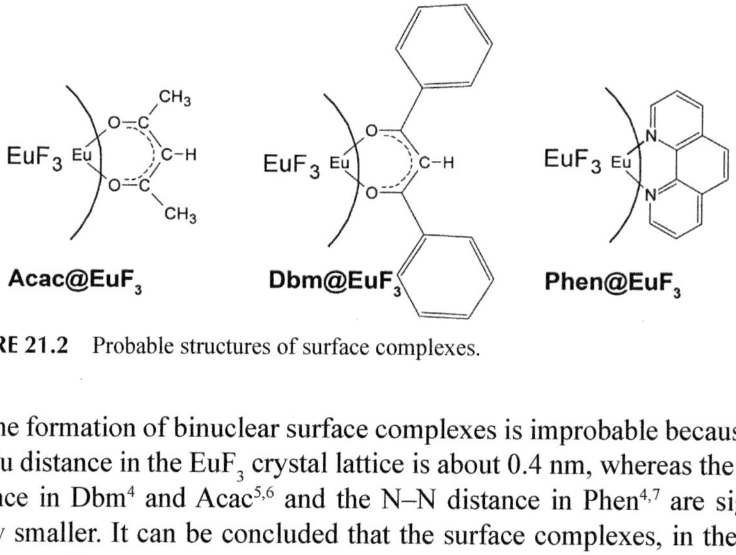

FIGURE 21.2 Probable structures of surface complexes.

The formation of binuclear surface complexes is improbable because the Eu–Eu distance in the EuF_3 crystal lattice is about 0.4 nm, whereas the O–O distance in Dbm[4] and Acac[5,6] and the N–N distance in Phen[4,7] are significantly smaller. It can be concluded that the surface complexes, in the case of Citr-modified powders, IR spectra have characteristic bands of COO^- and deformation bands of the bond between Eu^{3+} and O of COO^-. The difference between the frequencies of asymmetric and symmetric stretching vibrations of the coordinated COO^-,

$$\Delta = v_{as}(COO^-) - v_s(COO^-),$$

is equal to ~160 cm^{-1}. According to G. B. Deacon and R. J. Phillips' work,[8] we can conclude that COO^- groups are coordinated with Eu^{3+} on the surface with the formation, mainly, of bridging or bidentate structures.

Formation of the surface complexes follows also from the luminescence spectroscopy data (Fig. 21.3). Photoluminescence excitation (PLE) spectra have the characteristic bands attributed to the electronic transitions in Eu^{3+} ions. Appearance of additional bands in the region of 220–390 nm in the

PLE spectra of the modified samples corresponds to absorption by the organic ligands.

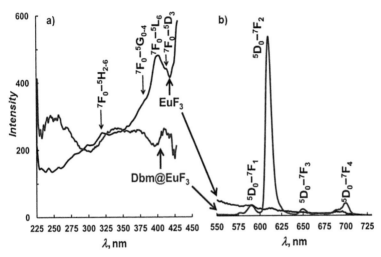

FIGURE 21.3 PLE spectra (a) and PL spectra measured at λ_{exc} = 377 nm (b) of unmodified EuF$_3$ and EuF$_3$ modified with Dbm by impregnation.

In photoluminescence (PL) spectra, surface modification results in increase of luminescence intensity (up to 20 times) and change in ratios of different bands intensities. The largest increase in intensity is observed for the $^5D_0 \rightarrow ^7F_2$ transition at 610–612 nm. This transition, being forbidden as electric dipole one, can be allowed as an induced electric dipole transition. This is hypersensitive to a local symmetry of the Eu^{3+} environment and Eu^{3+} ligand bond ionicity. For analysis of PL spectra of Eu^{3+} containing materials, ratio of the intensities,

$$R = I(^5D_0 \rightarrow ^7F_2)/I(^5D_0 \rightarrow ^7F_1)$$

is frequently used. Low R values (main emission at ~590 nm) correspond to high local symmetry of Eu^{3+} environment. R increase (increase of the emission at ~612 nm) can be caused by a ligand environment asymmetry and decrease of Eu^{3+} lig and bond ionicity. R values for EuF$_3$ and the surface modified samples are given in Table 21.1.

As follows from these data, when Eu^{3+} ions are incorporated into a host lattice, $R < 1$ (usually 0.1–0.3).[9] In the case of EuF$_3$ nanoparticles, a large portion of Eu^{3+} ions is located at the surface where local symmetry reduces due to surface defects and interaction with components of media. It results in the larger R values ($R \approx 1$). The samples modified with Dbm and Phen show $R >> 1$. It is evident that the reasons for the significant increase of R values after surface modification are changes in the surface structure:

1. reduction of luminescence quenching by OH^- and H_2O molecules due to ligand exchange on the surface;
2. energy transfer from the ligands to Eu^{3+} ions in the surface complexes, as shown in Figure 21.4.

TABLE 21.1 R Parameter for Different Samples Containing Eu^{3+}.

Sample	Excitation wavelength λ_{exc}, nm	R
Phen@EuF$_3$ (sorption)	396	7.6
	377	8.4
Dbm@EuF$_3$ (sorption)	396	2.1
Dbm@EuF$_3$ (impregnation)	409	13
	377	10.8
Acac@EuF$_3$ (mechanochemical treatment)	377, 399	0.7
EuF$_3$ doughnut-like aggregates	377, 396, 399, 409	0.6–0.8
La$_2$O$_3$:Eu^{3+} submicron particles	399	0.26 [9]

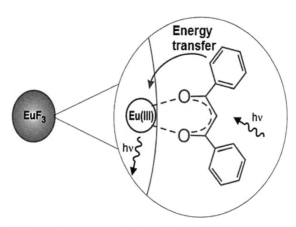

FIGURE 21.4 Scheme of energy transfer from Dbm to Eu^{3+} ion in the surface complex.

It seems that the energy transfer from the coordinated Phen and Dbm makes a main contribution to the luminescence enhancement effect. Acac, being not a sensitizer, does not enhance R values in modified samples.

Routes to Chemical Modification

In the case of in situ modification, the formation of nanosized RE fluoride particles is shown by XRD and TEM. IR spectra of the modified powders proved the formation of the surface complexes. It follows from the spectra that COO⁻ groups of Citr coordinate RE ions on the surface through the bidentate structure. From comparison of the citric acid distribution diagram and behavior of the Citr-modified nanoparticle colloids at different pH values, it is found that only one COO⁻ group of citric acid molecule is coordinated to the surface. Two other COO⁻ groups of the molecule are free and negatively charged that inhibits aggregation of nanoparticles and causes stabilization of these aqueous colloids (Fig. 21.5). Stability of the colloids, however, depends on pH value of the solutions. At pH < 3, the nanoparticles precipitate. Reverse transition of the precipitate into the solution takes place at pH > 4.5. The precipitation-peptization processes given in Figure 21.5 are reversible that can be used for purification and separation of nanoparticles synthesized.

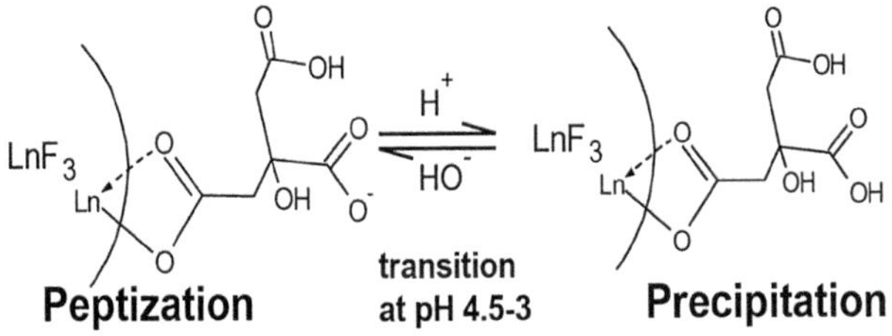

FIGURE 21.5 States of citric acid coordinated on the nanoparticle surface.

TEM revealed that polycarboxylic acids affect the growth of the LaF₃ particles (Fig. 21.6).

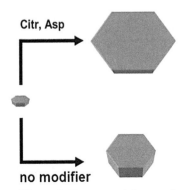

FIGURE 21.6 Effect of modifier on LaF₃ nanoparticle growth.

Citr and Asp lead to the formation of "flattened" nanoparticles—twice less in thickness and twice bigger in basal planes as compared with unmodified LaF_3 particles. This can be caused by selective sorption of these modifiers on basal (001) planes and inhibition of particle growth in [001] direction. Monocarboxylic acids (Gly and Trp) do not show the same effect.

From the results of elemental analysis and surface measurements, the surface density ρ, molecule nm^{-2} of modifiers was calculated (Table 21.2). As follows from Table 21.2, in situ modification gives higher ρ values in comparison with post-synthesis modification. Asp and Citr form dense monolayers on the particle surface. This is probably a reason for high stability of corresponding aqueous sols. The concentrations of RE fluoride nanoparticles in such sols reach several grams per liter. Among the methods of post-synthesis modification, the most effective one is sorption. This method gives relatively high surface density values.

TABLE 21.2 Surface Density (ρ) of Ligands in Modified RE Fluorides.

Modification method		Sample	ρ, molecule nm^{-2}
Post-synthesis modification	Impregnation	Dbm@EuF$_3$	0.3
	Mechanochemical treatment	Dbm@EuF$_3$	0.5
		Phen@EuF$_3$	0.5
	Sorption	Dbm@EuF$_3$	1.0
		Phen@EuF$_3$	1.5
		Citr@EuF$_3$	1.5
In situ modification	Double-jet precipitation	Citr@EuF$_3$	3.4
		Citr@LaF$_3$	3.9
		Asp@LaF$_3$	3.9
		Gly@LaF$_3$	1.1
		Trp@LaF$_3$	0.7

21.4 CONCLUSION

In conclusion, we have demonstrated that surface modification of RE fluoride nanoparticles can be carried out by post-synthesis methods (impregnation, sorption from solution, and mechanochemical treatment) as well as in situ methods (double-jet precipitation). The modification process goes through complexation of ligands with RE ions on the particle surface. An amount of the ligand coordinated on the surface depends on modification method.

Ligand assisted synthesis of RE fluorides makes it possible to obtain stable colloids of surface modified nanoparticles. This method gives products with higher values of coordinated ligand surface density and can be also used to control the nanoparticle size and shape. Formation of a chemical bond between the ligands and metal ions is shown by IR and luminescence spectroscopies. We demonstrated that surface modification is a way to enhance the luminescence. EuF_3 nanoparticles modified with Dbm and Phen have intense luminescence, with red emission at ~612 nm prevailed. It is shown that energy transfer from the coordinated Phen and Dbm to Eu^{3+} ions in the surface complexes makes a main contribution to the luminescence intensity increase. Such luminescent materials can have many applications as phosphors, components of optical devices, and luminescent labels.

KEYWORDS

- **surface modification**
- **lanthanides**
- **nanoparticles**
- **luminescence**
- **complexes**
- **IR spectroscopy**

REFERENCES

1. Safronikhin, A.; Ehrlich, H.; Lisichkin, G. Double-jet Precipitation Synthesis of CaF_2 Nanoparticles: The Effect of Temperature, Solvent, and Stabilizer on Size and Morphology. *J. Alloys Compd.* **2017,** *694,* 1182–1188.
2. Safronikhin, A.; Ehrlich, H.; Lisichkin, G. Chemical Modification of the Surface of Highly Dispersed Metal Salt Crystals. *Prot. Met. Phys. Chem. Surf.* 2014, *50* (5), 578–586.
3. Safronikhin, A.; Ehrlich, H., et al. Formation of Complexes on the Surface of Nanosized Europium Fluoride, Colloid. *Surf. A Physicochem. Eng. Aspect.* 2011, *377,* 367–373.
4. Ahmed, M. O.; Liao, J. L., et al. Anhydrous Tris(Dibenzoylmethanido)(o-phenanthroline) Europium(III), [Eu(DBM)$_3$(Phen)]. *Acta Cryst.* 2003, *59,* m29–m32.
5. Lingafelter, E. C.; Braun, R. L. Interatomic Distances and Angles in Metal Chelates of Acetylacetone and Salicylaldimine. *J. Am. Chem. Soc.* 1966, *88,* 2951–2956.
6. Cunningham, J. A.; Sands, D. E., et al. The Crystal and Molecular Structure of Ytterbium Acetylacetonate Monohydrate. *Inorg. Chem.* 1969, *8,* 22–28.

7. Tsaryuk, V.; Turowska-Tyrk, I., et al. Spectra and Details of the Structure of Europium Aliphatic Carboxylates with 1,10-Phenanthroline Derivatives. *J. Alloys Compd.* 2002, *341*, 323–332.
8. Deacon, G. B.; Phillips, R. J. Relationship Between the Carbon-oxygen Stretching Frequencies of Carboxylate Complexes and the Type of Carboxylate Coordination. *Coord. Chem. Rev.* 1980, *33*, 227–250.
9. Yu, L.; Song, H., et al. Fabrication and Photoluminescent Characteristics of $La_2O_3:Eu^{3+}$ Nanowires. *Phys. Chem. Chem. Phys.* 2006, *8* (2), 303–308.

CHAPTER 22

PROSPECTS FOR THE NANOSILICA POWDER AQUASPERSIONS IN FEED FOR FISH

A. A. LAPIN[1*], M. L. KALAYDA[1], V. V. POTAPOV[2], and V. N. ZELENKOV[3]

[1]*Department of Water Resource, Kazan State Power Engineering University, Krasnoselskaya St. 51, Kazan, The Republic of Tatarstan, Russia*

[2]*Scientific-Research Geotechnological Center of Russian Academy of Sciences, North East Highway, 30, Petropavlovsk-Kamchatsky, Russia*

[3]*FGBNU Institute of Vegetable Crops, Vereya, Ramensky District, Moscow, Russia*

Corresponding author. E-mail: lapinanatol@mail.ru

ABSTRACT

The results of testing of the influence exercised by aqueous dispersions of samples of experimental and industrial purpose use nanosilica powders (separated from hydrothermal heat-transfer agent solutions at the geothermal power plants) upon the death rate of crustaceans *Daphnia magna Straus* are discussed. In accordance with the attested procedure, the lack of their toxic effect in aqueous medium in concentrations of 4×10^{-2} g/dm^3 and less is established. Hydrothermal origin nano-dispersive silica (NDS) in plant growing for non-root treatments of plants at the open soil and in production of new types of feed for fish in interior and exterior reservoirs is proposed.

22.1 INTRODUCTION

Nanotechnology (NT) is one of the most important and fast developing spheres of modern-day scientific and practical activities. The modern NT knowledge rather significantly changes our understanding of interaction between materials and systems on the one hand, and man and environment on the other. Nanomaterials (NM) and systems may be of great benefit to the human society in such spheres as improvement of medical diagnostics, methods of medical treatment, better monitoring of water and air pollution, solar energy, improvement of water supply and purification systems, increasing reliability, and efficiency of technical systems—so the spheres of NM application in modern science and engineering are rather diverse.[1-3]

Nanostructures may exercise considerable influence upon the environment due to the following main causes:

- bioaccumulation, especially in case nanostructures absorb the contaminants, such as pesticides, cadmium, and organic substances, and transfer them up along the food chain;
- production of non-biodegradable contaminants (including nanoparticles (NPs) themselves) which are very hard to detect.

Utilization of NT may significantly change the operation of human and ecological biosystems. In particular, the usage of active nanoproducts may bring such consequences as:

- changes at the genetic level;
- creation of nanodevices that influence the functioning of brain and other human organs and parts of body; and
- changes in the environment, influence upon the safety, and quality of human life.

As there are not enough scientific research data on the nanoproducts properties and their influence upon biological, natural, and technical objects, it brings on high level of uncertainty for the decisions taken at different stages of analysis and estimation of risks.[4] There are plenty of scientific papers both in Russia and abroad on the toxicity of NM, but significant part of these are on research concerning the higher animals.[5,6]

For the estimation of NM toxicity, they also propose various test-systems: protozoa organisms, microorganisms, some cell and subcell elements, hydrobionts, plants, and insects. In ecological toxicology, the research is conducted upon the hydrobionts, with estimation of death rate of crustaceans daphnia, infusoria *Paramecium caudatum*, such protozoa as *Stylonychia mytilus* and *Tetrachymena pyriformis*.[7]

Quite simple and handy test for preliminary estimation of NM toxicity was suggested by the researchers at Biological Faculty of Moscow State University named after M. V. Lomonosov as the Biotester they used colon bacilli genetically modified with genes from luminescent sea bacteria. Luminescence of cells helped to estimate the level of cells' metabolism, which is selectively influenced by carbon nanotubes. The latter hinder the vital capacity of the above bacilli; destroy their cell walls and underlying membrane. The material itself, forming the nanotubes, does not influence the cells this way.[8]

NPs may exercise destructive effect upon the cells of animals and humans in vitro and are also toxic when inhaled, but being introduced into alimentary tract these are low-toxic.[9] There are not enough data on genes-toxicity, embryo-toxicity, and genes-mutation effect of NP over the living organisms, and also on the effect of NP upon the hormonal and immune status of these. NP properties very often differ from their micro- and macro-analogues, so NM constitutes in essence a new factor that influences a living organism. That is why the estimation of NM influence upon environment and living organisms is such a pressing problem of today.[10]

One of the most common NM, which happens to being introduced into a human organism on the ever-growing scale is nanostructural amorphous silicon dioxide (silica, $(SiO_2)_n$). This NM, known also under the "Aerosil" brand name, is widely used as sorbent and filling agent, also as food additive, as well as a component to the cosmetic products and medicinal preparations.[11]

Nonporous amorphous silicon dioxide is used as a food additive, registered under number E551 that prevents the main product from lumping and caking. Food quality silicon dioxide is used in pharmaceutical production as enterosorbent medicine and also in production of toothpaste. The above-named substance is come across in the composition of chips, dried-bread snacks, maize sticks, instant coffee, etc. It is officially confirmed that the substance, silicon dioxide passes through the alimentary tract and completely evacuated out of an organism, the structure of the substance not changed. According to the 15-year-long research by the French specialists, regular use of drinking water with high content of silicon dioxide lowers the risk of Alzheimer's disease by 10%. Thus, it seems that information on the harmful effects of silicon dioxide, which is an inert substance in a chemical sense, is false: food additive E551, when taken orally, is absolutely safe for health.[12] But at the same time, according to the results of some research, one may assume that under certain conditions NP of amorphous SiO_2 may exercise some toxic effect on an organism. These NP have characteristics of exciting catalytic generation of reactive forms of oxygen, which was revealed in non-cell system and in culture of human keratinocytes and alveolar epithelio-cytes.[9] Biological and medical aspects of silica usage, especially as a carrier for medicinal and other biologically active substances, as enterosorbents and adsorbents with a wide spectrum of useful qualities, are connected to the usage of pure and rather homogenous synthetic amorphous silica. Adsorption of biomolecules at the hard surface of enterosorbent may influence their bioactivity. The ions, present in water solution, influence the thickness of double electrical layer, which alters the adsorption characteristics due to shielding of surface charge at the enterosorbent. This process may be influenced by non-uniform ion distribution in the medium and at the silica surface under the different levels of pH and ion force of solutions. It should be noted, that living cells, being self-regulatory systems, can modify charge distribution at the membrane surface and thus influence the long-range interaction with the particles' surface.[13]

Interaction of microorganisms with hard materials exercises significant influence over the physiological activity of microbe populations by varying the rate and directivity of biochemical processes, rate of micro-organisms breeding, rate of biomass growth, and size and morphology of cells. At present, great number of papers is published concerning the effect of adhesion to the hard of microorganisms over their vital functions. In some cases, the reduction of metabolic activity of microorganisms under the interaction with materials of high degree of dispersion is highly important for biotechnology. This first of all concerns the production of bacterial preparations and preservation of animals' cells. Through blockading the cells' surface and creating certain micro-environment for bacteria or cells, the materials of high degree of dispersion assist the survival of these biological objects during their prolonged storage. And activity of bacteria associated with hard particles in water medium may more than 40 times exceed the activity of free-living microorganisms.[13]

Along with the natural adsorbents, also of great scientific and practical interest are synthetic materials of high degree of dispersion based on silicon dioxide. High-dispersive silicon dioxide (HDS) possesses several important properties, such as it is chemically pure, its chemical composition is homogenous, it possesses biological and thermal stability, it is harmless physiologically, and its particles have small dimensions (5–50 nm) and great specific surface area. There is not much data on the interaction of bacteria with HDS and its modified forms, and that mostly concerns yeasts, nitrogen-fixing, and methane-trophic bacteria. Adsorption properties of HDS in relation to normal alimentary tract micro-flora and a number of pathogenic organism's strains were also studied. HDS and its modified forms stimulate bacteria of *Azotobacter* genus on the synthesis of group B vitamins. Under the influence of HDS the production of thiamine (vitamin B1) by *Azotobacter vinelandii* 56 grew up 43%, and in the presence of HDS, modified with aluminum oxide, in the Ashby medium (with saccharose) the production of the above vitamin by *Axhroococcum* bacteria was higher by 116%. Synthesis of pyridoxine (vitamin B_6) by *Axhroococcum* 20 in similar conditions was higher by 29%. At the same time, nitrogen-fixing properties of these bacteria grew significantly.[13]

Toxins possess higher reactive properties than normally functioning proteins, so when NP of silica gel (possessing anti-oxidant properties) are introduced into the alimentary tract, they collect the active radicals-toxins and selectively bind pathogenic microorganisms, hindering their development. The above properties show that silica turns out to be a harmless preservative for concentrated animal feed. High-binding properties are due to agglutination of microorganisms by sorbent particles. The latter are considerably smaller in dimensions (3–30 nm) than the former, and it is exactly the sorbent particles that are adsorbed at the microbe cells and like a glue bind these into a conglomerate; it is estimated that specific surface area of the NDS pores is 417–418 m^2/g, micropores area not more than 28–29 m^2/g, volume of pores about 1.0 cm^3/g, volume of micropores 0.008 cm^3/g, medium pore diameter on the order of 8.55 nm.[14]

NM may reveal toxic properties when acting as catalysts activating redox reactions, inducing oxygen and water molecules accompanied by formation of peroxide radicals. Introduction of high-dispersive materials (HDM) to the water medium of culture liquids is followed by the higher oxygen mass-transfer therein, while during the stirring of liquid medium, there form highly turbulent micro-zones causing the above mass-transfer effect. As the oxygen supply rate significantly influences aerobe microorganisms vital functions, quite obvious is the probability of HDM stimulating the microorganisms physiology. Even more pronounced influence over oxygen mass-transfer is noticed in case of HDS, modified with aluminum oxide. When its concentration in liquid is 0.1%, the rate of oxygen mass-transfer becomes 117.9% compared to the controls, while raising the concentration up to 0.2%, the rate of oxygen mass-transfer grows by more than 120%.[15] In Russia, for the last 15 years, a new trend for production and usage of nano-dispersive silica (NDS) of hydrothermal origin was being developed. As raw materials, for production of NDS they use both water from deep wells of hydrothermal power plants (Pauzhet HPP and Mutnov HPP) and from natural hydrothermal springs of Kamchatka (Paratun and other springs). Application of modern ultrafiltration techniques not only created new technology for purification of thermal waters, but also in a highly effective way use the process of polycondensation of orthosilicic acid for production and concentration solid-phase NP of silica in sol form with high SiO_2 content in the solid-phase[16-18]. NDS, possessing highly-developed surface (specific area of NDS pores is 417–418 m^2/g) and due to the adsorption properties of surface hydroxyl groups (concentration at 200°C is 4.9 (OH/nm^2)), as a result of stirring and ultrasonic treatment raises the molecular mass of water nano associates while lowering the anti-oxidant activity of water NDS suspension. When hydrogen peroxide is added to the water NDS suspension, the above activity is lowered due to creation of hydroxyl radicals.[19]

Studies of toxic effects of the abovementioned form of nanostructural silicon dioxide, peroral administered into the organism, are of great interest both for solving fundamental problems of nanotoxicology and practice of hygienic standardization of nanoproducts.

The purpose of this chapter—studying of the influence, exercised by aqueous dispersions of the samples use nanosilica upon the death rate of crustaceans *Daphnia magna Straus* for experimental and industrial purposes.

22.2 EXPERIMENTAL

22.2.1 MATERIALS

Samples of experimental and industrial purposes use NDS were produced through drying of aqueous nanosilica soles obtained from hydrothermal solutions by applying baromembrane concentration to the aqueous phase of hydrothermal heat-transfer agent from the wells of Mutnov (Kamchatka) geothermal power plants. Specific surface area of the NDS pores is 417–418 m^2/g, micropores area not more than 28–29 m^2/g, volume of pores about 1.0 cm^3/g, volume of micropores 0.008 cm^3/g, and medium pore diameter on the order of 8.55 nm.

22.2.1 TECHNIQUES

Biotesting of samples, containing NDS, was performed in accordance with the MVI 4–2010 (МВИ 4-2010) while revealing the death rate of crustaceans

Daphnia magna Straus at the temperature of 25°C.[20] Statistical treatment of the research results was performed through application of Microsoft Excel computer program.[21]

22.3 RESULTS AND DISCUSSION

For testing the NDS, we chose the technique for estimation of acute toxicity of nanopowders, NM produce, nanocoatings, waste, and sewage sediments, that contains NPs. Research was performed in laboratory conditions with lower crustaceans *Daphnia magna Straus* as the test organism.[20]

The present technique is based on the estimation of death rate of Daphnia under the influence of aqueous dispersive nanomaterial systems, containing the NPs being tested (experimental group) in comparison with control culture in samples not containing the artificially introduced NPs (controls). Acute toxic influence over the daphnia in the tested samples was estimated out of the death rate during the 48 h of exposition. Quantitative characteristics of toxicity was estimated according to the generally accepted indicator: toxicity index (I)—relative (in %) meaning of death rate for each dispersion system (DS) variant in comparison with controls: $I = (Nc–Nds)/Nc \times 100\%$, where Nc and Nds are quantity of daphnia after exposition in controls and DS, correspondingly.

Criterion for high degree of acute toxicity is the death rate of 50% or more of daphnia ($I \geq 50$) in the tested sample under the condition that it controls all the daphnia retain their vital capacity. For the biotesting, we employed cultivators in complete with glass test tubes of 100 cm^3 volume, which provided for uniform and controlled conditions as for temperature, rate of lighting, aeration, stirring, NPs accessibility, and modeling of convection flows similar to those at natural water reservoirs.

In cultivation water (controls), survival of test culture of daphnia crustaceans through the 48 h period of biotesting equaled 100%. When potassium dichromate was introduced here in concentrations of 1.0–2.0 mg/dm^3, by the end of the 24 h exposition period was registered death rate of crustaceans in excess of 50%. Recording of daphnia death rate at experimental group and controls was conducted after 24 and 48 h, the results are listed in Table 22.1 on analog with paper.[22]

When studying the effect of NDS in concentration of 4×10^{-2} g/dm^3, the rate of toxicity was found to be in the range stipulated by the Norms and totaled less than 20% (Table 22.2), which demonstrates absence of general toxic effect of NDS particles at such a concentration.

TABLE 22.1 Toxicity Rate during Studying of General Toxic Effect of Nano-dispersive Silica over Daphnia.

Sample no.	Concentration of NDS, g/dm^3	Toxicity index (I), %
1	1×10^{-1}	73.30 ± 1.47
2	2×10^{-1}	82.50 ± 5.22
3	1×10^{-2}	0
4	4×10^{-2}	6.67 ± 2.85
5	1×10^{-3}	6.67 ± 1.43
6	1×10^{-4}	3.33 ± 0.72
7	1×10^{-5}	6.67 ± 1.43

When lower concentrations of 1×10^{-5}, 1×10^{-4}, 1×10^{-3}, and 1×10^{-2} g/dm^3 were studied, there was no marked toxic effect of NDS over daphnia. The experiment allowed to reveal the "dose–effect" relation. Thus, the highest death rate of daphnia, compared to controls, was induced by the maximal concentrations of 1×10^{-1}–2×10^{-1} g/dm^3, where toxicity rate constituted $73.30 \pm 11.47\%$ – $82.50 \pm 15.22\%$. The concentration of 1×10^{-2} g/dm^3 brought the toxicity rate equal to 0%, the concentration of 1×10^{-3} g/dm^3—$6.67 \pm 1.43\%$, the concentration of 1×10^{-4} g/dm^3—$3.33 \pm 0.72\%$ and the concentration of 1×10^{-5} g/dm^3—$6.67 \pm 1.43\%$.

Thus during experiments was revealed, the toxic effect of NDS particles over *Daphnia magna Straus* under rather low concentrations of less than 4×10^{-2} g/dm^3.

When estimating the toxicity rate of studied samples, we used the system of classification according to the value of toxicity index (Table 22.2). One of the most important factors that determine the test-reaction is the creation of adequate test-system: concentration of substances introduced for analysis to the test-system should correspond to the values possibly found under the natural conditions. In reality, aqueous dispersions of hydrothermal nanosilica concentrations used in the agriculture of Russia equal less than 50 g/hectare in plant growth, which corresponds to nanosilica concentrations by the factor of millions less than 4×10^{-2} in the water reservoirs.[23] Thus, it shows that concentrations of hydrothermal nanosilica powder aqueous dispersions used as feed additives for fish in interior (artificial) and exterior reservoirs are ecologically harmless for biots and microbiots of water reservoirs and rivers. The usage of hydrothermal nanosilicas within the modern techniques for agriculture industry at the open soil provides new perspectives not only for preservation of natural environment ecology around the agricultural fields but also for production of ecologically pure and safe food production.

TABLE 22.2 System of Classification of Analyzed Samples According to the Value of Toxicity Index.

Groups	Concentration of NDS, g/dm³	Toxicity index value (I)	Conclusion on the rate of sample toxicity
1	4×10^{-2}	Less than 20%	Permitted rate of toxicity
2	From 4×10^{-2} to 1.1×10^{-1}	20–50%	Sample is toxic
3	Over 1.1×10^{-1}	More than 50%	Sample is highly toxic

22.4 CONCLUSION

The results of testing of the influence exercised by aqueous dispersions of samples of experimental and industrial purpose use nanosilica powders (separated from hydrothermal heat-transfer agent solutions at the geothermal power plants) upon the death rate of crustaceans *Daphnia magna Straus* in accordance with the attested procedure show the lack of their toxic effect in aqueous medium in concentrations of 4×10^{-2} g/dm³ and less. Thus, being proved is the ecological safety of developed in Russia techniques for use of hydrothermal origin NDS in plant growing for nonroot treatments of plants at the open soil and in production of new types of feed for fish in interior and exterior reservoirs.

KEYWORDS

- nano-dispersive hydrothermal silica
- toxicity
- biotesting
- lower crustaceans *Daphnia magna Straus*
- ecology of natural environment
- plant growing
- feed for fish

REFERENCES

1. Durnev, A. D. 'Otsenka Genotoksichnosti Nanochastits Pri Ispolzovanii v Meditsine' [Estimation of Nanoparticles Genotoxicity Using in Medicine]. *Gigiyena i Sanitariya.* **2014,** *2,* 76–84.

2. Kurlyandsky, B. A. O Nanotekhnologii i Svyazannykh s Neyu Toksikologicheskikh Problemakh [About Nanotechnology and Related Toxicological Problems]. *Toksikologicheskiy vestnik.* **2007,** *6,* 2–4.
3. Khamidulina, Kh. K.; Davydova, Yu. O. Mezhdunarodnyye Podkhody k Otsenke Toksichnosti i Opasnosti Nanochastits i Nanomaterialov [International Approaches to Toxicity Estimation and Hazards of Nanoparticles and Nanomaterial]. *Toksikologicheskiy Vestnik.* **2011,** *6,* 53–57.
4. GOST R 54617.1-2011. *Menedzhment riska v nanoindustrii [State Standard R 54617.1-2011. Risk management in the nanoindustry];* Standartinform Publications: Moscow, 2013; p 19.
5. Rosenfeldt, R. R.; Seitz, F.; Schulz, R. Heavy Metal Uptake and Toxicity in the Presence of Titanium Dioxide Nanoparticles: A Factorial Approach Using Daphnia Magna. *Environ. Sci. Technol.* **2014,** *48* (12), 6965–6972.
6. Skjolding, L. M.; Kern, K.; Hjorth, R. Uptake and Depuration of Gold Nanoparticles in Daphnia Magna. *Ecotoxicology* **2014,** *23* (7), 1172–1183.
7. Potapov, V. V.; Sivashenko, V. A.; Zelenkov, V. N. Nanodispersnyy Dioksid Kremniya: Rasteniyevodstvo i Veterinariya [Nanodisperse Silica: Plant Growing and Veterinary]. *Nanoindustriya* **2013,** *4,* 18–25.
8. Kirpichnikov, M. P.; Shaitan, K. V. Cpetsialisty Biologicheskogo Fakulteta MGU im. M. V. Lomonosova Predlozhili Prostoy i Udobnyy Test Dlya Predvaritel'noy Otsenki Toksichnosti Nanomaterialov [Specialists of Lomonosov Moscow State University, of Biological faculty offer a simple and convenient test for a preliminary assessment of the toxicity of nanomaterials]. 2009. Available at: http://www.da-brand.com/?p=1422 (acceessed June 11, 2016).
9. Zaitseva, N. V.; Zemlyanova, M. A.; Zvezdin, V. N.; Dovbysh, A. A.; Gmoshinsky, I. V.; Khotimchenko, S. A.; Safenkova, I. V.; Akafyeva, T. I. Toksikologicheskaya Otsenka Nanostrukturnogo Dioksida Kremniya. Parametry Ostroy Toksichnosti [Toxicological Estimation of Nanostructured Silica. Acute Toxicity Parameters]. *Voprosy Pitaniya* **2014,** *83* (2), 42–49.
10. Onishchenko, G. G. Obespecheniye Sanitarno-epidemiologicheskogo Blagopoluchiya Naseleniya v Usloviyakh Rasshirennogo Ispolzovaniya Nanomaterialov i Nanotekhnologiy [Ensuring Sanitary and Epidemiological Welfare of Population by Expanded Using of Nanomaterials and Nanotechnologies]. *Gigiyena i Sanitariya* **2010,** (2), 4–7.
11. Ebbesen, M.; Jensen, T. G. Nanomedicine: Techniques, Potentials, and Ethical Implications. *J. Biomed. Biotechnol.* **2006,** *1,* 1–11.
12. Liu, R.; Zhang, X.; Pu, Y., et al. Small-sized Titanium Dioxide Nanoparticles Mediate Immune Toxicity in Rat Pulmonary Alveolar Macrophages in Vivo. *J. Nanosci Nanotechnol.* **2010,** *10,* 5161–5169.
13. Mwaanga, P.; Carraway, E. R.; van den Hurk, P. The Induction of Biochemical Changes in Daphnia Magna by CuO and ZnO Nanoparticles. *Aquat Toxicol.* **2014,** *150,* 201–209.
14. Potapov, V. V.; Muradov, S. V.; Sivashenko, V. A. Rezultaty Eksperimentov po Ispytaniyu Nanodispersnogo Kremnezema v Kachestve Kormovoy Dobavki v Selskom Khozyaystve [The Results of Experiments on the Testing of Nanodispersed Silica as a Feed Additive in Agriculture] Collection of Scientific Papers "Unconventional Natural Resources, Innovative Technologies and Products". Publishing House of the Academy of Natural Sciences: Moscow, **2010,** *18,* 82–88.
15. Seifulla, R. D.; Kim, E. K. Problemy Toksichnosti Nanofarmakologicheskikh Preparatov [The Toxicity Problems of Nanopharmacological Drugs]. *Exp. Clin. Pharm.* **2013,** *76* (2), 44.

16. Potapov, V. V. *Kolloidnyy Kremnezem v Vysokotemperaturnom Gidrotermalnom Rastvore [Colloidal Silica in the High-temperature Hydrothermal Solution];* Dalnauka: Vladivostok, 2003; pp 8–33.
17. Potapov, V. V.; Zelenkov, V. N.; Gorbach, V. A.; Kashpura, V. N.; Min, G. M. *Izvlecheniye Kolloidnogo Kremnezema iz Gidrotermalnykh Rastvorov Membrannymi Metodami [Extraction of Colloidal Silica from Hydrothermal Solutions by Membrane Methods];* RANS: Moscow, 2006; p 228.
18. Potapov, V. V.; Zelenkov, V. N.; Kashpura, V. N.; Gorbach, V. A.; Muradov, S. N. *Polucheniye Materialov na Osnove Nanodispersnogo Kremnezema Gidroterma'lnykh Rastvorov [Obtaining Materials Based on Nanodispersed Silica of Hydrothermal Solutions];* RANS: Moscow, 2010; p 296.
19. Lapin, A. A.; Kalayda, M. L.; Potapov, V. V.; Zelenkov, V. N. *[Nanodispersed Silica Using in Fish Farming]. Ot Nanostruktur, Nanomaterialov i Nanotekhnologiy k Nanoindustrii: Tezisy Dokl. Shestoy Mezhdunar. Konf. (Rossiya, Izhevsk, 4–6 aprelya 2017 g.)/Pod Obshch. Red. Prof. V. I. Kodolova [From Nanostructures, Nanomaterials and Nanotechnologies to the Nanoindustry: Abstracts,* The Sixth International Conference, Russia, Izhevsk, April 4–6, 2017; Kodolov, V. I., Ed.; IzhSTU: Izhevsk, Russia, 2017, pp 50–54.
20. Potapov, V. V.; Serdan, A. A.; Kashutin, A. N. Amorfnyy Nanorazmernyy Kremnezem Gidrotermalnogo Proiskhozhdeniya: Tekhnologii Polucheniya, Fiziko-khimicheskiye Kharakteristiki, Opyt i Perspektivy Innovatsionnogo Primeneniya Dlya Neorganicheskogo Materialovedeniya [Amorphous Nano-sized Silica of Hydrothermal Origin: Producing Technology, Physical and Chemical Characteristics, Experience and Prospect for Innovative Application of Inorganic Material Authority]. *Butlerovskiye Soobshcheniya [Butlerov Commun.]* **2015,** *43 (9),* 74–77.
21. Potapov, V. V.; Shitikov, E. S.; Tatarinov, S. A.; Portnyagin, V. N.; Zelenkov, V. N.; Lapin, A. A. Primeneniye Nanodispersnogo Kremnezema Dlya Povysheniya Prochnosti Tsementnykh Obraztsov [Application of Nanodispersed Silica for Increasing the Strength of Cement Samples]. *Butlerovskiye Soobshcheniya [Butlerov Commun.]* **2010,** *21 (9),* 50–58.
22. Morgalev, Yu. N.; Morgaleva, T. G.; Grigoriev, Yu. S. Metodika Opredeleniya Indeksa Toksichnosti Nanoporoshkov, Izdeliy iz Nanomaterialov. Nanopokrytia, Otkhodov i Osadkov Stochnykh Vod, Soderzhashchikh Nanochastitsy, po Smertnosti Test-organizma *Daphnia magna Straus* [Method for Determine the Toxicity Index of Nanopowders, Products from Nanomaterials. Nanocoatings, Wastes and Sewage Sludge Containing Nanoparticles, According to the Mortality of the Test Organism *Daphnia Magna Straus*]. MVI 4-2010. Svidetel'stvo ob Attestatsii no 4–10 ot 27.07.2010. Gosudarst-vennoye Obrazovatel'noye Uchrezhdeniye Vysshego Professional'nogo Obrazovaniya «Tomskiy Gosudarstvennyy Universitet». [Certificate of Attestation no. 4–10 dated 27.07.2010. State Educational Establishment of Higher Professional Education "Tomsk State University". 27-07-2010]. Tomsk, 2010; p 34.
23. Berk, K.; Qeyri, P. *Analiz Dannykh s Pomoshyu Microsoft Excel: Per. S angl. [Data Analysis Using Microsoft Excel: Translated from English];* Vilyams Publications: Moscow, 2005; p 560.

PART V
Selected Short Notes, Abstracts, and Conference Communications

CHAPTER 23

CARBON NANOPARTICLES BASED ON GRAPHITE NITRATE COINTERCALATION COMPOUNDS

E. V. RAKSHA[1*], YU. V. BERESTNEVA[1], O. M. PADUN[1],
A. N. VDOVICHENKO[1], SAVOSKIN M. V.[1], and G. E. ZAIKOV[2]

[1]*L. M. Litvinenko Institute of Physical Organic and Coal Chemistry,
R. Luxemburg St. 24, Donetsk 83114, Ukraine*

[2]*Department of Bio-Chemistry, N. M. Emanuel Institute of
Biochemical Physics, Russian Academy of Sciences, Kosygin St. 4,
Moscow 119991, Russia*

Corresponding author. E-mail: elenaraksha411@gmail.com

ABSTRACT

A number of new graphite nitrate cointercalation compounds with organic substances such as esters, ethers, ketones, carboxylic acids, and others were synthesized. A characteristic of these compounds stability was suggested. New low-temperature wet chemistry method for producing carbon nanoscrolls was proposed. The method is based on the use of readily available acceptor-type graphite intercalation compounds. Initial graphite intercalation compound is first exfoliated to produce a suspension of separate graphene monolayers, which is subsequently sonicated yielding a final product—carbon nanoscrolls.

23.1 INTRODUCTION

Acceptor-type graphite intercalation compounds are appropriate for the transformation of the intercalation/exfoliation technique into a large-scale synthesis of graphene and nanoscrolls. Acceptor-type graphite intercalation compounds such as graphite nitrate and hydrosulfate are much more stable substances produced in large quantities for many years. Here, we present our results on synthesis of carbon nanoparticles from the acceptor-type graphite intercalation compounds.

A number of new graphite nitrate cointercalation compounds with organic substances, such as esters, ethers, ketones, carboxylic acids, and others were synthesized. A characteristic of these compounds stability—retention of the ability to thermal expansion after standard thermal treatment (K_{st})—was proposed. For the cointercalants of different chemical nature, the K_{st} value was experimentally shown to vary from zero to one. It was found that there is a symbiosis between stability indices and the calculated proton affinities for the most of cointercalants[1] despite a wide range of the used modifiers chemical nature.

23.2 EXPERIMENTAL

A technique for carbon nanoparticles (few-layer graphene and nanoscrolls) producing from the graphite nitrate cointercalation compounds via their sonication was proposed.[2,3] Recently, we obtained new ternary graphite cointercalation compounds via additional intercalation of different organic compounds into the graphite nitrate. It was revealed that variation in the organic cointercalants nature and composition allows one to control the stability/lability of the resulting compounds within a wide range.

Graphite nitrate cointercalation compounds were synthesized in a thermostatic reactor at 25°C. Nitric acid with a density of 1.502 g/cm³ was added to natural flake graphite GT-1 (Zavalie Graphite Works, Kirovograd Region, Ukraine) sample and the mixture was then stirred. Then co-intercalants (individual compounds or their binary mixture (1:1)) were added and the system was stirred again (Fig. 23.1).

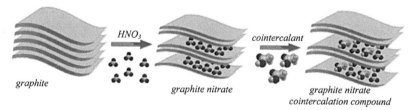

FIGURE 23.1 Scheme of the graphite nitrate cointercalation compounds synthesis.

23.3 RESULTS AND DISCUSSION

The new cointercalation compounds based on graphite nitrate are black-gray powders, resembling the original graphite. The morphology of the resulting

was studied by scanning electron microscopy (SEM). Figure 23.2a presents SEM image of the compound based on the graphite nitrate cointercalated with ethyl formate and acetic acid binary mixture. The primary analysis of the powder X-ray diffraction (PXRD) patterns shown in Figure 23.2b, confirmed the formation of binary and ternary compounds during the modification of graphite nitrate by individual cointercalants or their binary mixtures.

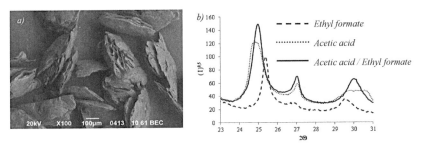

FIGURE 23.2 SEM image of the graphite nitrate cointercalated with ethyl formate and acetic acid (a); PXRD patterns of the graphite nitrate cointercalated compounds (b).

By ultrasound exfoliation in various liquid media of thermo-expanded graphite obtained from graphite nitrate cointercalated with ethyl formate and acetic acid, suspensions of few-layer graphene particles also containing single-layered grapheme, were obtained. On TEM image (Fig. 23.3), graphene-like, generally planar, nanoparticles, yielding clear reflexes of the hexagonal structure, are visible. Structure of these nanoparticles does not change after drying on an amorphous carbon substrate for 40 days. From the electron diffraction pattern (Fig. 23.3), and Raman spectra (containing G ($\sim$1572 cm^{-1}), 2D ($\sim$2667 cm^{-1}), and D ($\sim$1327 cm^{-1}) lines), the thickness of such particles is from 1 to 4 and above atomic layers.

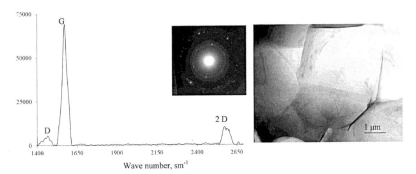

FIGURE 23.3 Raman spectra, electron diffraction pattern, and TEM image of the graphene-like particles obtained from graphite nitrate cointercalated with ethyl formate and acetic acid.

23.4 CONCLUSIONS

The presented approach can be the basis of pilot as well as industrial technologies of the carbon nanoparticles production for the development of new materials with improved properties.

KEYWORDS

- **graphite nitrate**
- **morphology**
- **ultrasound exfoliation**
- **grapheme-like nanoparticles**
- **Raman spectroscopy**
- **electron diffraction**
- **scanning electron microscopy**

REFERENCES

1. Savoskin, M. V.; Yaroshenko, A. P.; Whyman, G. E.; Mestechkin, M. M.; Mysyk, R. D.; Mochalin, V. N. Theoretical Study of Stability of Graphite Intercalation Compounds with Brønsted Acids. *Carbon* **2003,** *41,* 2757–2760.
2. Savoskin, M. V.; Mochalin, V. N.; Yaroshenko, A. P.; Lazareva, N. I.; Konstantinova, T. E.; Barsukov, I. V.; Prokofiev, L. G. Carbon Nanoscrolls Produced from Acceptor-type Graphite Intercalation Compounds. *Carbon* **2007,** *45,* 2797–2800.
3. Savoskin, M. V.; Vdovichenko, A. N.; Raksha, E. V.; Berestneva, Yu. V.; Vishnevskij, V. Yu.; Padun, O. M.; Pavlenko, A. V.; Yurasov, Yu. I. In *Production of Carbon Nanoparticles by Exfoliation of Graphite Nitrate and Its Co-Intercalation Compounds,* Proceedings of the 2016 International Conference on Physics, Mechanics of New Materials and Their Applications, Ivan, A., Parinov, S. H. C., Muaffaq, A. J., Eds.; Nova Science Publishers: Hauppauge, NY, In Press.

CHAPTER 24

SELECTED COMMUNICATIONS, SHORT NOTES, AND ABSTRACTS

24.1 A STUDY OF THE DEPENDENCE OF THE FORMATION OF A DIAMOND FILM ON NUCLEATION CENTERS IN THE PRESENCE OF A COPPER SUBLAYER

D. S. VOKHMYANIN[*]

Faculty of Mechanical Engineering, Perm National Research Polytechnic University, Perm Krai, Russia

[*]*Corresponding author. E-mail: dima5907@bk.ru*

This chapter considers study of the dependence of the formation of the morphology of a diamond film on the surface of carbide plates WC-Co from the nucleation centers of diamond crystallites in the presence of a copper sublayer. Removal of cobalt ligament and forming a copper layer in a surface sublayer carried out by chemical vapor deposition from aqueous solutions of $CuSO_4$. The centers of nucleation of diamond were additionally applied from aqueous suspensions during the formation of a sublayer. It is found that the globular structure of the film formed by using only the copper underlayer, ultradispersed diamonds, and with coprecipitation of diamond/copper. If deposition was carried out after removal of the sublayer, as well as the double layer of copper with a diamond, the diamond film is preferably oriented in the (100) direction. X-ray diffraction analysis and Raman spectroscopy establish the phase composition of the coating obtained. On the X-ray patterns, a diamond reflex is identified at 2Θ 44.0. The Raman spectrum has high luminescence. In the spectrum, a diamond line is identified. Its position is shifted toward large angles when using a diamond/copper sublayer. When using only diamond seeding in the direction of smaller angles, in addition to the diamond line, the spectrum contains lines 1150, 1230, 1350, 1500, and 1580 cm^{-1}. According to the shift of the Raman diamond line, the bi-axial stresses arising in the obtained coating are calculated. If only diamonds are used as a seed, tensile stresses of about 1.1 GPa arise in the coating, in other cases, compressive stresses from 1.3 to 2.6 GPa the value of which depends on the method of surface preparation.

KEYWORDS

- **diamond films**
- **tungsten alloy**
- **copper sublayer**
- **diamond powder**

24.2 THE "RELEASE FORMS" OF THE METAL/CARBON NANOCOMPOSITES

Y. V. PERSHIN[1,2], N. M. KARAVAEVA[1], V. V. TRINEEVA[2,3], and V. I. KODOLOV[2,4*]

[1] *Nanostructures Laboratory, Izhevsk Electromechanical Plant "KUPOL", Izhevsk, Russia*

[2] *Department of Chemistry and Chemical Technology, Kalashnikov Izhevsk State Technical University, Studencheskaya St. 7, Izhevsk, Russia*

[3] *Institute of Mechanics, Ural Division, Russian Academy of Sciences, T. Baramzinoy 34, Izhevsk, Russia*

[4] *Basic Research-High Educational Centre of Chemical Physics and Mesoscopy, Ural Division of Russian Academy of Sciences, Izhevsk, Udmurt Republic, Russia*

Corresponding author. E-mail: kodol@istu.ru

Metal/carbon nanocomposites are metal-containing clusters associated with a shell from carbon fibers. Their activity as well as the activity of stable free radicals is very high. The ability to self-organize nanoparticles is realized due to their energy saturation, which manifests itself in their oscillations and electronic state. The "release forms" of a metal/carbon nanocomposite is required to ensure the safety of the activity of the obtaining nanocomposites. As a shell, into which fractional charge will be partially transmitted, linear functional polymers can be used for nanocomposite particles, which readily dissolve in organic solvents, in water and easily combined with components of most polymer compositions. In the method for obtaining the "final form" of a metal/carbon nanocomposite, it is proposed to use shells of polyethylene glycol (PEG) and polycarbonate (PC). The X-ray diffraction spectra of the "release form" of the Cu/C nanocomposite in PEG are investigated, and the morphology of the Cu/C nanocomposite in the PC envelope was analyzed.

KEYWORDS

- **metal/carbon nanocomposites**
- **polar polymers**
- **synergetics**
- **mesoscopic physics**

24.3 THE SURFACTANT EFFECT ON THE DISTRIBUTION OF CARBON NANOTUBES IN AQUEOUS DISPERSIONS DURING PREPARATION OF A NANOMODIFYING ADDITIVE TO CONSTRUCTION COMPOSITE MATERIALS

YU. TOLCHKOV, T. PANINA, Z. MIKHALEVA, E. GALUNIN*, N. MEMETOV, and A. TKACHEV

Department of Technology and Methods of Nanoproducts Manufacturing, Tambov State Technical University, Tambov, Russia

Corresponding author. E-mail: evgeny.galunin@gmail.com

This chapter describes the results of a study on the homogeneity of suspensions considering the effect of different surfactants and concentrations thereof on uniform distribution of carbon nanotubes (CNTs)-based modifying additive to construction composite materials in aqueous dispersions. The uniformity of the CNTs distribution in these matrices was achieved by using ultrasound, which makes it possible to disperse globules of CNTs particle agglomerates and gain a 15–20-fold decrease in their average size, thereby efficiently using CNTs as cement modifier. The experimental results revealed the surfactant type and its concentration that can promote uniform distribution of the CNTs in the bulk of the suspension and, correspondingly, in the construction composite matrix. Besides, it was shown that the introduction of the CNTs-based additive improves physical, mechanical, and operational characteristics of the original construction material.

KEYWORDS

- **homogeneity of suspensions**
- **surfactants**
- **carbon nanotubes**
- **modifying additive**
- **cement modifier**
- **construction material**

24.4 ELECTROCHEMICAL SYNTHESIS OF HIGHLY DISPERSED SYSTEMS BASED ON ALUMINA AND CALCIUM OXIDE

A. I. KHAYRULLINA[*], E. V. PETROVA, and A. F. DRESVYANNIKOV

Department of Analytical Chemistry, Kazan National Research Technological University, Kazan, Russia

[*]Corresponding author. E-mail: alina17xaj@mail.ru

Physicochemical properties of dispersed systems based on alumina by introducing a small amount of modifier can be improved. They form thermally stable compounds having higher thermo-mechanical and catalytic properties. Highly dispersed alumina systems have been obtained by the electrochemical method. The possibility of modification of alumina by calcium ions during electrolysis was shown. It was found that the variation of conditions of electrolysis allows us to control particle size and phase compounds of oxide systems. The properties of synthesized oxide systems have been studied by various physical methods: dynamic light scattering, XRD, and electron microscopy.

KEYWORDS

- alumina oxide
- electrochemical synthesis
- modification
- phase composition
- morphology
- calcium oxide
- properties
- physical methods

24.5 SYNTHESIS OF PRECURSORS OF COMPLEX IRON AND TITANIUM OXIDES BY THE ELECTROCHEMICAL METHOD

L. R. KHAYRULLINA*, I. O. GRIGORYEVA, and
A. F. DRESVYANNIKOV

Department of Analytical Chemistry, Kazan National Research Technological University, Kazan, Russia

Corresponding author. E-mail: leniza_rinatovna@mail.ru

Highly dispersed complex iron and titanium oxides were obtained by simultaneous electrochemical dissolution of the same metals in halide-containing solutions. The possibility of the rate control of anodic dissolution of iron and titanium by means of changing the electrolyte composition, anodic current density, and ratio of the metal surfaces was shown. Phase and elemental composition, morphology of the surface, and the shape of the particles of obtained product were studied by XRD and scanning electron microscopy.

KEYWORDS

- electrochemical synthesis
- phase composition
- anodic dissolution
- combined electrode
- highly disperse system
- goethite
- ulvospinel
- hematite
- pseudobrookite

24.6 THE FORMATION OF DISPERSED STRUCTURAL COMPONENTS OF Ti IN ALLOYING LAYER OF CASTINGS THROUGH THE OCCURRENCE OF SELF-PROPAGATION HIGH-TEMPERATURE SYNTHESIS

V. B. DEMENT'YEV*, P. G. OVCHARENKO, A. Y. LESHCHEV, and K. E. CHEKMYSHEV

Institute of Mechanics, Udmurt Federal Research Center, Ural Branch of the Russian Academy of Sciences, T. Baramzinoy 34, Izhevsk, Russia

Corresponding author. Email: demen@udman.ru

Selected Communications, Short Notes, and Abstracts

The nanostructured components obtaining method in alloying layers on the surface of castings, which made of iron–carbon alloys is presented. Castings are produced by casting on consumable pattern through the occurrence of self-propagation high-temperature synthesis in alloying compositions. The behavior of synthesis is achieved by using compositions containing, as one component a titanium and a ferrotitanium, and initiation of synthesis is made at the moment of castings shaping due to thermal energy of the melt. Principle stages of the observed method and regularities of synthesis behavior in are reflected in this chapter.

KEYWORDS

- **casting on consumable pattern**
- **surface alloying synthesis**
- **alloying compositions**
- **iron–carbon alloys**
- **borides**
- **carbides**

24.7 HIGH-SPEED LASER SYNTHESIS OF NANOSCALE COMPOSITE LAYERS TO IMPROVE THE CORROSION RESISTANCE OF METALLIC MATERIALS

S. M. RESHETNIKOV[*]

Department of Physical and Organic Chemistry, Udmurt State University, Universitetskaya 1, Izhevsk, Russia

[*]*E-mail: smr41@mail.ru*

In the report compiled the studies, which performed under the supervision of the author in the field of short-pulse laser processing for the synthesis of functional layers that increase the corrosion resistance of metallic materials on the example of non-alloy steels and iron.

KEYWORDS

- laser processing
- corrosion properties
- surface phase
- nanoscale layers

24.8 SYNTHESIS OF NANOCRYSTALLINE Bi_2Se_3 FILM BY VACUUM-THERMAL ANNEALING OF Se/Bi HETEROSTRUCTURE

V. YA. KOGAI*, K. G. MIKHEEV, and G. M. MIKHEEV

Laser Laboratory, Institute of Mechanics, Ural Branch of the Russian Academy of Sciences, T. Baramzinoy 34, Izhevsk, Russia

*E-mail: vkogai@udman.ru

The phase transformations in Se/Bi heterostructure during vacuum-thermal annealing are studied by methods of XRD and Raman scattering. It is established for the first time that stoichiometric Bi_2Se_3 film forms only at certain ratio of Se and Bi layers thicknesses equal to 3.13. The average crystallite size of stoichiometric Bi_2Se_3 film is 30 nm.

KEYWORDS

- phase transformations
- Se/Bi heterostructure
- vacuum-thermal annealing
- X-ray diffraction
- Raman scattering
- stoichiometric Bi_2Se_3 film

Selected Communications, Short Notes, and Abstracts

24.9 CARBON NANOPARTICLES BASED ON NEW TRIPLE GRAPHITE COINTERCALATION COMPOUNDS

E. V. RAKSHA[1*], YU. V. BERESTNEVA[1], VISHNEVSKIJ V. YU.[1], MAYDANIK A. A.[1], GLAZUNOVA V. A.[2], BURKHOVETSKIJ V. V.[2], A. N. VDOVICHENKO[1], and SAVOSKIN M. V.[1]

[1]L. M. Litvinenko Institute of Physical Organic and Coal Chemistry, R. Luxemburg St. 24, Donetsk 83114, Ukraine

[2]Public Institution Donetsk Institute for Physics and Engineering, Donetsk, Ukraine

*Corresponding author. E-mail: elenaraksha411@gmail.com

It has been shown that sonication of new triple graphite cointercalation compounds in ethyl alcohol gives carbon nanoparticles. Samples of graphite nitrate sequentially cointercalated with ethyl formate and 1,4-dioxane, ethyl formate and acetic acid as well as ethyl formate and ethyl acetate were selected as precursors of carbon nanoparticles. The microstructure of the obtained carbon nanoparticles was investigated by the transmission electron microscopy method.

KEYWORDS

- **carbon nanoparticles**
- **triple graphite cointercalation compounds**
- **scanning electron microscopy**
- **transmission electron microscopy**

REFERENCES

1. Dresselhaus, M. S.; Dresselhaus, G. Lattice Mode Structure of Graphite Itercalation Compounds. In *Intercalation Layered Materials*; Springer: Berlin, Germany, 1979; Vol. 6, pp 422–480.

310 Nanoscience and Nanoengineering: Novel Applications

2. Gubin, S. P.; Tkachev, S. V. Graphen i Materialy Na Ego Osnove [Graphene and Materials Based on It]. Radioelektronika. Nanosistemy. Informatsionnye Tekhnologii – Radioelectronics Nanosystems. *Inf. Technol.* **2010,** *10* (1–2) 99–137.
3. Savoskin, M. V.; Yaroshenko, A. P.; Whyman, G. E.; Mysyk, R. D. New Graphite Nitrate Derived Intercalation Compounds of Higher Thermal Stability. *J. Phys. Chem. Sol.* **2006,** *67,* 1127–1131.
4. Savoskin, M. V.; Mochalin, V. N.; Yaroshenko, A. P.; Lazareva, N. I.; Konstantinova, T. E.; Barsukov, I. V.; Prokofiev, I. G. Carbon Nanoscrolls Produced from Acceptor-type Graphite Intercalation Compounds. *Carbon* **2007,** *45,* 2797–2800.
5. Savoskin, M. V.; Yaroshenko, A. P.; Whyman, G. E.; Mestechkin, M. M.; Mysyk, R. D.; Mochalin, V. N. Theoretical Study of Stability of Graphite Intercalation Compounds with Brønsted Acids. *Carbon* **2003,** *41,* 2757–2760.
6. Saidaminov, M. I.; Maksimova, N. V.; Zatonskih, P. V.; Komarov, A. D.; Lutfullin, M. A.; Sorokina, N. E.; Avdeev, V. V. Thermal Decomposition of Graphite Nitrate. *Carbon* **2013,** *59,* 337–343.
7. Savoskin, M. V.; Yaroshenko, A. P.; Mysyk, R. D.; Whyman, G. E.; Vovchenko, L. L.; Popov, A. F. Stabilization of Graphite Nitrate by Intercalation of Organic Compounds. *Theor. Exp. Chem.* **2004,** *40* (2), 92–97.
8. Rao, C. N. R.; Sood, A. K.; Voggu, R.; Subrahmanyam, K. S. Some Novel Attributes of Graphene. *J. Phys. Chem. Lett.* **2010,** *1,* 572–580.
9. Lee, J. H.; Shin, D. W.; Makotchenko, V. G.; Nazarov, A. S.; Fedorov, V. E.; Yoo, J. H.; Yu, S. M.; Choi, J. Y.; Kim, J. M.; Yoo, J. B. The Superior Dispersion of Easily Soluble Graphite. *Small* **2010,** *6,* 58–62.
10. Grayfer, E. D.; Nazarov, A. S.; Makotchenko, V. G.; Kim, S. J.; Fedorov, V. E. Chemically Modified Graphene Sheets by Functionalization of Highly Exfoliated Graphite. *J. Mater. Chem.* **2011,** *21,* 3410–3414.

24.10 SOME QUESTIONS OF FORECASTING THE DEVELOPMENT OF NANOTECHNOLOGY

M. R. MOSKALENKO*

History of Science and Technology Department, Institute of Humanities and Arts, Ural Federal University, Ekaterinburg, Russia

*Corresponding author. E-mail: max.rus.76@mail.ru

ABSTRACT

This paper discusses the scientific and technological forecasting. Characteristics studied for scientific and technological forecasts applicable to the development of nanotechnology.

Selected Communications, Short Notes, and Abstracts

INTRODUCTION

Nowadays, the intermittent development of the sphere of scientific researches, designing and engineering on an atomic scale, which is called nanotechnology, can be observed. Of course, tempestuous development of nanotechnology gives life to the most incredible and bold forecasts of its usage in everyday life and industry, which are stirred up by sensation-seeking journalists. It is necessary to know the general specifications of scientific–technical forecasts in order to estimate all the perspectives of the development of nanotechnology[1,2].

Usually, these standard methods are used in scientific–technical forecasts: analysis of trends and tendencies of development; extrapolation of existing tendencies; elicitation of the processes' cyclicality; studying of principles and regularities of the process or phenomenon; and making of scenarios. Sometimes science fiction is related to a specific type of forecasting where the method of intuitional foreseeing and artistic imagination of artists are used. Several issues should be considered when we speak about scientific–technical forecasting in general and especially when we speak about forecasting of nanotechnology.

HOW DO SOCIAL CONDITIONS OF ANY SPECIFIC SOCIETY CONTRIBUTE TO INNOVATIONS?

In other words, how do social conditions of any specific society contribute to the development of high-technological economics and integration? For example, a well-developed system of education and the cult of scientific knowledge in the late USSR contributed to the creation of many original development projects and inventions[3,4].

However, economic system was not able to realize them; no required mechanisms and institutions were made to provide the integration. Also, the phenomenon of "resource curse" can also take place—when the excessive amount of natural resources and corresponding export profits turn all other branches of economics (including high-tech ones) to be secondary and subordinated to the import.

ABOUT VALUE OF NEW BREAKTHROUGH TECHNOLOGIES AND INVENTIONS

It can be quite difficult to value new breakthrough technologies and inventions. For example, airplane appeared later then dirigibles did, despite the fact that before the First World War dirigibles had been predicted to be the main striking air force; the attitude toward first machine-guns was skeptical among soldiers; first cars and vacuum cleaners often caused nothing but skeptical smile, etc.

There could be some engineering designs that seem to be quite realistic and perspective, but then appear being hardly possible to built-in practice: nuclear aircraft in the 1950s; enormous cargo and passenger seaplanes (some development projects could weight-up to 1000 tons); hovercraft tanks, etc.

Tempestuous development of any technology can crucially change industry and everyday life within a short period of time (automotive industry in 1900 s, personal computers in 1980–1990 s, cellular network in 1990– early 2000 s, etc.).

NONLINEAR DEVELOPMENT OF TECHNOLOGY AND TECHNIQUES

The development of technology and technique is often nonlinear: tempestuous, intermittent development can slow down and become more fluent (classic example—the history of tanks and aviation).

Tempestuous development of any sphere of science and technology always goes with excessively optimistic forecasts, like how this thing will change the face of humanity; then comes a period of more deliberated valuation and normally forecasts become more restrained.

CONCLUSION

These peculiarities of scientific–technical forecasting should be considered while forecasting and planning the development of nanotechnology in various branches of industry.

KEYWORDS

- **scientific forecasting**
- **technological forecasting**
- **nanotechnology**

Selected Communications, Short Notes, and Abstracts 313

REFERENCES

1. Canton, J. The Strategic Impact of Nanotechnology on the Future of Business and Economics. URL:http://www.globalfuturist.com/dr-james-canton/insights-and-future-forecasts/stratigic-impact-of-nanotechnology-onbusiness-and-economics.html.
2. Nanotechnology. A Technology Forecast. http://www.tfi.com/pressroom/pr/nano.html.
3. Toffler, A. *Future Shock;* Random House: New York, NY, 1970; p 505.
4. Zigunenko, S. N. Hundred-of Great Records for Military Technique. Veche: Moscow, 2012; p 432.

24.11 ANALYSIS OF TiC-TiAl AND TiB$_2$-Ti-Al SYSTEMS ELECTROEXPLOSION COATING STRUCTURE BY TRANSMISSION ELECTRON MICROSCOPY

D. A. ROMANOV[*], V. E. GROMOV, M. A. STEPICOV, E. A. GAEVOI, E. A. GRIGORIEVA, and V. O. APANINA

Siberian State Industrial University, Novokuznetsk, Russia

[*]*Corresponding author. E-mail: romanov_da@physics.sibsiu.ru*

Composite coating TiC-TiAl system[1,2] was applied to the die tooling of die steels and H12MF 5HNM, by spraying on electroexplosive installation EVU 60/10 M with aluminum foil, as well as titanium, titanium diboride, and titanium carbide powders with a particle size of 0.1... 1, 0 μm. Coatings are applied during thermal exposure, causing heating of the substrate surface to the melting temperature. Pulse remelting of the surface layer carried by high electron beams providing controlled heating to a temperature above the melting temperature of the self-hardening and melt at a cooling speed to 106 K/s. These conditions allow the irradiation surface to form alloys with nano and submicrocrystalline of high physical and mechanical properties. TEM was investigated TiC-TiAl system electro-explosion coating structure after electron-beam processing[3]. It was found that the structure of the coating is a multi-phase and contains titanium carbide, solid solution on the basis of α-iron, and a solid solution based on γ-Fe. The structure of the surface layer of the coating composition TiB$_2$-Ti-Al, the inclusions formed plate-shaped, the transverse dimensions of which vary in the range 0.7–1.0 μm; the longitudinal dimensions vary between 0.7 and 3.5 μm. Inclusions plate shape rounded inclusions coexist (globular) shape, the dimensions of which vary in a very wide range from 50 to 200 nm. Grains of α-phase (solid solution based on iron with a bcc lattice); grain size from 0.5 μm to 1.5 km. As grain boundaries are located α-phase switching $Ti_3.3Al$; particle sizes vary from 30 to 50 nm.

KEYWORDS

- **electroexplosive sputtering**
- **electron beam processing**
- **nanostructuring surface**
- **structure nonporous coating**
- **titanium carbide**
- **titanium**
- **aluminum**

REFERENCES

1. Wang, J.; Song, R. G.; Lin, X.; Huang, W. D. Microstructure and Properties of Laser Cladding TiC/TiAl Composite Coatings on γ-TiAl in-termetallic Alloy. *Surf. Eng.* **2009,** *25* (6), 196–200
2. Mahmoodiana, R.; Hassana, M. A.; Hamdi, M., et al. In Situ TiC–Fe–Al2O3–TiAl/Ti3Al Composite Coating Processing Using Centrifugal Assisted Combustion ynthesis. *Compos. B Eng.* **2014,** *59,* 279–284.
3. Romanov, D. A.; Budovskih, E. A.; Gromov, V. E.; Ivanov Ju, F. *Jelektrovzryvnoe Napylenie Iznoso- i Jelektrojerozionnostojkih Pokrytij;* Novokuzneck Publ.: OOO "Poligra-fist," 2014; p 203.

24.12 EFFECTIVENESS OF MICROFERTILIZERS IN NANOFORM IN OAT GROWING TECHNOLOGY

I. SH. FATYKHOV[*], A. I. KADYROVA, V. G. KOLESNIKOVA, and T. N. RYABOVA

Department of Plant-Growing, Izhevsk State Agricultural Academy, Izhevsk, Russia

[*]*Corresponding author. E-mail: info@izhgsha.ru*

Selected Communications, Short Notes, and Abstracts

The reaction of oats varieties Ulov and Gunter to pre-seeding treatment with micronutrient fertilizers in various forms is studied. The purpose of research is to determine the relative effectiveness of presowing seed treatment with various forms of micronutrient fertilizers in the cultivation process of grain varieties of oats Ulov and Gunter in the sod-podzolic soil. Presowing seed treatment of oats grain Ulov with micronutrient fertilizers provides an increase in grain yield by 12–25% and the variety Gunter by 12–19%. The average crop yield of oat varieties in versions with nanometal of copper and zinc was on the same level as in the versions with cobalt and copper sulfates. The productivity increase was formed due to the increase of plant density of productive stems to the harvest and grain content of panicles. Micronutrient fertilizers contributed to the rise of the protein content of the grain oats Ulov by 0.5–1.9% and Gunter by 0.6–2.1%. The highest protein content in the grain of 12.1–12.4% was observed with presowing seed treatment with solutions of nanometals on average. Presowing seed treatment of oats varieties with different forms of fertilizers raised the fat content in the grain yield by 0.2–0.7% on average. With the application of micronutrient fertilizer ZhUSS in grain oats Ulov the methionine content was higher by 0.12%, valine by 0.08%, and threonine by 0.25% as compared to similar values in the control variant. The application of micronutrient fertilizer in Gunter grain increased the methionine content by 0.06%, valine by 0.32%, and threonine by 0.1%.

KEYWORDS

- oat
- variety
- grain yield
- presowing seed treatment
- micronutrient fertilizers
- nanometals
- ZhUSS
- protein
- fat
- amino acids

24.13 THE POLARIZATION-ORIENTATION SENSITIVE PHOTOCURRENT IN SILVER–PALLADIUM NANOCOMPOSITES

A. S. SAUSHIN* and G. M. MIKHEEV

Laser Laboratory, Institute of Mechanics, Udmurt Federal Research Center, Ural Branch of the Russian Academy of Sciences, Baramzinoy 34, Izhevsk, Russia

Corresponding author. E-mail: alex@udman.ru

The polarization-orientation sensitive photocurrent in silver–palladium (Ag/Pd) nanocomposite films induced by pulse laser radiation was studied. The photocurrent was measured in two directions: perpendicular (transverse photocurrent) and parallel (longitudinal photocurrent) to incidence plane. We showed that any photo-induced current is a linear function of the incident beam power. The longitudinal photocurrent contains constant contribution and does not depend on polarization. The transverse photocurrent consists of linear and circular contributions. The photocurrent can be induced in a wide spectral range from ultraviolet to infrared radiation. The photocurrent origin in Ag/Pd nanocomposite films is suggested to be photon drag effect. The phenomenological theory was developed. An experimental approach[1–12] for the direct measurement of the circular contribution has been developed.

KEYWORDS

- **polarization-orientation sensitive photocurrent**
- **Ag/Pd nanocomposite films**
- **pulse laser radiation**
- **incident beam power**
- **photon drag effect**

Selected Communications, Short Notes, and Abstracts 317

REFERENCES

1. Zhevandrov, I. D. *Primemenie Polyarizovannogo Sveta* (*Polarized Light Application*); Nauka Publications: Moscow, 1978; p 176.
2. Mikheev, G. M.; Saushin, A. S.; Goncharov, O. Y.; Dorofeev, G. A.; Gil'mutdinov, F. Z.; Zonov, R. G. Effect of the Burning Temperature on the Phase Composition, Photovoltaic Response, and Electrical Properties of Ag/Pd Resistive Films. *Phys. Solid State* **2014**, *56* (11), 2286–2293.
3. Mikheev, G. M.; Aleksandrov, V. A.; Saushin, A. S. Circular Photogalvanic Effect Observed in Silver-palladium Film Resistors. *Tech. Phys. Lett.* **2011**, *37* (6), 551–556.
4. Saushin, A. S.; Mikheev, G. M. Vliyanie Polyarizacii Izlucheniya na Parametric Fotovoltaicheslikh Impulsov v Nanostrukturirovanikh Serebro-palladievikh Rezistivnikh Plyonkakh [Influence of Radiation Polarization on Parameters of Photovoltaic Pulses in Silver–Palladium Resistive Films]. *Chem. Phys. Mesoscopy* **2013**, *15* (1), 127–137.
5. Mikheev, G. M.; Saushin, A. S.; Zonov, R. G.; Styapshin, V. M. Spectral Dependence of Circular Photocurrent in Silver–Palladium Resistive Films. *Tech. Phys. Lett.* **2014**, *40* (5), 424–428.
6. Mikheev, G. M.; Saushin, A. S.; Vanyukov, V. V. Helicity-dependent Photocurrent in the Resistive Ag/Pd Films Excited by IR Laser Radiation. *Quantum Electron.* **2015**, *45* (7), 635–639.
7. Mikheev, G. M.; Saushin, A. S.; Vanyukov, V. V.; Mikheev, K. G.; Svirko, Y. P. Circular Photocurrent in Ag/Pd Resistive Films Upon Excitation by Femtosecond Laser Pulses. *Phys. Solid State.* **2016**, *58* (11), 2345–2352.
8. Ivchenko, E. L. Circular Photogalvanic Effect in Nanostructures. *Phys. Usp.* **2002**, *45* (12), 1299–1303.
9. Mikheev, G. M.; Saushin, A. S.; Vanyukov, V. V.; Mikheev, K. G.; Svirko, Y. P. Femtosecond Circular Photon Drag Effect in the Ag/Pd Nanocomposite. *Nanoscale Res. Lett.* **2017**, *12* (1), 1–7.
10. Saushin, A. S.; Zonov, R. G.; Mikheev, K. G.; Shamshetdinov, R. R.; Mikheev, G. M. Polarization-sensitive Photocurrent in the Resistive Ag/Pd Films. *J. Phys. Conf. Ser.* **2016**, *741*, 012093-1-012093-6.
11. Saushin, A. S.; Zonov, R. G.; Mikheev, K. G.; Aleksandrovich, E. V.; Mikheev, G. M. The Influence of PDO Content on Circular Photocurrent in Resistive Ag/Pd Films. *Tech. Phys. Lett.* **2016**, *42* (9), 963–966.
12. Saushin, A. S.; Mikheev, K. G.; Styapshin, V. M.; Mikheev, G. M. Direct Measurement of the Circular Photocurrent in the Ag/Pd Nanocomposites. *J. Nanophotonic.* **2017**, *11* (3), 032508-1-032508-8.

24.14 INVESTIGATION OF THE OPTICAL PROPERTIES AND SIZE DEPENDENCE OF RUTILE NANOCRYSTALS

E. V. POPOVA*, A. N. LATYSHEV, and O. V. OVCHINNIKOV

Faculty of Physics, Voronezh State University, Voronezh, Russia

*Corresponding author. E-mail: elina.vacheslavovna@gmail.ru

Special interest represent are nanoparticles of titanium dioxide in size less than 10 nm since their properties differ significantly from bulk samples due to the fact that as the size decreases, the edge of the optical absorption band shifts toward short waves. Most authors[1-5] conclude that the quantum-size effect is absent for nanoparticles of titanium dioxide, whose size exceeds 2 nm. However, some authors claim that there is a size effect for particles in the range from 1 to 20 nm, while the magnitude of the dimensional shift varies from 0.1 to 1.26 eV. Such contradictory results, not a random statistical error, but a manifestation of the method of nanoparticle formation and the manifestation of the additional influence of the modeling matrix in which nanoparticles are placed. All this affects the shape of the electronic absorption spectrum. To solve the problem of the presence of a size effect in nanoparticles with dimensions of several nanometers, there is the problem of choosing a single criterion by which their absorption spectra can be compared. In this chapter, we propose a method for determining the short-wavelength shift of absorption bands of nanoparticles of ultra-small size, by converting the spectral curves to the unit area of one nanoparticle. To determine the effect of the shift of the absorption bands of nanoparticles of different sizes, a comparison is made of their spectra, which are reduced to an absorption factor. For example, of nanoparticles of titanium dioxide obtained by mechanically crushing a polycrystalline titanium dioxide powder. A process for the preparation of small particles of rutile is also shown. The proposed method shows the presence of a quantum-size effect. A shift of the bands was observed with a decrease in the nanoparticle size from 4 to 2 nm by 0.3 eV in the direction of higher energy.

KEYWORDS

- **titanium dioxide**
- **mechanical grinding**
- **rutile**
- **absorption efficiency factor**
- **quantum-size effect**

Selected Communications, Short Notes, and Abstracts

REFERENCES

1. Gupta, S. M.; Tripathi, M. A Review of TiO_2 Nanoparticles. *Chinese Sci. Bull.* **2011,** *56* (16), 1639−1657.
2. Serpone, N.; Lawless, D.; Khairutdinovt, R. Size Effects on the Photophysical Properties of Colloidal Anatase TiO_2 Particles: Size Quantization or Direct Transitions in This Indirect Semiconductor? *J. Phys. Chem.* **1995,** *99,* 16646–16654.
3. Monticone, S.; Tufeu, R.; Kanaev, A. V.; Scolan, E.; Sanchez, C. Quantum Size Effect in TiO Nanoparticles: Does It Exist? *Appl. Surf. Sci.* **2000,** *162–163,* 565–570.
4. Satoh, N.; Nakashima, T.; Kamikura, K.; Yamamoto, K. Quantum Size Effect in TiO_2 Nanoparticles Prepared by Finely Controlled Metal Assembly on Dendrimer Templates. *Nat. Nanotechnol.* **2008,** *3,* 106–111.
5. Bohren, C. F.; Huffman, D. R. *Absorption and Scattering of Light by Small Particles;* John Wiley & Sons: Hoboken, NJ, 1983.

24.15 ANTIBACTERIAL COATING BASED ON NANOSTRUCTURED TITANIUM DIOXIDE

A. A. GUROV[1*], S. E. POROZOVA[1], A. M. KHANOV[1], O. YU. KAMENSCHIKOV[2], and O. A. SHULYATNIKOVA[3]

[1]*Faculty of Mechanical Engineering, Perm National Research Polytechnic University, Perm Krai, Russia*

[2]*Perm State National Research University, Perm, Russia*

[3]*Perm State Medical Academy, Perm, Russia*

Corresponding author. E-mail: gurov5991@yandex.ru

Polyphasic materials based on titanium dioxide and its structural modifications have long been used in the manufacture of catalysts and microelectronics. Low-temperature form of titanium dioxide anatase is characterized by the presence of anti-bacterial properties, which is of considerable interest in the creation of implants for various purposes. The goal of this research was obtaining of antibacterial coatings based on nanostructured titanium dioxide (anatase phase). Studied the surface structure of titanium dioxide (anatase phase) on the surface of ceramic samples of nanosized titanium dioxide (rutile phase), obtained by sol–gel technology. It is shown that the formation of the coating occurs on an intermittent mechanism. The phase composition and the composition of the surface samples was determined by spectroscopy Raman scattering on Fourier-spectrometer Senterra (Bruker, Germany) at a wavelength of the emitting laser of 532 nm. Definitely, when you use this technique on the surface, a solid coating of different thickness, it completely or partly overlaps the high-temperature modification of titanium dioxide. Image of the coating on the fractured sample, obtained using a scanning electron microscope, Hitachi (Japan). Coating thickness is 60 ± 15 μm. Coating placer was generated from layered structures, reminiscent of Schiller layers. Experiments conducted on laboratory rats showed good interaction of the coating with living tissues and absence of rejection reactions.

KEYWORDS

- **titanium dioxide**
- **anatase phase**
- **antibacterial coating**
- **spectroscopy**
- **microscope**
- **sol–gel technology**
- **Shiller layers**

24.16 PHOTOSENSITIZATION OF THIONINE DYE LUMINESCENCE BY Mn-DOPED MIXED ZnS-CdS QUANTUM DOTS

D. V. VOLYKHIN* and V. G. KLYUEV

Optics and Spectroscopy Department, Voronezh State University, Voronezh, Russia

Corresponding author. E-mail: volykhin.d@ya.ru

Manganese-doped mixed CdZnS quantum dots were successfully prepared by aqueous synthesis in photographic inert gelatin. Quantum dots were formed with a cubic crystal lattice (zinc-blender) with an average particle size of about 2 nm. Spectroscopic manifestation of association of CdZnS:Mn quantum dots with dye molecules was observed[1]. The possibility of photosensitization of thionine emission by pure and Mn-doped CdZnS quantum dots is shown.

Selected Communications, Short Notes, and Abstracts

KEYWORDS

- aqueous synthesis
- inert gelatin
- CdZnS:Mn quantum dots
- Mn-doped CdZnS quantum dots

REFERENCE

1. Klyuev, V. G.; Volykhin, D. V.; Ovchinnikov, O. V.; Pokutniy, S. I. Relationship Between Structural and Optical Properties of Colloidal $Cd_xZn_{1-x}S$ Quantum Dots in Gelatin. *J. Nanophotonics* **2016,** *10* (3), 033507.

24.17 PHOTOINDUCED TRANSFORMATIONS IN DARK-RED GLASSY SELENIUM FILMS AT LOW-ENERGY QUANTUM EFFECTS

E. V. ALEKSANDROVICH[1]*, R. A. GAISIN[2], K. G. MIKHEEV[1] and G. M. MIKHEEV[1]

[1]Laser laboratory, Institute of Mechanics Ural Branch of Russian Academy of Sciences, T. Baramzinoy 34, Izhevsk, Russia

[2]Kalashnikov Izhevsk State Technical University, Studencheskaya St. 7, Izhevsk, Russia

*Corresponding author. E-mail: evalex@udman.ru

The thermally deposited films of glassy dark-red selenium are studied by X-ray diffraction, spectrophotometry, and Raman scattering[1–5]. Based on the results of optical transmittance measuring in the region of Urbach edge, the absorption factors of the films before and after ~520–610 nm exposure with different intensities are calculated. It is established that deposited films are direct bandgap semiconductors with energy of the optical gap equal to ~2.07 eV.

Polymorphoid induces photo-transformations, leading to changes of their optical properties (reflection, optical gap width, refraction, and absorption, etc.). Either photobleaching occurs due to the concentration ratio of polymorphides of high- or low-temperature polymorphic modifications in a glassy matrix Se.

KEYWORDS

- optical gap width
- polymorphoid
- irradiation

REFERENCES

1. Mooradian, A.; Wright, G. B. The Raman Spectrum of Trigonal, α-Monoclinic and Amorphous Selenium. In *The Physics of Selenium and Tellurium;* Charles Cooper, W.; Ed.; Pergamon Press: Oxford, London, 1969; pp 269–276.
2. Minaev, V. S.; Timoshenkov, S. P.; Kalugin, V. V. Structural and Phase Transformations in Condensed Selenium. *J. Optoelectron. Adv. Mater.* **2005,** *7* (4), 1717–1741.
3. Swanepoel, R. Determination of the Thickness and Optical Constants of Amorphous Silicon. *J. Phys. E Sci. Instrum.* **1983,** *16,* 1214–1221.
4. Felts, A. *Amorphous and Glass Similar Inorganic Solids;* Mir: Moscow, 1986; p 558.
5. Minaev, V. S.; Timoshenkov, S. P.; Vassiliev, V. P.; Aleksandrovich, E. V.; Kalugin, V. V.; Korobova, N. E. The Concept of Polymer Nano-heteromorphic Structure and Relaxation of the Glass-forming Substance by Chalcogenides, Oxides and Halides Example. Some Results and Perspective. *J. Optoelectron. Adv. Mater.* **2016,** *18,* 10–23.

24.18 HYDROGEN EFFECT ON ANTIOXIDANT ACTIVITY OF AQUEOUS SYSTEMS

A. A. LAPIN[1*] and S. D. FILIPPOV[2]

[1]Department of Water Resources, Kazan State Power Engineering University, Krasnoselskaya St. 51, Kazan, The Republic of Tatarstan, Russia

[2]Akvalon Ltd, Chelyabinsk, Russia

[*]Corresponding author. E-mail: lapinanatol@mail.ru

Activated water is the kind of water where hydrogen bonds are loosened, which makes molecules more mobile and, therefore, easier for living organisms to consume and also facilitates biological waste removal. The principle of water activation lies in destroying cluster structures to saturate the water with monomolecules. An activated water cluster contains 5–6 molecules (13–16 molecules in regular water), which makes it easier for cells of living organisms to consume such activated water and get rid of biological waste. Such water is considered more active as indicated by biological and biophysical parameters. Water activated via any method has a higher liquidity (due to lower surface tension) and a higher dissolving capacity. Water with smaller clusters have improved reactivity, better penetrates biological membranes, and is quicker extracted out of body. It efficiently substitutes and complements all kinds of cleansing supplements and physiotherapeutic procedures. Activated water is used for all-around detoxification, neutralization, and removal of waste, toxins carcinogens, and radionuclides. This chapter deals with the experimental data to determine the total antioxidant activity of aqueous systems, saturated with hydrogen using a coulometric analysis method, wherein water is observed on the highest increase of its activity up to 20 times. We analyzed the Arkhyz + antioxidant = Zhivitsa functional drink, where all the ingredients are of natural origin. Saturating the drink by hydrogen leads to an insignificant increase (16.5% rel.) of total antioxidant activity. The presence in the water of active organic compounds reduces the influence of hydrogen on the total antioxidant activity, probably consisting of different numbers of molecules.

KEYWORDS

- hydrogen
- antioxidant activity
- water
- activation of the redox potential
- total antioxidant activity

24.19 HEATING INFLUENCE ON THE OPTICAL LIMITING IN ZINC PHTHALOCYANINE SOLUTION

R. Y. KRIVENKOV[1], E. M. KHUSNUTDINOV[1], T. N. MOGILEVA[1], I. P. ANGELOV[2], and G. M. MIKHEEV[1*]

[1]*Laser Laboratory, Institute of Mechanics, Udmurt Federal Research Center, Ural Branch of the Russian Academy of Sciences, T. Baramzinoy 34, Izhevsk, Russia*

[2]*Institute of Organic Chemistry with Centre of Phytochemistry, BAS, Sofia 1113, Bulgaria*

Corresponding author. E-mail: mikheev@udman.ru

The results of the optical limiting (OL) and investigation in the solution of zinc phthalocyanine (ZnPc) dye in dimethylsulfoxide at the temperature range of 25–70°C by z-scan method are presented[1-6]. It is revealed that underexposure of the 532 nm laser radiation, the ZnPc solution exhibits a significant OL effect. Enhance the OL effect with the temperature increase is discovered. The character of the OL threshold dependence on temperature is determined.

KEYWORDS

- **nonlinear optical properties**
- **optical limiting**
- **z-scan**
- **laser radiation**
- **elevated temperature**

Selected Communications, Short Notes, and Abstracts

REFERENCES

1. McKeown, N. B. *Phthalocyanine Materials: Synthesis, Structure and Function;* Cambridge University Press: Cambridge, England, 1998; Vol. 6, p 193.
2. Mikheev, G. M.; Angelov, I. P.; Mantareva, V. N.; Mogileva, T. N.; Mikheev, K. G. Thresholds of Optical Limiting in Solutions of Nanoscale Compounds of Zinc Phthalocyanine with Galactopyranosyl Radicals. *Tech. Phys. Lett.* **2013,** *39* (7), 664–668.
3. Hyojung, Yu.; Sok Won Kim. Temperature Effects in an Optical Limiter Using Carbon Nanotube Suspensions. *J. Korean Phys. Soc.* **2005,** *47* (4), 610–614.
4. Chapple, P. B.; Staromlynska, J.; Hermann, J. A., et al. Single-beam Z-Scan: Measurement Techniques and Analysis. *J. Nonlinear Opt. Phys. Mater.* **1997,** *6* (3), 251–293.
5. Mikheev, G. M.; Krivenkov, R. Yu.; Mikheev, K. G.; Okotrub, A. V.; Mogileva, T. N. Z-scanning Under Monochromatic Laser Pumping: A Study of Saturatable Absorption in a Suspension of Multiwalled Carbon Nanotubes. *Quantum Electron.* **2016,** *46* (8), 719–725.
6. Mikheev, G. M.; Mogileva, T. N.; Popov, Y. A.; Kaluzhny, D. G. A Computer-Based Laser System for the Diagnostics of Hydrogen in Gas Mixtures. *Instrum. Exper. Tech.* **2003,** *46* (2), 233–239.

24.20 GENERATION OF POLARIZATION-SENSITIVE PHOTOCURRENT IN NANOCRYSTALLINE COPPER SELENIDE FILMS

Yu. A. OSTROUH*, V. YA. KOGAI, K. G. MIKHEEV, R. G. ZONOV, and G. M. MIKHEEV

Laser Laboratory, Institute of Mechanics, Ural Branch of the Russian Academy of Sciences, T. Baramzinoy 34, Izhevsk, Russia

Corresponding author. E-mail: ostrouh@udman.ru

Nanocrystalline copper selenide (CuSe) with 280 nm-thickness films are synthesized. It was found out that the photocurrent pulse arises in the CuSe filmed under the 1064 nm pulsed nanosecond laser radiation. It is shown that the photocurrent polarity depends on the incidence angle sign and the photocurrent amplitude depends both on the incidence angle sign and incident radiation polarization. The mechanism of the observed phenomenon is suggested[1–9].

KEYWORDS

- copper selenide films
- photocurrent pulse generation
- laser radiation
- polarization
- incidence angle sign
- photocurrent pulse polarity

REFERENCES

1. Mikheev, G. M.; Zonov, R. G.; Obraztsov, A. N.; Volkov, A. P.; Svirko, Y. P. Spectral Dependence of the Optical Rectification Effect in Nanographite Films. *Tech. Phys. Lett.* **2005,** *31* 94–96.
2. Mikheev, G. M.; Aleksandrov, V. A.; Saushin, A. S. Circular Photogalvanic Effect Observed in Silver–Palladium Film Resistors. *Tech. Phys. Lett.* **2011,** *37* (6), 551–556.
3. Mikheev, G. M.; Nasibulin, A. G.; Zonov, R. G., et al. Photon-drag Effect in Single-walled Carbon Nanotube Films. *Nano Lett.* **2012,** *12* (1), 77–83.
4. Akbari, M.; Onoda, M.; Ishihara, T. Photo-induced Voltage in Nano-porous Gold Thin Film. *Opt. Express.* **2015,** *23* (2), 823–832.
5. Mikheev, G. M.; Saushin, A. S.; Vanyukov, V. V., et al. Femtosecond Circular Photon Drag Effect in the Ag/Pd Nanocomposite. *Nanoscale Res. Lett.* **2017,** *12* (1), 39.
6. Mikheev, G. M.; Styapshin, V. M. Nanographite Analyzer of Laser Polarization. *Instruments Exp. Tech.* **2012,** *55* (1), 85–89.
7. Kogai, V. Y.; Vakhrushev, A. V.; Fedotov, A. Y. Spontaneous Explosive Crystallization and Phase Transformations in a Selenium/Copper Bilayer Nanofilm. *JETP Lett.* **2012,** *95* (9), 454–456.
8. Sakr, G. B.; Yahia, I. S.; Fadel, M., et al. Optical Spectroscopy, Optical Conductivity, Dielectric Properties and New Methods for Determining the Gap States of CuSe Thin Films. *J. Alloys Compd.* **2010,** *507* (2), 557–562.
9. Obraztsov, P. A.; Mikheev, G. M.; Garnov, S. V., et al. Polarization-sensitive Photoresponse of Nanographite. *Appl. Phys. Lett.* **2011,** *98* (9), 91903.

24.21 PECULIARITIES OF NANOSTRUCTURED STEELS TRANSFORMATIONS

T. M. MAKHNEVA*, A. A. SUKHIKH, and V. B. DEMENT'YEV

Selected Communications, Short Notes, and Abstracts

Laser Laboratory, Institute of Mechanics, Ural Branch of the Russian Academy of Sciences, T. Baramzinoy 34, Izhevsk, Russia

*Corresponding author. E-mail: mah@udman.ru

The investigation results are presented for the phase $\alpha \rightarrow \gamma$-transformation in two chromium-nickel nanostructured steels. Specific features of the process have been revealed, the main of which is the decrease of the critical temperature A_n of the $\alpha \rightarrow \gamma$-transformation in the temperature range of aging; it creates premises for the radical change of the mechanical properties of the studied steels (08Cr15Ni5Cu2Ti and 03Ni18Co9Mo5Ti).

KEYWORDS

- **chromium-nickel nanostructured steels**
- **phase $\alpha \rightarrow \gamma$-transformation**
- **critical temperature**
- **temperature range of aging**
- **mechanical properties**

24.22 INVESTIGATION OF THERMOPHYSICAL PROPERTIES OF METALLIC NANOMATERIALS USING MATHEMATICAL MODELING METHODS

I. V. PAVLOV[1] and A. YU. FEDOTOV[2*]

[1]Kalashnikov Izhevsk State Technical University, Studencheskaya 7, Izhevsk, Russia

[2]Department of Mechanics of Nanostructure, Institute of Mechanics, Udmurt Federal Research Center, Ural Branch of the Russian Academy of Science, T. Baramzinoy 34, Izhevsk, Russia

*Corresponding author. E-mail: alezfed@gmail.com

Some questions concerning the thermal conductivity of homogeneous nano-materials are considered. The study of the thermophysical characteristics of the material and the determination of its thermal conductivity coefficient, those tasks to which a large number of numerical and practical experiments, are devoted[1–5].

KEYWORDS

- **nanomaterials**
- **thermal conductivity**
- **phonons**
- **modeling**
- **LAMMPS**
- **crystal structure**
- **molecular dynamics**

REFERENCES

1. Vakhrushev, A. V.; Severyukhin, A. V.; Fedotov Yu, A.; Valeev, R. G. Issledovanie Protsessov Osazhdeniya Nanoplenok na Podlozhku iz Poristogo Oksida Alyuminiya Metodami Matematicheskogo Modelirovaniya [Research Nanofilms Deposition Processes on a Substrate of Porous Aluminum Oxide by Means of Mathematical Simulation]. *Vychislitel'naya Mekhanika Sploshnykh sred [Comput. Continuum Mech.]*. **2016,** *9* (1), 59–72.
2. Vakhrushev, A. V.; Fedotov, A. Y.; Vakhrushev, A. A.; Golubchikov, V. B.; Givotkov, A. V. Multilevel Simulation of the Processes of Nanoaerosol Formation. Part 1. Theory Foundations. *Int. J. Nanomech. Sci. Technol.* **2011,** *2* (2), 105–132.
3. Vakhrushev, A. V.; Fedotov, A. Yu.; Severyukhin, A. V.; Suvorov, S. V. Modelirovanie Protsessov Polucheniya Spetsial'nykh Nanostrukturnykh Sloev v Epitaksial'nykh Strukturakh Dlya Utonchennykh Fotoelektricheskikh Preobrazovateley [Modelling of Processes of Special Nanostructured Layers of Epitaxial Structures for Sophisticated Photovoltaic Cells]. *Khimicheskaya fizika i mezoskopiya [Chem. Phys. Mesoscopics]*. **2014,** *16* (3), 364–380.
4. VMD, Visual Molecular Dynamics. Theoretical and Computational Biophysics Group [Electronic Resource]. https://www.ks.uiuc.edu/Research/vmd (accessed March 01, 2017).
5. LAMMPS Molecular Dynamics Simulator [Electronic resource]. http://lammps.sandia. gov (accessed March 01, 2017).

24.23 COMPOSITION, STRUCTURE AND ELECTROCHEMICAL PROPERTIES OF THE SURFACE NANO-SCALE LAYERS OF ARMCO–IRON, DOPED CARBON

E. M. BORISOVA[1*], S. M. RESHETNIKOV[1], O. R. BAKIEVA[2], F. Z. GILMUTDINOV[2], and V. Y. BAYANKIN[2]

[1]*Department of Physical and Organic Chemistry, Udmurt State University, Universitetskaya St. 1, Izhevsk, Russia*

[2]*Physical-Technical Institute, Udmurt Fedral Research Center, Ural Division, Russian Academy of Science, Kirova St. 132, Izhevsk 426000, Russia*

*Corresponding author. E-mail: borisovayelena@mail.ru

By method X-ray photoelectron spectroscopy, X-ray diffraction, EELFS, atomic force microscopy, and nanoindentation studied the chemical composition, structure, topography, and hardness of thin carbon coating on iron, obtained by magnetron sputtering and their subsequent processing flow of argon ions. The possibility of increasing the corrosion resistance of the treated this way iron samples.

KEYWORDS

- **carbon films**
- **ion–beam mixing**
- **chemical composition**
- **structure**
- **anodic behavior**

24.24 THE NANOSTRUCTURED FIBRE OF POLYETHYLENE TEREPHTHALATE AND ITS PROPERTIES

YU. M. VASILCHENKO[1*], M. YU VASILCHENKO[2], V. A. BAZHENOV[2], and K. P. SHIROBOKOV[1]

[1]*Department of Chemistry and Chemical Technology, Kalashnikov Izhevsk State Technical University, Studencheskaya St. 7, Izhevsk, Russia*

[2]*Izhevsk State Agricultural Academy, Izhevsk, Russia*

Corresponding author. E-mail: cct@istu.ru

A perspective way of utilization of polymeric waste is their processing in fibrous material as a result of inflation of a stream of the melted raw materials the air stream directed via the inflation device. In this case, all production operations are carried out on one unit with the minimum harm to the environment. For improvement of operational characteristics of fibrous materials (sorption ability and durability), giving of new properties to them (an electromagnetic susceptibility and assignment of a static charge), and also the stage of modification of intermediate fusion by nanostructures is offered to include expansions of field of industrial use of this technology.

KEYWORDS

- polyethylene terephthalate
- modification
- metal/carbon nanocomposite

Selected Communications, Short Notes, and Abstracts 331

24.25 THE USE OF PHOSPHORUS-CONTAINING COPPER/CARBON NANOCOMPOSITES TO OBTAIN MODIFIED INTUMESCENT FIREPROOF EPOXY COATINGS

R. V. MUSTAKIMOV[1,2] and V. I. KODOLOV[1,2*]

[1]*Basic Research-High Educational Centre of Chemical Physics and Mesoscopy, Ural Branch of the Russian Academy of Science, Izhevsk, Udmurt Republic, Russia*

[2]*Department of Chemistry and Chemical Technology, Kalashnikov Izhevsk State Technical University, Studencheskaya St. 7, Izhevsk, Russia*

Corresponding author. E-mail: kodol@istu.ru

2In this chapter, a study of an intumescent fireproof epoxy coating modified with phosphorus-containing copper/carbon nanocomposites was carried out. It has been determined that an epoxy coating containing a copper/carbon nanocomposite in an amount of 0.003% and ammonium polyphosphate in an amount of 10% has been found to have the best fireproof properties. The foam coating of the epoxy coating was examined by SEM. The introduction of the nanocomposite led to the formation of pores with the more rounded shape, whereas in the foam without pores are more like pentagons and hexagons. Such a change in the geometric shape should lead to an increase in the strength of the foam box.

KEYWORDS

- **copper/carbon nanocomposites**
- **epoxy resin**
- **intumescent fireproof coating**
- **foam coke**
- **microstructure**

24.26 INTRAMEDULLAR INTRODUCTION OF NANOSIZED MATERIAL "LITAR" AS ALTERNATIVE OF MODERN CELLULAR TECHNOLOGY

S. D. LITVINOV* and I. I. MARKOV

Medical University "Reaviz", Tchapaevskaya St. 227, Samara, Russia

Corresponding author. E-mail: LSD888@ramber.ru

The report presents the main cases of material "LitAr" to restore (replenish) of different tissues. The possibility of stem cells stimulation for regeneration without their removal from the body is discussed[1]. Intramedullary proposed applications of the product as an alternative to the modern technologies of cellular regenerative process are proposed[2] school.

KEYWORDS

- material "LitAr"
- stem cells stimulation
- regeneration
- intramedullary
- modern technologies
- cellular regenerative process

REFERENCES

1. Levard, F.; Laurent-Maquin, D.; Guillaume, C., et al. Zink Doped Hydroxiapatite as Immunomodulatori Biomaterials. *Int. J. Artif. Organs.* **2012,** *35* (8), 569.
2. Litvinov, S. D.; Markov, I. I.; Van'kov, V. A. Intramedullary Introduction of the Material LitAr Stimulates Posttraumatic Cell Regeneration. *Int. J. Artif. Organs.* **2012,** *35* (8), p 569.

Selected Communications, Short Notes, and Abstracts

24.27 INVESTIGATIONS REVIEW IN TREND OF POLYMER NANOCOMPOSITES AND NANOSTRUCTURED MATERIALS

SABU THOMAS*

School of Chemical Sciences, International and Inter University Centre for Nanoscience and Nanotechnology, Mahatma Gandhi University, Priyadarshini Hills P.O., Kottayam 686560, Kerala, India

Corresponding author. E-mail: sabuchathukulam@yahoo.co.uk

ABSTRACT 1. POLYMER NANOCOMPOSITES FOR DENTAL APPLICATIONS

Prosthodontics is the branch of dental medicine dealing with restoration and rehabilitation of lost oral and maxillofacial structures of human body. Among the various available materials for rehabilitation, polymethyl methacrylate (PMMA) is the most universally accepted one. It is a biocompatible and cost-effective material. It has got some demerits like decreased retention and increased staining of the prosthesis leading to discoloration. It is also ineffective in resisting plaque and microbial accumulation. In an attempt to rectify the existing problems, nano metal and metal oxides are added to the polymer matrix. Zinc oxide nanoparticles and silver nanoparticles of various weight percentages were incorporated into the PMMA matrix. Scanning electron microscopy studies were carried out for the morphological analysis. Mechanical properties were investigated using universal testing machine. Biofilm formations of the samples were assessed for evaluating the anti-Candidal property. Water sorption and salivary sorption studies were carried out along with contact angle measurements. Observations revealed that polymer nanocomposites show reduced biofilm formation, improvement in wettability, and retention of the polymer system compared to pure PMMA.

ABSTRACT 2. NANOSTRUCTURED FILTERS BASED ON CARBOHYDRATE POLYMERS

Water is the essential lifesaving materials. According to the United Nations, around 18% of the world population (almost 1.1 billion people) cannot obtain safe water at this time. Due to remote access of safe drinking water, in most developing countries waterborne diseases, such as cholera, dysentery, enteric fever, and hepatitis A are very common. In an estimate, the World Health Organization states that more than 1.6 million individuals mostly children die annually due to waterborne diseases. So, there is an urgent need for developing an economic and easy to use water treatment process. There are numerous water treatment technologies studied till date and the research is still going on to develop a perfect water filter. Nanomaterials are emerging very fast as new water purification materials, due to their high surface area, higher removal capacities, and reusability. But main challenges with almost all the nanomaterials are their high cost. This problem can be addressed by using inexpensive carbohydrate polymers. Mother Nature gifted us a vast amount of carbohydrate polymer-based biomasses which have variable surface sites and are used to capture different pollutants from water. But this biopolymer too has some disadvantages like low removal efficiency and low specific surface area. Thus modification of carbohydrate polymeric materials to nanoscale or modified them with other nanomaterials could solve the problems. For enhancing the specific surface area nanocellulose, nanochitin, and graphene oxides could be prepared from different biomass sources. Also, several different functional groups can be introduced to these nanobiopolymers to enhance their removal efficacy and reusability. These materials could be very good filter materials for water purifications.

ABSTRACT 3.   ENGINEERING AT THE NANOSCALE: A STRATEGY FOR DEVELOPING HIGH-PERFORMANCE FUNCTIONAL MATERIALS

The talk will concentrate on various approaches being used to engineer materials at the nanoscale for various applications in future technologies. In particular, the case of clay, carbon nanostructures (e.g., nanotubes and graphene), metal oxides, and bionanomaterials (cellulose, starch, and chitin) will be used to highlight the challenges and progress. Several polymer systems will be considered, such as rubbers, thermoplastics, thermoetts, and their blends for the fabrication of functional polymer nanocomposites. The interfacial activity of nanomaterials in compatibilizing binary polymer blends will also be discussed. Various self-assembled architectures of hybrid nanostructures can be made using relatively simple processes. Some of these structures offer excellent opportunity to probe novel nanoscale behavior and can impart unusual macroscopic end properties. I will talk

Selected Communications, Short Notes, and Abstracts

about various applications of these materials, taking into account their multifunctional properties. Some of the promising applications of clay, metal oxides, nanocellulose, chitin, carbon nanomaterials, and their hybrids will be reviewed. Finally, the effect of dewetting upon solvent rinsing on nanoscale thin films will also be discussed.

REFERENCES

1. Thomas, S., et al. *Langmuir* 2016.
2. Thomas, S., et al. *Macromolecules* **2016,** 5b0243.
3. Thomas, S., et al. *Prog. Polym. Sci.* **2014,** *39* (4), 749–780.
4. Thomas, S., et al, *Soft Matter Accep.*
5. Thomas, S., *J. Phys. Chem. B* **2010,** *114,* 13271–13281.
6. Thomas, S., *J. Phys. Chem. B* **2009,** *113,* 5418-5430.
7. Thomas, S. *J. Phys. Chem. B* **2008,** *112,* 14793–14803.
8. Thomas, S., et al. *Appl. Clay Sci.* **2016,** *123,* 1–10.
9. Thomas, S., et al. *Rubber Chem. Technol.* 2016.
10. Thomas, S., *et al. Comp. Sci. Technol.* 116, 9–17.
11. Thomas, S., et al. *Phys. Chem. Chem. Phys.* **2015,** *17* (29), 19527–19537.
12. Thomas, S., et al. *J. Mater. Chem. C* **2014,** *2* (40), 8446–8485

ABSTRACT 4. ENGINEERING AT THE NANOSCALE: STATE OF THE ART, CHALLENGES, AND NEW OPPORTUNITIES

The talk will concentrate on various approaches being used to engineer materials at the nanoscale for various applications in future technologies. In particular, the case of clay, carbon nanostructures (e.g., nanotubes and graphene), metal oxides, bionanomaterials (cellulose, starch, and chitin) will be used to highlight the challenges and progress. Several polymer systems will be considered, such as rubbers, thermoplastics, thermosets, and their blends for the fabrication of functional polymer nanocomposites. The interfacial activity of nanomaterials in compatibilizing binary polymer blends will also be discussed. Various self-assembled architectures of hybrid nanostructures can be made using relatively simple processes. Some of these structures offer excellent opportunity to probe novel nanoscale behavior and can impart unusual macroscopic end properties. I will talk about various applications of these materials, taking into account their multifunctional properties. Some of the promising applications of clay, metal oxides, nanocellulose, chitin, carbon nanomaterials, and their hybrids will be reviewed.

ABSTRACT 5.   PRODUCTION OF NANOMATERIALS FROM BIOWASTES

Green chemistry started for the search of benign methods for the development of nanoparticles from nature and their use in the field of antibacterial, antioxidant, and antitumor applications. Biowastes are eco-friendly starting materials to produce typical nanoparticles with well-defined chemical composition, size, and morphology. Cellulose, starch, chitin, and chitosan are the most abundant biopolymers around the world. All are under the polysaccharides family in which cellulose is one of the important structural components of the primary cell wall of green plants. Cellulose nanoparticles (*fibers, crystals, and whiskers*) can be extracted from agro waste resources, such as jute, coir, bamboo, pineapple leaves, coir, etc. Chitin is the second most abundant biopolymer after cellulose, it is a characteristic component of the cell walls of fungi, the exoskeletons of arthropods, and of chitin (*fibers and whiskers*) can be extracted from shrimp and crab shells. Chitosan is the derivative of chitin, prepared by the removal of acetyl group from chitin (deacetylation). Starch nanoparticles can be extracted from tapioca and potato wastes. These nanoparticles can be converted into smart and functional biomaterials by functionalization through chemical modifications (*esterification, etherification, TEMPO oxidation, carboxylation, hydroxylation, etc.*) due to presence of large amount of hydroxyl group on the surface. The preparation of these nanoparticles includes both series of chemical and mechanical treatments; crushing, grinding, alkali, bleaching, and acid treatments. Transmission electron microscopy, scanning electron microscopy, and atomic force microscopy are used to investigate the morphology of nanoscale biopolymers. Fourier-transform infrared spectroscopy, and X-ray diffraction are being used to study the functional group changes, crystallographic texture of nanoscale biopolymers, respectively. Since large quantities of biowastes are produced annually, further utilization of cellulose, starch, and chitins as functionalized materials is very much desired. The cellulose, starch, and chitin nanoparticles are currently obtained as aqueous suspensions which are used as reinforcing additives for high performance environment-friendly biodegradable polymer materials. These nanocomposites are being used as biomedical composites for drug/gene delivery, nano scaffolds in tissue engineering, and cosmetic orthodontics. The reinforcing effect of these nanoparticles results from the formation of a percolating network based on hydrogen bonding forces. The incorporation of these nanoparticles in several bio-based polymers has been discussed.

Selected Communications, Short Notes, and Abstracts 337

The role of nanoparticle dispersion, distribution, interfacial adhesion, and orientation on the properties of the eco-friendly bionanocomposites has been carefully evaluated.

ABSTRACT 6. MANUFACTURING OF NANOMATERIALS FROM BIOWASTES AND PRODUCTION OF ECO-FRIENDLY BIONANOCOMPOSITES

Green chemistry started for the search of benign methods for the development of nanoparticles from nature and their use in the field of antibacterial, antioxidant, and antitumor applications. Biowastes are eco-friendly starting materials to produce typical nanoparticles with well-defined chemical composition, size, and morphology. Cellulose, starch, chitin, and chitosan are the most abundant biopolymers around the world. All are under the polysaccharides family in which cellulose is one of the important structural components of the primary cell wall of green plants. Cellulose nanoparticles (*fibers, crystals, and whiskers*) can be extracted from agro waste resources, such as jute, coir, bamboo, pineapple leaves, coir, etc. Chitin is the second most abundant biopolymer after cellulose, it is a characteristic component of the cell walls of fungi, the exoskeletons of arthropods, and nanoparticles of chitin (*fibers and whiskers*) can be extracted from shrimp and crab shells. Chitosan is the derivative of chitin, prepared by the removal of acetyl group from chitin (Deacetylation). Starch nanoparticles can be extracted from tapioca and potato wastes. These nanoparticles can be converted into smart and functional biomaterials by functionalization through chemical modifications (*esterification, etherification, TEMPO oxidation, carboxylation, hydroxylation, etc.*) due to presence of large amount of hydroxyl group on the surface. The preparation of these nanoparticles includes both series of chemical and mechanical treatments; crushing, grinding, alkali, bleaching, and acid treatments. Transmission electron microscopy, scanning electron microscopy, and atomic force microscopy are used to investigate the morphology of nanoscale biopolymers. Fourier-transform infra-red spectroscopy and X-ray diffraction are being used to study the functional group changes, crystallographic texture of nanoscale biopolymers, respectively. Since large quantities of biowastes are produced annually, further utilization of cellulose, starch, and chitins as functionalized materials is very much desired. The cellulose, starch, and chitin nanoparticles are currently obtained as aqueous suspensions which are used as reinforcing additives for high performance environment-friendly biodegradable polymer materials. These nanocomposites are being used as biomedical composites for drug/gene delivery, nano scaffolds in tissue engineering, and cosmetic orthodontics. The reinforcing effect of these nanoparticles results from the formation of a percolating network based on hydrogen bonding forces. The incorporation of these nanoparticles in several bio-based polymers has been discussed. The role of nanoparticle dispersion, distribution, interfacial adhesion, and orientation on the properties of the eco-friendly bionanocomposites has been carefully evaluated.

ABSTRACT 7. CHEMISTRY OF NANOMATERIALS AND THEIR POLYMER NANOCOMPOSITES

The historical development of the science and technology of nanomaterials will be presented from the very early stages. The major contributors to the field of nanoscience and nanotechnology will be reviewed. The basic definitions and concepts will be discussed. The applications of nanomaterials in all the major fields will be presented. The field of polymer nanocomposites is stimulating both fundamental and applied research because these nanoscale materials often exhibit physical and chemical properties that are dramatically different from conventional microcomposites. A large number of nanoparticles, layered silicates, and polymeric whiskers are being used for the preparation of nanocomposites. Since the Toyota research group's pioneering work on nylon6/layered silicate nanocomposites, polymer nanocomposites containing layered silicates have attracted much attention. The polymer/layered nanocomposites can exhibit increased modulus, decreased thermal expansion coefficient, reduced gas permeability, increased solvent resistance, and enhanced ionic conductivity when compared to the polymer alone. In the proposed talk, the different preparation techniques for polymer nanocomposites will be discussed. The role of various surfactants in improving the polymer/filler interaction will be reviewed. The various characterization techniques for nanocomposites will be addressed. In the case of semi-crystalline polymers, the role of crystallization on the intercalation and exfoliation will be discussed. The important properties of nanocomposites will also be presented. I will also present recent developments in cellulose nanocomposites and bionanocomposites. Finally, the new developments in the field of nanomedicine (drug delivery and scaffolds) will be discussed in detail.

Selected Communications, Short Notes, and Abstracts 339

ABSTRACT 8. COMPATIBILIZATION OF POLYMER BLENDS

Most pairs of high molecular weight polymers are immiscible and incompatible. These materials are characterized by a two-phase morphology, a narrow interface and poor physical and chemical interactions across the phase boundaries. As a consequence of this, incompatible blends often exhibit poor mechanical properties and highly unstable morphologies. This problem can be alleviated by the addition or in-situ formation of compatibilizers, also called emulsifying agent or interfacial agents. The addition of suitably selected compatibilizer to an immiscible binary blend should reduce the interfacial energy between the phases, permit finer dispersion during mixing, provide stability against gross segregation, and result in improved interfacial adhesion. In recent years, more attention has been focused on reactive compatibilization of immiscible polymer blends. This is a very fast, easy, and cost-effective alternative. This method has been applied to a large number of blend systems. Different types of chemical reactions can be used in compatibilizing polymer blends; reactions as imidization, esterification, amidation, aminolysis ester–ester interchange, amide–ester interchange, ring opening, and ionic bonding can occur rapidly at elevated processing temperatures. In this chapter, physical and reactive compatibilization of several polymer pairs will be discussed. The effects of structure and molecular weight of homo and block copolymers concentration of the compatibilizer mode of addition of the compatibilizer blend ratio mode of mixing, etc., will be analyzed. The role of compatibilizers on the morphology development, rheology, morphology stability and interface adhesion, and mechanical properties have been investigated. Attempts have been made to compare the experimental results with compatibilization theories.

ABSTRACT 9. NATURAL FIBER REINFORCED GREEN POLYMER COMPOSITES: MACRO TO NANOSCALES

Composites from polymers (rubbers and plastics) and reinforcing fibers provide best properties of each. They replace conventional materials in many structural and non-structural applications. Both fibers and plastics are light, in combination they give composites of very high strength to weight ratio. In recent years, composites made from natural (cellulosic) fibers and organic polymers have gained a lot of interest in construction and automobile industry. Unlike synthetic fibers, natural fibers are abundant, renewable, cheap, and of low density. Composites made from natural fibers are cost effective and environment friendly. However, lack of interfacial adhesion and poor resistance to moisture absorption makes the use of natural fibers less attractive for critical applications. However, these problems can be successfully alleviated by suitable chemical treatments. This presentation deals with the use of natural fibers, such as pineapple leaf fiber, coir fiber, sisal fiber, oil palm fiber, and banana fiber as reinforcing material for various thermoplastics, thermosets, and rubbers. The fiber surface modifications via various chemical treatments to improve the fiber-matrix interface adhesion on mechanical, viscoelastic, dielectric rheological aging, and thermal properties will also be discussed. Experimental results will be compared with theoretical predictions. The advantages of hybridizing natural and glass fibers also will be scanned briefly. The use of these composites as building materials will be discussed. Finally, recent developments in cellulose nanocomposites will also be presented.

ABSTRACT 10. BIOINSPIRED GREEN MICRO AND NANOCOMPOSITES FOR THE FUTURE

Micro and nano bioinspired composite materials are the best future materials for the coming millennium. Cellulose fibers, chitin, and starch in different length scales offer outstanding properties like stiffness, toughness, and other mechanical properties. Composites from polymers (rubbers and plastics) and reinforcing fibers provide best properties of each. They replace conventional materials in many structural and non-structural applications. Both natural fibers and polymers are light, in combination they give composites of very high strength to weight ratio. In recent years, composites made from natural (cellulosic) fibers and organic polymers have gained a lot of interest in construction and automobile industry. Unlike synthetic fibers, natural fibers are abundant, renewable, cheap, and of low density. Composites made from natural fibers are cost effective and environment friendly. However, lack of interfacial adhesion and poor resistance to moisture absorption makes the use of natural fibers less attractive for critical applications. However, these problems can be successfully alleviated by suitable chemical treatments. This presentation deals with the use of natural fibers, such as pineapple leaf fiber, coir fiber, sisal fiber, oil palm fiber, and banana fiber as reinforcing material for various thermoplastics, thermosets, and rubbers. The fiber surface modifications via various chemical treatments to improve the fiber-matrix interface adhesion on mechanical, viscoelastic, dielectric rheological aging, and thermal properties will also be discussed. Experimental results

Selected Communications, Short Notes, and Abstracts

will be compared with theoretical predictions. The advantages of hybridizing natural and glass fibers also will be scanned briefly. The use of these composites as building materials will be discussed. Finally, recent developments in cellulose nanocomposites, chitin nanocomposites, and starch nanocomposites will also be presented.

ABSTRACT 11. INVESTIGATION OF POLYMER BLEND MISCIBILITY ON A NANOSCALE BY FLUORESCENCE SPECTROSCOPY

Miscibility and phase behavior are of crucial importance for the final properties of polymer blends. Many experimental techniques have been developed for the characterization of polymer blend miscibility. However, the most fundamental question that still remains in the characterization of polymer blend miscibility is the scale over which the components are molecularly mixed. It is well accepted that the measured miscibility is strongly dependent upon the techniques used. Therefore, it is important to develop more sensitive technique for the study of miscibility. In recent years, excimer fluorescence spectroscopy has been proven to be a highly promising technique. The important advantages of this technique that distinguish it from other techniques are the high sensitivity of detecting phase separation and the ability to analyze phase behavior of blends which contain very small concentration (even less than 1%) of one of the components. The recent studies on excimer fluorescence spectroscopy to investigate miscibility of polymer blends will be briefly reviewed. Comparison will be made between fluorescence, light scattering, microscopy, and differential scanning calorimetry. The use of this technique to study the effects of casting solvents, molecular weight, etc. on miscibility will be discussed. More attention will be given to the miscibility, phase separation, and kinetics of phase separation of SAN/PMMA and PS/PMMA systems.

ABSTRACT 12. POLYMER NANOCOMPOSITES: RECENT ADVANCES

The field of polymer nanocomposites is stimulating both fundamental and applied research because these nanoscale materials often exhibit physical and chemical properties that are dramatically different from conventional microcomposites. A large number of nano particles, layers silicates, and polymeric whiskers are being used of the preparation of nanocomposites.

Since the Toyota research group's pioneering work on nylon6/layered silicate nanocomposites, polymer nanocomposites containing layered silicates have attracted much attention. The polymer/layered nanocomposites can exhibit increased modulus, decreased thermal expansion coefficient, reduced gas permeability, increased solvent resistance, and enhanced ionic conductivity when compared to the polymer alone. In the proposed talk, the different preparation techniques for polymer nanocomposites will be discussed. The role of various surfactants in improving the polymer/filler interaction will be reviewed. The various characterization techniques for nanocomposites will be addressed. In the case of semi-crystalline polymers, the role of crystallization on the intercalation and exfoliation will be discussed. The important properties of nanocomposites will also be presented. Finally, recent developments in cellulose nanocomposites and bionanocomposites will also be described.

ABSTRACT 13. MICRO AND NANOSTRUCTURED EPOXY RESIN-BASED POLYMER BLENDS

Epoxy-based blends have generated a lot interest recently due to their increasing commercial importance. Epoxy resin is often blended with rubbers and thermoplastics to generate micromorphologies for the better impact performance. Very recently, nanostructured blends based on epoxy resin/block copolymers systems have appeared in literature. These nanostructured blends have the potential to show super toughness. In the proposed talk, miscibility, phase separation, morphology, mechanical, and viscoelastic properties of a series of epoxy-based blends will be discussed. The phase separation, gelation, and vitrification will be investigated detail. More attention will be given to the development of nanomorphologies. Techniques, such as scanning electron microscopy, transmission electron microscopy, OM, laser light scattering, and rheology will be made use of the characterization of the morphologies. Finally, the role of the nano and micromorphologies on the mechanical and viscoelastic properties will be disused.

ABSTRACT 14. RECYCLING OF POLYMER BLENDS

The end-of-life of all polymeric materials (neat polymers, blends, and composites) has become a great challenge, due to environmental concern, and to international and national regulations. Among all the possible ways to manage polymer waste, a hierarchy could be established. The most preferred option is the minimization of waste, followed by reuse of materials in the same application, recycling in another application (including recovery of monomers or low-weight molecules), incineration with energy recovery and finally incineration without energy recovery, or landfilling. Recycling of polymers that used to end up only at city *landfills* or *incinerators* is increasing around the world. As with any technological trend, the engineering profession plays an important role. Discarded polymer products and packaging make up a growing portion of municipal solid waste. The environmental protection agency (EPA) estimates that by the year 2010, the amount of plastics throw away will be 50% greater than at the beginning of the 1990s. EPA also says that polymer waste accounts about one-fifth of all waste in the waste stream. Over the past two decades, recycling of polymers has dramatically increased. In the proposed talk, all the recent trends in polymer recycling will be discussed. This will include melt recycling, monomer recycling, thermal depolymerization, heat compression process, and more efficient solid-state pulverization process. Very specific cases of polymer blend recycling will be discussed which include thermoplastic–thermoplastic, rubber–rubber, rubber thermoplastic, etc. Finally, attention will be given to specific polymer product recycling.

ABSTRACT 15.   POLYMER LATEX NANOCOMPOSITES

The field of polymer latex nanocomposites is stimulating both fundamental and applied research because these nanoscale materials often exhibit physical and chemical properties that are dramatically different from conventional microcomposites. A large number of nanoparticles, layers silicates, CNT, and polymeric nanowhiskers are being used for the preparation of latex stage nanocomposites. Since the Toyota research group's pioneering work on nylon6/layered silicate nanocomposites, rubber latex nanocomposites containing nanofillers have attracted much attention. The latex nanocomposites can exhibit increased modulus, decreased thermal expansion coefficient, reduced gas permeability, increased solvent resistance, and enhanced ionic conductivity when compared to the neat latex alone. In the proposed talk, the different preparation techniques for latex nanocomposites will be discussed. The role of various surfactants in improving the latex/filler interaction will be reviewed. The various characterization techniques for latex nanocomposites will be addressed. In the case of semicrystalline lattices, the role of crystallization on the intercalation and exfoliation will be discussed. The dynamic properties of latex nanocomposites, such as Payne effect and Mullins effect will also be presented. Finally, recent developments in the applications of latex nanocomposites will also be described.

ABSTRACT 16. VISCOELASTIC PHASE SEPARATION PROCESS AND THE DEVELOPMENT OF MICRO AND NANO MORPHOLOGIES IN EPOXY-BASED BLENDS FOR SUPER TOUGHNESS

Phase separation in general could be either by diffusion or by diffusion and hydrodynamic flow. A new model has been suggested recently to follow the phase separation process in dynamically asymmetric mixtures composed of fast and slow components. This new model is often called the *viscoelastic phase separation process* due to the fact that viscoelastic effects play a dominant role in the phase separation process.[1-6] The dynamic asymmetry can be induced by either the large size difference (molecular weight) or the difference in glass-transition temperature between the components of a mixture or blend. The molecular weight difference often exists in complex fluids, such as polymer solutions, polymer blends, micellar solutions, colloidal suspensions, emulsions, and protein solutions. The Tg differences can exist in any mixtures. In dynamically asymmetric mixtures, phase separation generally leads to the formation of a long-lived "interaction network" (a transient gel) of slow-component molecules (or particles), if the attractive interactions between them are strong enough. Because of its long relaxation time, it cannot catch up with the deformation rate of the phase separation itself and as a result the stress is asymmetrically divided between the components. This leads to the transient formation of network-like or sponge-like structures of a slow-component-rich phase and its volume shrinking.[4] In this talk, we present our new results on the viscoelastic phase separation process in epoxy/SAN and epoxy/ABS blends.[1-4] Epoxy resin is often blended with high molecular weight thermoplastics to generate microstructured morphologies for the better impact performance. These systems are very ideal to follow the viscoelastic phase separation process on account of their molecular weight and Tg differences. We have looked at the phase separation process in these blends by various techniques, such as optical microscopy, scanning electron microscopy, transmission electron microscopy, atomic force microscopy, and small angle laser light scattering. The dynamics of phase separation has been carefully followed by optical microscopy and laser light scattering. In most cases, the system

Selected Communications, Short Notes, and Abstracts 345

undergoes spinodal decomposition and the viscoelastic phase separation was prominent at higher concentration of the thermoplastic phase where phase inversion occurs.[1-6] The particle in particle morphology (secondary, ternary, and quaternary phases), IPN type of structures, and unusual shrinkage have been examined as a result of the viscoelastic phase separation process.[1,4] All these phenomena have been carefully quantified and co-related with the viscoelastic phase separation process.

REFERENCES

1. Jyotishkumar, P.; Pionteck, J.; Ozdilek, C.; Moldenaers, P.; Cvelbar, U.; Mozetic, M.; Thomas, S. *Soft Matter Accep.*
2. Jyotishkumar, P.; Ozdilek, C.; Moldenaers, P.; Sinturel, C.; Janke, A.; Pionteck, J.; Thomas, S. *J. Phys. Chem. B* **2010,** *114,* 13271–13281.
3. Jyotishkumar, P.; Koetz, J.; Tierisch, B.; Strehmel, V.; Häßler, R.; Ozdilek, C.; Moldenaers, P.; Thomas, S. *J. Phys. Chem. B* **2009,** *113,* 5418–5430.
4. Jose, J.; Joseph, K.; Pionteck, J.; Thomas, S. *J. Phys. Chem. B* **2008,** *112,* 14793–14803.
5. Zhong, X.; Liu, Y.; Su, H.; Zhan, G.; Yu, Y.; Gan, W. *Soft Matter.* **2011,** *7,* 3642–3650.
6. Tanaka, H. *Adv Mater.* **2009,** *21,* 1872–1880.

ABSTRACT 17. DEVELOPMENT OF POLYMER NANOCOMPOSITE SCAFFOLDS FOR TISSUE ENGINEERING

Biodegradable polymer scaffolds are useful materials to integrate the femoral part of the implant with the bone and provide a matrix for cellular growth. Synthetic biodegradable polymers can provide temporary scaffold for cell adhesion and expansion both in vitro and in vivo and guide tissue regeneration with defined sizes and shapes. The fibrillar structure is important for cell attachment, proliferation, and differentiated function in tissue engineering. The structure allows for growth and is convenient for transport of nutrients. The synthetic polymers, such as polycaprolactone, poly l-lactic acid, and their copolymers have attracted wide attention for their biodegradation in the human body and are used for tissue engineering. Several methods have been practiced to create highly porous scaffold including fiber bonding, solvent casting/salt leaching, gas foaming, phase separation, and electrospinning. Out of which electrospinning is the simple and cost-effective technique for producing nanofibers from polymer solution. Introduction of organically modified clay in polymers leads to different types of structures which include intercalated or exfoliated morphology. The nano reinforcement increases the mechanical rigidity, mobility, stiffness, and

346

biodegradability in biodegradable polymers. Moreover, it also increases the porosity of the polymer nanocomposite. Nanoparticle reinforced scaffolds are yet to achieve importance. In fact, they have wide range of interest in tissue engineering. Literature reports regarding nanoparticle reinforced scaffolds are very scant. Hence, the present investigation will be interesting and will find application in tissue engineering in the foreseeable future. In this talk, the state of the art on the synthesis, morphology, structure, properties, and applications of dual porous nanocomposite scaffolds will be presented.

ABSTRACT 18.   NONLINEAR VISCOELASTICITY OF RUBBER NANOCOMPOSITES

Nonlinear viscoelasticity of rubber nanocomposites (Payne effect and Mullin's effect) has received a lot of attention during the last few years since rubber nanocomposites are being used in a variety of dynamic applications. The Payne effect is a very interesting feature of the dynamic behavior of rubber composites, especially rubber compounds containing fillers, such as carbon black, carbon nanotubes (CNTs), silica, etc., from macro- to nanoscales. The Payne effect is named after the British rubber scientist A. R. Payne, who made extensive studies of the effect. The effect is sometimes also known as the Fletcher–Gent effect, after the authors of the first study of the phenomenon (Fletcher & Gent, 1953). The effect is observed under cyclic loading conditions with small strain amplitudes and is manifest as a dependence of the viscoelastic storage modulus on the amplitude of the applied strain. Above approximately 0.1% strain amplitude, the storage modulus decreases rapidly with increasing amplitude. At sufficiently large strain amplitudes (roughly 20%), the storage modulus approaches a lower bound. In that region where the storage modulus decreases the loss modulus shows a maximum. The Payne effect depends on the filler content of the material and vanishes for unfilled elastomers. Physically, the Payne effect can be attributed to deformation-induced changes in the material's microstructure, that is, to breakage and recovery of weak physical bonds linking adjacent filler clusters. Since the Payne effect is essential for the frequency and amplitude-dependent dynamic stiffness and damping behavior of rubber bushings, automotive tires, and other products, constitutive models to represent it have been developed in the past (e.g., Lion et al., 2003). Similar to the Payne effect under small deformations is the Mullins effect that is observed under large deformations. In this talk, the Payne effect and Mullins effect of nanocomposites will be presented. Fillers, such as nano silica, carbon

Selected Communications, Short Notes, and Abstracts 347

black, CNTs, nanoclay, etc., show strong Payne effect in rubbers. The effect of coupling agent and temperature on the Payne effect has been carefully studied. The variable density model of Goritiz and Mayer has been utilized to model the behavior of the nanocomposites. The model gave an excellent fit with the experimental.

ABSTRACT 19. AYURVEDIC "BHASMS" AS ETHNO-NANOMEDICINE

Nanoscience and nanotechnology are considered to be one of the biggest technological innovations since the industrial revolution. This new technology promises to re-engineer the man-made world, molecule by molecule, sparking a wave of novel revolutionary commercial products from machines to medicine. The origin of nanoscience and nanotechnology is often attributed to a concept and novel ideas proposed by the Nobel laureate Richard P. Feynman. The word nano derived from the Greek word means dwarf. A nanometer is one billionth of meter or length of 10 hydrogen atom placed side by side or 1/80,000 of thickness of human hair. The particle size of nanoparticles in medicine ranges from 5 to 100 nm. These are produced by various chemical or physical processes and having specific properties. Traditional medicine systems, such as Ayurveda and Unani can serve as an excellent template for the development of nanomedicine for human theragnostics. Recent studies show that traditional medicines, such as Ayurvedic Bhasms may hold strong relevance in the emerging era of nanomedicine. There is an urgent need for amalgamation of traditional medicine system involving Ayurvedic Bhasms, with the evolving field of metal-based nanomedicine. Recent reports also support the view that Ayurvedic Bhasms as nanomedicine resemble nanocrystalline materials and are similar in their physico–chemical properties. The studies show that the Bhasms can be employed for targeted drug delivery as they are non-toxic, biocompatible, and non-antigenic in nature.

ABSTRACT 20. SELF-HEALING EPOXY COATINGS

The self-healing coating has attracted increasing attention because it automatically repairs and prevents corrosion of the underlying substrate. It responds without external interference to environmental stimuli in an effective pattern. The corrosion growth relates to the oxidation and the practical loss of material. The protection of materials can be accomplished by coatings. It is estimated that approximately 90% of corrosion defensive methods and service costs are related to protective coatings. Among the extensively used protective coatings, organic coatings can provide favorable corrosion protection for different materials.

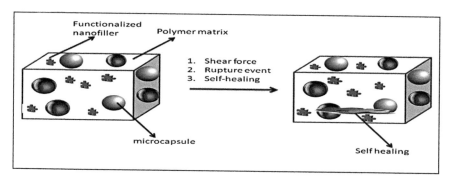

In recent years, the self-healing process based on microcapsule consists of reactive core materials got great consideration, because the microcapsules can be comfortably synthesized and integrated into the matrix materials. The microcapsules enclosed in the coatings will liberate the embedded active and repairing material when the coating integrity is broken by the outer impact and heal the damaged region through the polymerization reactions.

Among diverse core materials, epoxy resin has the healing capability without the need of additional catalysts and should be a valuable option as healing agent for the fabrication of self-healing composites, generally epoxy-amine or epoxy-amide systems which are slightly alkaline. In these systems, the polymerization processes related to the epoxy discharged from fractured microcapsules would take place to heal microcracks generated by external stress. The healing process will be accomplished through the polymerization reactions. One of the ways is the curing reaction of epoxy with an amine-curing agent, which is based on the polymerization of epoxy groups and amino groups, and the other is the polymerization reaction between epoxy molecules under the alkaline condition.

PART 21. PROCESSING OF POLYMER COMPOSITES

In this talk, general aspects about composites will be discussed in the beginning. This will be followed by an elaborate discussion polymer processing techniques. The details are given below:

Thermoset Composites

- Hand lay-up and Spray lay-up
- Vacuum bag molding
- Compression molding
- Filament winding
- Pultrusion
- Resin transfer molding
- Structural reaction injection molding

Thermoplastic Composites

- Calendaring
- Sheet molding
- Film casting
- Injection molding
- Extrusion
- Blow molding
- Rotational molding
- Thermoforming

24.28 A DEVICE FOR STUDYING OPTICAL LIMITING IN NANOMATERIALS SUSPENSIONS AT LOW TEMPERATURES

D. L. BULATOV[*], K. G. MIKHEEV, YU. A. OSTROUH, and G. M. MIKHEEV

Laser Laboratory, Institute of Mechanics, Ural Branch of the Russian Academy of Sciences, T. Baramzinoy 34, Izhevsk, Russia

[*]*Corresponding author. E-mail: dlbulatov@udman.ru*

A device based on a Peltier module for studying nonlinear optical properties of various materials at negative temperatures by z-scan technique is developed.[1-4] The device is meant for temperatures as low as $-50°C$ and keeping a preset temperature constant during the experiment. The performance of the device is described.

KEYWORDS

- **nonlinear optical properties**
- **optical limiting**
- **z-scan**
- **laser radiation**
- **low temperature**
- **Peltier module**

REFERENCES

1. Chen, Y.; Lin, Y.; Liu, Y., et al. Carbon Nanotube-based Functional Materials for Optical Limiting. *J. Nanosci. Nanotechnol.* **2007,** *7* (4–5), 1268–1283.
2. Wang, Y.; Lv, M.; Guo, J., et al. Carbon-based Optical Limiting Materials. *Sci. China Chem.* **2015,** *58* (12), 1782–1791.
3. Chen, Y.; Bai, T.; Dong, N., et al. Graphene and Its Derivatives for Laser Protection. *Prog. Mater. Sci.* **2016,** *84,* 118–157.
4. Dini, D.; Calvete, M. J. F.; Hanack, M. Nonlinear Optical Materials for the Smart Filtering of Optical Radiation. *Chem. Rev.* **2016,** *116* (22), 13043–3233.

24.29 X-RAY PHOTOELECTRON STUDY OF THE MECHANISM OF THE FUNCTIONALIZATION OF THE CARBON METAL-CONTAINING NANOSTRUCTURE SURFACE WITH P-ELEMENTS

I. N. SHABANOVA[1*], TEREBOVA N. S.[1], and SAPOZHNIKOV G. V.[1,2]

[1]Laboratory of Nanostructures, Physico-Technical Institute, Udmurt Federal Research Center, Ural Branch of the Russian Academy of Sciences, Kirov St. 132, Izhevsk 426000, Russia

[2]FGBOU GOU VPO "Udmurtski Gosudarstvenni Universitet", Universitetskaya St. 1, Izhevsk 426034, Russia

[*]Corresponding author. E-mail: xps@ftiudm.ru

In this chapter, the X-ray photoelectron spectroscopy method has been used for studying the chemical bond of atoms on the surface of the carbon metal-containing nanostructures functionalized with the atoms of sp-elements, such as silicon, phosphorus, sulfur, nitrogen, fluorine and iodine, and the influence of the functionalization on the variation of the metal (Fe, Ni, and Cu) atomic magnetic moment. It is shown that on the surface of the nanostructures, strong covalent bonds between d-metal atoms and atoms of silicon, phosphorus, and sulfur and between carbon atoms and atoms of fluorine, nitrogen, and iodine are formed. It is also shown that at the functionalization of carbon metal-containing nanostructures with silicon, sulfur, and phosphorus, the atomic magnetic moment on metal increases.

KEYWORDS

- **X-ray photoelectron spectroscopy**
- **satellite structure of C1s- and Cu3s-spectra**
- **carbon metal-containing nanostructures**
- **functionalization**
- **atomic magnetic moment**

INDEX

A

Acceptor-type graphite intercalation compounds, 297–298

Activated water, 323
   cluster, 323

Adding energy characteristics
   two principles, 44–8, 59–63

Ammonium polyphosphate (APP), 164
   experimental technique, 164–165
   investigation methods
     IR-spectroscopic, 165
     XPS studies, 165
   results and discussion
     IR-spectroscopy, 165–168
     XPS-study, 168–71

Anodic aluminum oxide (AAO), 134

ASNOM probe, 221

B

Basic energy atomic characteristics
   comparison of, 70–71

Bio-structural SEP, 90

Biodegradable polymer scaffolds, 345

Biophysical processes
   nomograms of, 77–78

Biosystems, 85

Biowastes, 336
   manufacturing of nanomaterials from, 337–338

Boltzmann's constant, 73

Business and nature
   entropic criteria in, 79–80

C

Carbon copper-containing nanostructures
   characteristics of, 163
   experimental technique, 164–165
   investigation methods
     IR-spectroscopic, 165
     XPS studies, 165

results and discussion
   IR-spectroscopy, 165–168
   XPS-study, 168–171

Castings, 307

Cellular systems
   conformations of hexagonal structures in, 91–92

Cellulose nanoparticles, 336

Charge transfer process
   admixture particles in polymer between, 265–266
   current density, for, 261–264
   quantitative estimation for, 267–270
   rate factor of, 266

Chemical mesoscopics, 4
   conditions for process, 4–7
     diffract grams of, 5, 6
     PVA marks, influence of, 6
     quantity metallic nickel, influence of, 6
   perspectives of, 14–15

Chitin, 336

Chitosan, 336

Composite coating TiC-TiAl system, 313

Controlled rocket solid propellant engines
   dependence of pressure and temperature, 38

Copper/carbon nanocomposite
   diffract grams of, A
   IR spectra, F, G
   peaks intensity changes in IR spectra, E

D

Depth of mass center of atoms, D

Diamond film
   dependence of formation of morphology of, 301
   Raman spectrum, 301–2
   X-ray patterns, 301

Dispersed systems
   physicochemical properties of, 305

DNA molecules

354   Index

conformation of, 92–3
numerical and energy composition of, 93

E

Electrical conducting polymer composites (ECPC)
  growth of conductivity of, 260
Electron winding
  angle of, 50–5
Engineering at nanoscale
  high-performance functional materials, developing, 334–345
  state of art and, 335
Epoxy coating, 331
  foam, 331
Epoxy-based blends, 342

F

Fibrillar structure, 345
Fully automated topography imaging, 220
  AFM, 220–1
  height images of, 220
Fundamental interactions
  evaluation of intensity, 125
  types of, 121–4

G

Graphite nitrate cointercalation compounds, 297, 298
  electron diffraction pattern, 299
  experimental, 298
  Raman spectra, 299
  results and discussion, 298–299
  synthesis of, 298
  TEM image, 299
Green chemistry, 336

H

High-dispersive silicon dioxide (HDS), 286
  adsorption properties, 287
Highly dispersed complex iron and titanium oxides, 306
Homogeneity of suspensions, 304

I

Innovations

social conditions of society, contribution of, 20, 311
Interaction processes
  directedness of, 46, 62
  energy characteristics on two principles of, 44–48
Inventions
  new breakthrough technologies and, 311–312
  value of, 20
Iron and manganese complexes with Pcs, 190
  experimental, 190–191
  results and discussion, 191–195

L

Large-scale atomic/molecular massively parallel simulator (LAMMPS), 135
Ligands in modified RE fluorides
  surface density of, 280
Lilies, 178
  diameter of open flower, 185
  height of flowering shoot, 182
  height of stem, G
  length of bud, 183
  number of buds, 183
  peat
    agrochemical characteristics of, 181
    characteristics during storage of, 180
    chemical analysis of, 181
  width of leaf, 184
Lorentz curve
  space-time graphic dependence, of, 78

M

Manganese-doped mixed CdZnS quantum dots, 320
Mathematical model
  problem setting and, 134–136
  processes, 38
    intrachamber, 39
  results and analysis, 136–9
Mechanical motions
  functionally interconnected, 97
  initial criteria, 98–99
Metal/carbon nanocomposites (NCs), 144, 149–150
  application in lily growing, 177

Index

cost-effectiveness of, 186
experimental researches, 179
results and discussion, 179–186
functional materials modification
processes by, 152–157
AFM images, 157
C1s spectrum of, 152
charge (electron) quantization, 154
composition polarization, 152
peak intensity change, 153, 154
x-ray photoelectron C1s spectra, 156
methods of estimation, 150–151
release forms of, 303
synthesis in nanoreactors of polymeric
matrices, 145
Metal-containing nanostructures (NCs)
RedOx synthesis of, 146
nanofilms images, 147
structure and characteristics of, 146–149
atomic magnetic moments, 149
C1s spectrum, 149
experimental EPR data, 149
metal/carbon nanocomposite, 149–150
microphotography, 148
nanofilms images, 147
nickel nanocrystal, 147
Micro and nano bioinspired composite
materials, 340–341
Micromechanics manufacturing
cartridge, 219
hybrid mode, 218
multi-probe cartridge and SPM head
parts, 219
progress in, 218–219
Modified method of the immersed atom
(MEAM), 134
Molecular dynamics (MD) method, 134

N

Nanobiobodies
estimation and prognosis methods, 27–29
Nanocrystalline copper selenide (CuSe),
325
Nanodispersive silica (NDS), 283, 287
experimental
materials, 288
results and discussion, 289–91

classification of analyzed samples,
system, 291
toxicity rate, 290
Nanoelements, 179
Nanofilm coatings formation
Problem solving, steps, 135
Nanomaterials (NM), 24
nonporous amorphous silicon dioxide,
285–286
polymer nanocomposites
chemistry between, 339
toxicity
estimation of, 284–285
nanostructural amorphous silicon
dioxide (silica, $(SiO2)n$), 285
Nanoparticle surface
states of citric acid coordinated, 279
Nanoreactors
cluster, 8–10
cluster-dimensional, 145
nanostructures formation within, 146
one-dimensional, 7, 145
two-dimensional, 8, 145
Nanoscience and nanotechnology, 347
Nanosized rare earth (RE) fluorides, 274
experimental researches, 275
results and discussion, 275–80
Nanostructured fibre of polyethylene terephthalate, 330
Nanostructured filters
carbohydrate polymers, based on,
333–334
Nanostructured steels transformations
peculiarities of, 327
Nanostructured titanium dioxide
antibacterial coating based on, 319–320
Nanostructures
media and compositions, influence on,
11–14
Nanotechnology (NT), 19, 284
utilization, 284
Nickel/carbon nanocomposite
diffract grams of, B
Non-stationary intra chamber processes, 31
Non-traditional electrical conducting materials, 259–260
Nonlinear development of technology and
techniques, 312

Nonlinear viscoelasticity of rubber nano-composites, 346–347

O

Optical limiting (OL), 324

P

Peltier module
  device based on, 349
Percentage of gallium atoms, E
Percentage of precipitated atoms, C
Phase separation, 344
Phase transformations, 308
Photoluminescence (PL) spectra, 277
Photoluminescence excitation (PLE)
  spectra, 276, 277
Photometric measuring equipment, 25
Photosynthesis, 24
  maxima and molar absorption coeffi-
    cients, 27
  photometric measuring equipment, 25
  role of, 25
  specific absorption coefficients, 27
Photosynthetic pigments, 24
Phthalocyanine (Pcs)–polymer complexes
  biocidal activity of, 189, 190
Physicochemical analysis, 26
Physiological systems
  entropic transitions in, 81–82
    by time t, 83
Pigment systems
  characteristics of, 26
  content in cucumber leaves, 29
Plant diagnostics, 25
Polarization-orientation sensitive photocurrent, 316
Polycarbonate
  XPS C1s-spectra of, 229
Polymer blends
  recycling of, 342–343
Polymer composites
  processing of, 348
    thermoplastic composites, 349
    thermoset composites, 349
Polymer films, 207–8
  basic part, 208–15
  dependence of local resistances of,
    212–214

dependences of oscillation frequency, 215
ferromagnetic film, 211
magnetization changes, 210
measuring of magnetic characteristics,
  209
rectangle and trapezoidal shape, 209
specific volumetric electric resistance,
  208
water solution, 208
Polymer latex nanocomposites, 343–344
Polymer materials, 162
Polymer materials modified with C/M
  nanostructures
  x-ray photoelectron investigation
    experiment, 226
    results and discussion, 226–232
    XPS C1s-spectrum of, 227
Polymer nanocomposites
  advances, 341–2
  dental applications for, 333
Polymethyl methacrylate (PMMA), 333
  C1s-spectra of, 227
  XPS C1s-spectra of, 228
Polymorphoid induces photo-transforma-
  tions, 321–2
Polypeptide chain
  formation of, 86, 88–9, 91
Polyvinyl alcohol
  modified with C/M nanostructures
    XPS C1s-spectra of, 231
  XPS C1s-spectra of, 230
Precipitation methods, 139
Prosthodontics, 333

Q

Quantum action
  act of, 48–9
Quantum transitions
  energies, 52
  parameters of, 50
Quark screw model
  calculation
    bond energy of deuteron, 128–9
    energy mass of free nuclide, 127–8
  electric charge, 126
  number of, 126
  structural scheme of, 127
  structure, 126

Index

R

Rocket solid propellant engines, 37
Rotation angles of planets, 102–4
  ratio of, 104
Rotational and orbital motion of planets
  characteristics of, 102
  equation of dependence, 99–100, 102
Rutile nanocrystals
  optical properties and size dependence
    of, 318

S

S-curves ("life lines"), 80
  dependence of, 81
Self-healing coating, 347–8
Silicon semiconductor structure chip, 222
Siltation
  porous template, 137
Silver nanoparticles, 200
  experimental, 200–1
  results and discussion, 201–5
Small-sized technological electron
    spectrometer
  magnetic energy-analyzer, with, 237–8
    results, 239–41
    technical characteristics, 242
Solid and liquid–gaseous planets
  ratio of dimensionless moments, 102
Solver-NANO, 221
Spatial-energy interactions
  entropic nomogram
    degree of, 74–5
    surface-diffusion processes, 75
  entropy of, 72
Spatial-energy parameter (SEP or P-param-
    eter), 58, 63–5, 85–6, 107, 108–11
  atoms calculated *via* electron bond
    energy, 85–6
  Lagrange and Hamilton functions, 66–9
  structural exchange of, 69, 72, 117–21
  wave
    equation of, 65–6, 114–15
    principles of, 115–17
    properties, 115–117
State-of-the-art chemistry
  objective of, 273

Structural PC-parameters
  calculate via electron bond energy, 87
Structural transitions
  geometry of quantization of, 53
Substrate diagnostics, 25–6
Surface complexes
  energy transfer from Dbm to Eu3+ ion
    in, 278
  formation of, 276–7
  probable structures of, 276
Surface modification, 274
Surface-diffusion processes
  entropic nomogram, 75
Surface-modified particle
  structure of, 274
System hamiltonian, 261

T

Technology and techniques
  nonlinear development of, 20–1
Tert-butyl hydroperoxide ((CH3)3COOH)
  NMR 1H and 13C spectroscopy, 248
    change of, 252, 253
    experimental, 248–9
    experimental parameters of, 250
    magnetic shielding tensors, 249
    molecular geometry parameters of, 251
    parameters of, 255
    results and discussion, 249–56
    structural model of, 252, 254
Thermodynamic probability, 73
Triple graphite cointercalation compounds
  sonication of, 309

U

Ulov and Gunter
  reaction to pre-seeding treatment, 315

V

Viscoelastic phase separation process, 344

X

X-ray photoelectron spectroscopy (XPS),
    238
  method, 351

9781771886963